Results and Problems in Cell Differentiation

A Series of Topical Volumes in Developmental Biology

16

Editors

W. Hennig, Nijmegen, L. Nover, Halle and U. Scheer, Würzburg

Results and Problems in Cell Differentiation

Volume 1 · H. Ursprung (Ed)
The Stability of the Differentiated State

Volume 2 · J. Reinert; H. Ursprung (Eds)
Origin and Continuity of Cell Organelles

Volume 3 · H. Ursprung (Ed)
Nucleic Acid Hybridization in the Study of Cell Differentiation

Volume 4 · W. Beermann (Ed)
Developmental Studies on Giant Chromosomes

Volume 5 · H. Ursprung; R. Nöthiger (Eds)
The Biology of Imaginal Disks

Volume 6 · W. J. Dickinson; D. T. Sullivan
Gene-Enzyme Systems in Drosophila

Volume 7 · J. Reinert; H. Holtzer (Eds)
Cell Cycle and Cell Differentiation

Volume 8 · W. Beermann (Ed)
Biochemical Differentiation in Insect Glands

Volume 9 · W. J. Gehring (Ed)
Genetic Mosaics and Cell Differentiation

Volume 10 · J. Reinert (Ed)
Chloroplasts

Volume 11 · R. G. McKinnell; M. A. DiBerardino
M. Blumenfeld; R. D. Bergad (Eds)
Differentiation and Neoplasia

Volume 12 · J. Reinert; H. Binding (Eds)
Differentiation of Protoplasts and of Transformed Plant Cells

Volume 13 · W. Hennig (Ed)
Germ Line—Soma Differentiation

Volume 14 · W. Hennig (Ed)
Structure and Function of Eukaryotic Chromosomes

Volume 15 · W. Hennig (Ed)
Spermatogenesis: Genetic Aspects

Volume 16 · L. Nover, D. Neumann, K.-D. Scharf (Eds)
Heat Shock and Other Stress Response Systems of Plants

Heat Shock and Other Stress Response Systems of Plants

Edited by Lutz Nover, Dieter Neumann, Klaus-Dieter Scharf

With 38 figures

With contributions by

Dieter Neumann[a], Lutz Nover[a], Benno Parthier[a], Rigomar Rieger[b], Klaus-Dieter Scharf[a], Reinhold Wollgiehn[a], Uta zur Nieden[a]

[a] Institute of Plant Biochemistry, Acad. Sci. of the G.D.R., Weinberg 3, DDR-4050 Halle

[b] Central Institute for Genetics and Cultural Plants, Acad. Sci. of the G.D.R., Correnssstr. 3, DDR-4325 Gatersleben

Springer-Verlag
Berlin Heidelberg New York
London Paris Tokyo

Lutz Nover
Dieter Neumann
Klaus-Dieter Scharf
Institute of Plant Biochemistry
Academy of Sciences of the G.D.R.
Weinberg 3
DDR-4050 Halle

Sole distribution rights for all non-socialist countries
Springer-Verlag Berlin Heidelberg New York London Paris Tokyo

ISBN 3-540-51837-1 Springer-Verlag Berlin Heidelberg New York
ISBN 0-387-51837-1 Springer-Verlag New York Berlin Heidelberg

Printed in the German Democratic Republic

Contents

Heat Shock and Other Stress Response Systems of Plants

By Dieter Neumann, Lutz Nover, Benno Parthier, Rigomar Rieger, Klaus-Dieter Scharf, Reinhold Wollgiehn, Uta zur Nieden

With 38 Figures, 8 Tables

A. Introductory part

A.1. Historical roots and scientific scope

Research on plant stress physiology goes far back into the last century (for references see ALEXANDROV, 1977, 1979; ALEXANDROV et al., 1970; LEVITT, 1958, 1980a, b; SACHS, 1864). The wealth of information collected in many thousands of papers documents the remarkable capability of plants to survive and grow under a variety of unfavourable conditions (for summaries see ALEXANDROV, 1977; ATKINSON and WALDEN, 1985; AYRES, 1984; KEY and KOSUGE, 1985; LEVITT, 1958, 1980a, b; MATTERS and SCANDALIOS, 1986b; RHODES, 1987; SACHS and HO, 1986). The particular interest in plant stress response systems stem from the following facts: (1) Due to their fixation to the natural habitat, plants are much less capable to escape stressful situations. (2) Severe anthropogenic changes of the environment considerably increased the number and/or amount of stressors. (3) Frequently, high-bred strains of cultural plants with a maximum productivity are particularly sensitive to growth and development under suboptimal conditions. The average losses of major crop plants due to the stressful environment may amount to 50–80% of their genetically determined productivity (BOYER, 1982; ORT and BOYER, 1985).

The general phenomenon of induced stress tolerance and the question of its underlying mechanisms are in the center of experiments in this field. Evidence for functions of stress metabolites and proteins in this respect are dealt with in many parts of this survey. We concentrated on the cell biological aspects of the stress response of plants, but included many key references to bacteria, eukaryotic microorganisms and animals as far as similar mechanisms of response are known. To a large extent this is true for heat shock, but also for other general stress situations, e.g. oxidation stress, anaerobiosis or mutagenic stress. For different reasons other stressors act more selectively, e.g. low temperature, starvation, wounding, water deficiency connected with salt or osmotic stress and biotic stress by pathogen infection.

Several factors contribute to the broad range intensivation of experimental efforts to characterize the molecular basis of stress response systems:

1. In 1962 RITOSSA reported on the heat shock-induced differential gene activation in *Drosophila*, and TISSIERES et al. (1974) described for the first time formation of the corresponding heat shock proteins (HSPs). With the advent of methods for gene cloning and sequencing, isolation and analysis of the *Drosophila* hs genes rapidly followed (see summaries by ASHBURNER and BONNER, 1979; NOVER, 1984, SCHLESINGER et al., 1982). In addition, induction of heat stress proteins was soon recognized as a conserved trait in all kinds of organisms, including plants (ALTSCHULER and MASCARENHAS, 1982; BARNETT et al., 1980; KEY et al., 1981, 1983; SCHARF and NOVER, 1982). The remarkably successful development of hs research in all kinds of organisms (part B of this review) is documented in numerous reviews and books (ASHBURNER and BONNER, 1979; ATKINSON and WALDEN, 1985; BIENZ, 1985; BIENZ and PELHAM, 1987; CRAIG, 1985; KEY et al., 1983, 1985a, b; LINDQUIST, 1986; LINDQUIST and CRAIG, 1988; NEIDHARDT et al., 1984; NOVER, 1984, 1987, 1990; PARDUE et al., 1989; PELHAM, 1986, 1988; SCHLESINGER, 1986; SCHLESINGER et al., 1982; SCHÖFFL et al., 1984b, 1986, 1988; TANGUAY, 1988). It has certainly influenced similar investigations on other stress response systems discussed in part C of this review.
2. It turned out that many stress-inducible genes are members of smaller or larger multigene families, which may be expressed also as part of developmental programs. Hence, the relatively easy detection and isolation of stress-inducible genes provides convenient probes for developmental genes and, thus, facilitates characterization of corresponding regulatory sequences connected with them. Moreover, stress situations may trigger or be inherent parts of developmental programs in microorganisms and plants. The intricate role of phytohormones or other modulators of the stress response in this respect is slowly emerging (Sect. C.2.).
3. Since the first reports on hyperthermic killing of cancer cells in the last century (BRUNS, 1887; BUSCH, 1866; COLEY, 1893; see review by NOVER, 1989) attempts to explore the cellular basis of this effect and to improve the clinical application in combination with radiotherapy and chemotherapy have provided deep insights into primary sites of heat shock damage and repair mechan-

isms as well as into the molecular basis of induced thermotolerance (for reviews and books see ANGHILERI and ROBERT, 1986; ARCANGELI et al., 1987; HAHN and LI, 1982; HENLE and DETHLEFSEN, 1978; NOVER, 1990 and OVERGAARD, 1985). Contributions from the experimental medicine to the cellular stress biology extend from heat shock to chemical stressors and modulators of the stress response. Many results obtained are generally valid and can help to understand similar reactions in plants as well.

Based on mostly unpublished material from our laboratories characteristic aspects of the plant heat shock response are reviewed in part B and compared or complemented with experimental data from animal and bacterial systems. To facilitate understanding of the material in figures and tables, essential experimental and technical remarks are summarized in the following section A.2. and in the legends to the figures. On the basis of the survey on the hs response in part B, other plant stress response systems are discussed in part C. Despite the considerable complexity of the plant stress response systems, even if reduced to the cellular level, our understanding is rapidly progressing and a general concept of plant/stress interactions becomes visible.

A.2. Materials, methods and scientific terms

Heat shock regimes

Plant material (leaf discs, root tips, hypocotyl segments etc.) or suspension culture cells in nutrient solutions were either heated immediately to the indicated temperature (regime I) or adapted slowly by about 2 °C/hr (regime II). Alternatively, a preinduction protocol was followed, i.e. after 15 min. heating to 40 °C, the plant material was allowed to recover for 2-3 hrs at 25 °C (preinduced material) and then subjected to a second, long-term hs at 40-42 °C (regime III).

Plant material

The following seedlings and plants respectively were cultivated in a greenhouse: *Arabidopsis thaliana, Catharanthus roseus, Cucurbita pepo, Glycine max, Hordeum vulgare, Lycopersicon esculentum, L. peruvianum, Peganum harmala, Nicotiana rustica, Pisum sativum, Vicia faba* and *Zea mays.* Sterile cultures of microorganisms were gifts of S. REINBOTHE (heterotrophic *Euglena gracilis*), E. OHMANN (autotrophic *Chlorella pyrenoidosa*), D. GRÖGER *(Claviceps purpurea)* and J. SCHIEMANN *(Aspergillus nidulans)*.
Plant cell suspension cultures were obtained from the following sources: *Catharanthus roseus* and *Peganum harmala* from G. KRAUSS and M. HIEKE (Halle), *Digitalis lanata, Lycopersicon peruvianum* and *L. esculentum* from our laboratories (see TEWES et al., 1982 and NOVER et al., 1982). *Petroselinum crispum* from K. HAHLBROCK and D. SCHEEL (MPI für Züchtungsforschung, Köln) and *Neobesseya similis* (cactus) callus was from S. LOPATO and J. GLEBA (Botanical Institute, Kiev).

Abbreviations and scientific terms

Amino acid residues in Figs. 6-13 are indicated by the one letter code: **A**, alanine, **C**, cysteine, **D**, aspartic acid, **E**, glutamic acid, **F**, phenylalanine, **G**, glycine, **H**, histidine, **I**, isoleucine, **K**, lysine, **L**, leucine, **M**, methionine, **P**, proline, **Q**, glutamine, **R**, arginine, **S**, serine, **T**, threonine, **V**, valine, **W**, tryptophan, **X**, unknown residue or either one of the 20 residues, **Y**, tyrosine.
Heat shock **(hs)**: induced genes and proteins are identified by **hsp** (for the genes) and **HSP** (for the proteins) plus numbers indicating the relative molecular weights (Mr) of the proteins in kilodalton (kDa). Thus, hsp70 refers to a gene coding for a HSP of the 70 kDa family. For a definition of the HSP families see Sect. B.1.2. The terms **hsc/HSC** are applied to constitutively expressed members of the eukaryotic HSP families. Organisms in connection with derived genes or proteins are denoted by their Latin names in a two letter code: **Bm**, *Bacillus megaterium* (bacilli); **Bo**, *Brassica oleracea* (cabbage); **Ce**, *Caenorhabditis elegans* (nematode); **Dd**, *Dictyostelium discoideum* (slime mold); **Dm**, *Drosphila melanogaster* (fruit fly); **Ec**, *Escherichia coli* (coli bacterium), **Gd**, *Gallus domesticus* (chicken); **Gm**, *Glycine max* (soybean); **Hs**, *Homo sapiens* (man); **Le**, *Lycopersicon esculentum* (tomato); **Ml**, *Mycobacterium leprae* (bacterium); **Mm**, *Mus musculus* (mouse); **Nc**, *Neurospora crassa* (fungus); **Pc**, *Petroselinum crispum* (parsley); **Ph**, *Petunia hybrida* (petunia); **Rn**, *Rattus norvegicus* (rat); **Sc**, *Saccharomyces cere-*

visiae (yeast); **Ta**, *Triticum aestivum* (wheat), **Tb, Tc**, *Trypanosoma brucei, T. cruzi* (trypanosomes); **Xl**, *Xenopus laevis* (claw-toed frog); **Zm**, Zea mays (corn).
Other abreviations: **ACC**, amino-cyclopropane carboxylic acid (precursor of ethylene biosynthesis); **CHS**, chalcone synthase; **PAL**, phenylalanine-ammonia lyase. Separations of proteins were either done by one-dimensional **(1D)** electrophoresis in sodium dodecyl sulfate containing polyacrylamide gradient gels (Fig. 1, SDS-PAGE) or two-dimensionally **(2D)** by combination with isoelectric focusing (Figs. 2–5, see NOVER and SCHARF, 1984; SCHARF and NOVER, 1982). Explanations and abbreviations for electron-microscopic data are summarized in the legend to Fig. 18.

Acknowledgements:

We gratefully acknowledge the kind supply of unpublished manuscripts and sequences, of clones and many other materials provided in support of the scientific work in our laboratories by N. AMRHEIN (Zurich), E. BAULIEU (Bicetre), L. BJÖRK (Uppsala), ELIZABETH CRAIG (Madison, Wisc.), K. HAHLBROCK (Cologne), B. KORANT (Wilmington, Del.), SUSAN LINDQUIST (Chicago, Ill.), H. LODISH (Boston, MA), F. SCHÖFFL (Bielefeld). R SINIBALDI (Palo Alto, CA), I. TÖRÖK (Szeged), ELIZABETH VIERLING (Tucson, Ariz.), JILL WINTER (Palo Alto, CA), R. YOUNG (Boston, MA) and M. H. ZENK (Munich).

B. Heat shock response

B.1. Plant heat shock proteins and genes

B.1.1. Induced HSP synthesis in microorganisms and higher plants

If proper conditions are applied, the high selectivity of gene expression during hs allows the easy detection of heat shock proteins even by one-dimensional electrophoresis of labeled proteins (Fig. 1). Dominant representatives in plants are the 'large' HSPs, e.g. HSP95 (no. 0), HSP80 (no. 1) and HSP70 (no. 3) and the 'small' HSPs, e.g. HSP21 (no. 18) and the HSP17 complex (no. 22). They represent three dominant families of hs genes (see below). HSP80 belongs to the HSP90 family, HSP70 belongs to the HSP70 family and HSPs17-21 belong to the HSP20 family. HSP95 is short lived and synthesized only transiently in the early phase of the hs response. It is unclear, whether it belongs to the HSP90 family. Representatives of two additional hs protein families are not detected on these 1D gels. The HSP60 family comprises organellar proteins (see Sect. B.1.2.4.), and the HSP8.5 (ubiquitin family) runs out of the range of most gels.
Induction of HSP synthesis by hs or some chemical stressors was reported for all types of green algae and higher plants (Table 1). Whenever analyzed by two-dimensional electrophoresis, the pattern of HSPs appears much more complex, especially with respect to the HSP20 family with 15-20 major and minor representatives (Nover and Scharf, 1984; Key et al., 1985b; Mansfield and Key, 1987). Illustrative examples are given in Figs. 2 to 5 showing autoradiographs of HSP patterns of 16 experimental objects. Similar to the 1D-electropherograms, hs-induced proteins are labeled with dots. Mrs and numbers of prominent HSPs are given in the legends.

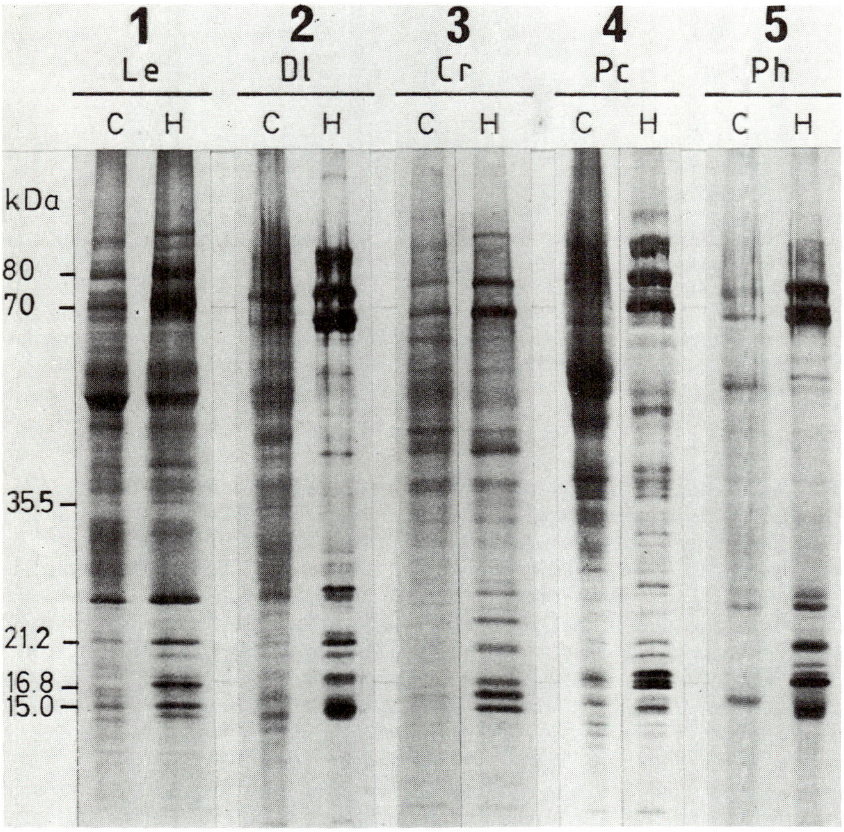

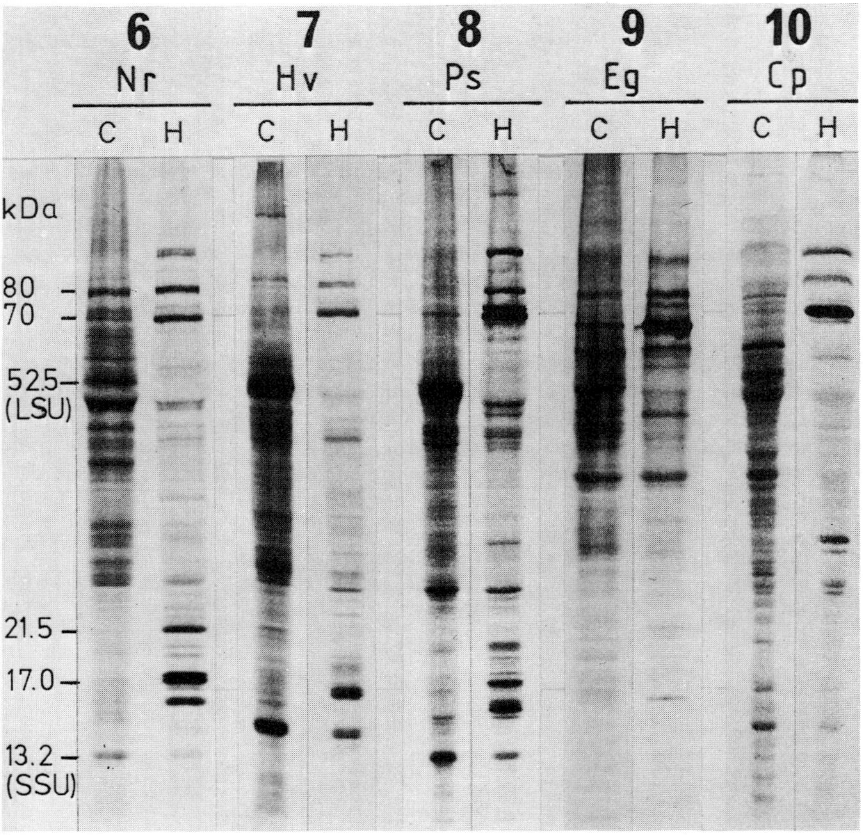

Fig. 1. Patterns of induced heat shock proteins in plants.
The figure represents autoradiographs of total cell proteins from the indicated plant materials after labeling with ^{35}S-methionine under control (C) and hs conditions (H) respectively. Proteins were separated by SDS-PAA gradient gel electrophoresis (for details see SCHARF and NOVER, 1982; NOVER and SCHARF, 1984). **1–5**, protein samples from cell cultures of *Lycopersicon esculentum, Digitalis lanata, Catharanthus roseus, Petroselinum crispum* and *Peganum harmala*. **6–8**, samples from leaves of *Nicotiana rustica, Hordeum vulgare* and *Pisum sativum*. **9**, proteins from *Euglena gracilis* (unicellular green flagellate) and **10**, *Claviceps purpurea* (ergot fungus). For heat shock conditions see legends to Figs. 2–5

Microorganisms (Fig. 2): When inspecting patterns of labeled proteins for two unicellular green algae, *Euglena gracilis* (A) and *Chlorella pyrenoidosa* (B), and two fungi, *Claviceps purpurea* (C), and *Aspergillus nidulans* (D), the 'large' HSPs belonging to the HSP90 and HSP70 families are observed in electrophoretic positions similar to those found in higher plants (Figs. 3 to 5). However, composition and abundance of small HSPs is highly divergent. There is one small HSP in *Euglena* (see Fig. 1, sample 9), which is not detectable in the 2D-pattern (Fig. 2A). In contrast there is a group of 15 not very abundant HSPs with Mr 15–30 kDa in *Chlorella* (B). Similar to yeast with only one HSP26 (PETKO and LINDQUIST, 1986), *Claviceps* produces a single highly abundant HSP24 (Fig. 2C), whereas in *Aspergillus* three minor HSPs are marked with Mrs 22, 28 and 33, respectively (Fig. 1D). It can be speculated that one or two of them correspond to mitochondrial HSPs30 and 34 reported for the related ascomycete *Neurospora crassa* (PLESOFSKY-VIG and BRAMBL, 1985).
Plants: The typical complexity of plant HSPs is demonstrated in three parts: Fig. 3 represents pictures obtained from labeling experiments with six heterotrophically grown cell suspension cultures,

Table 1. Induced heat shock protein synthesis in algae and plants

Organism/plant organ	References
Green algae	
Acetabularia mediterranea	Kloppstech et al., 1986
Euglena gracilis	This review, Figs. 1 and 2
Chlamydomonas reinhardii	Kloppstech et al., 1985
Chlorella pyrenoidosa	This review, Fig. 2
Chlorella protothecoides	Valliammai et al., 1987a, b
Volvox carteri	Kirk & Kirk, 1986
Monocotyledonous plants	
Tradescantia paludosa (spider wort), shoot apex	Xiao & Mascarenhas, 1985
Lilium longiflorum (lily), germinating pollen	Hong-Qi et al., 1984
Sorghum bicolor, seedling	Ougham & Stoddart, 1986
Gladiolus × *gandavensis,* cormel	Ginzburg & Salomon, 1986
Pennisetum americanum (pearl millet), seedling	Key et al., 1983; Mansfield & Key, 1987
Panicum miliaceum (millet)	Mansfield & Key, 1987
Hordeum vulgare (barley),	
Aleurone layer	Belanger et al., 1986
Leaf	This review, Figs. 1 and 5
Root, coleoptile	Necchi et al., 1987; Marmiroli et al., 1986
Triticum aestivum (wheat),	
Seedling	Mansfield & Key 1987; Zivy, 1987
Root, coleoptile	Necchi et al., 1987
Oryza sativa	Mansfield & Key 1987; Mocquot et al., 1987
Zea mays (corn),	
Seedling	Altschuler & Mascarenhas, 1982; Key et al., 1983; Mansfield & Key, 1987; Rees et al., 1987; Sinibaldi & Turpen, 1985; Yacoob & Filion, 1986
Mesocotyl	Bewley et al., 1983
Coleoptile	This review, Fig. 5
Plumule, radicle, mesocotyl, leaf	Baszczynski et al., 1982, 1983; Baszczynski, 1986; Cooper et al., 1984
Dicotyledonous plants	
Glycine max (soybean),	
Seedling root	This review, Fig. 5
Seedling	Altschuler & Mascarenhas, 1982; Key et al., 1981; Mansfield & Key, 1987
Suspension culture	Altschuler & Mascarenhas, 1982; Barnett et al., 1980; Vierling & Key, 1985
Phaseolus vulgaris (bean), developing seeds	Chrispeels & Greenwood, 1987
Pisum sativum (pea), seedling leaves and shoot apex	Key et al., 1983; Mansfield & Key, 1987; this review, Figs. 1, 5, 34
Vicia faba (broad bean), cotyledon	This review, Fig. 5

Organism/plant organ	References
Vigna unguiculata (cowpea), leaves	DAWSON & GRANTHAM, 1981
Cajanus cajan (pigeon pea)	KISHORE & UPADHYAYA, 1988
Peganum harmala (Syrian rue), cell cultures	This review, Fig. 3
Daucus carota (carrot), callus, protoplasts, embryos, plantlets	PITTO et al., 1983
Petroselinum crispum (parsley), suspension culture	This review, Fig. 3, WALTER, 1989; WALTER & HAHLBROCK, 1985
Arabidopsis thaliana (plantlets)	KURKELA et al., 1988; WU et al., 1988a
Brassica oleracea (broccoli), leaves	FABIJANSKI et al., 1987
Gossypium hirsutum (cotton),	
Leaves	BURKE et al., 1985
Seedlings	KEY et al., 1983
Spinacia oleracea (spinach)	GUY et al., 1985
Catharanthus roseus (periwincle), cell cultures	This review, Figs. 1 and 3
Lycopersicon esculentum (tomato),	
Suspension culture	This review, Figs. 1 and 3
Roots, leaves, ovaries, sepals	This review, Fig. 4
Lycopersicon peruvianum (wild tomato),	
Suspension culture	SCHARF & NOVER, 1982; NOVER & SCHARF, 1984, this review Figs. 3 and 30
Protoplasts	NEUMANN et al., 1987
Nicotiana tabacum (tobacco),	
Mesophyll protoplasts and 2 d old protopl. cultures	MEYER & CHARTIER, 1983
Suspension culture	BARNETT et al., 1980
Leaves	DAWSON & GRANTHAM, 1981; this review, Figs. 1 and 5
Nicotiana plumbaginifolia, protoplasts	RESTIVO et al., 1986
Digitalis lanata (fox glove), cell cultures	This review, Figs. 1 and 3
Helianthus annuus (sunflower), seedling	MANSFIELD & KEY, 1987

i.e. of two tomato species (*Lycopersicon peruvianum* and *L. esculentum*), of *Digitalis lanata, Catharanthus roseus, Petroselinum crispum* (parsley) and *Peganum harmala.* In Fig. 4 HSP patterns of different parts of a tomato plant are illustrated (*L. esculentum* roots, leaves, petals and ovaries). Finally, a collection of six other plants with their HSPs is shown in Fig. 5 including soybean roots *(Glycine max)*, maize coleoptiles *(Zea mays)*, and leaves of tobacco, barley, pea and broad bean.

For better orientation and reference to the 1D gels major HSPs are numbered in Fig. 3 A in correspondence with our original nomenclature given in NOVER and SCHARF (1984). The region of large HSPs is characterized by the HSP80 (pI 5.4–6.2) and the HSP70 (pI 5.0–5.7) complex. The presence of a third HSP95–100 (pI 6.5) evidently depends on the labeling conditions. At least in tomato cell cultures it is only synthesized shortly after the onset of the heat shock. Contrasting to the small

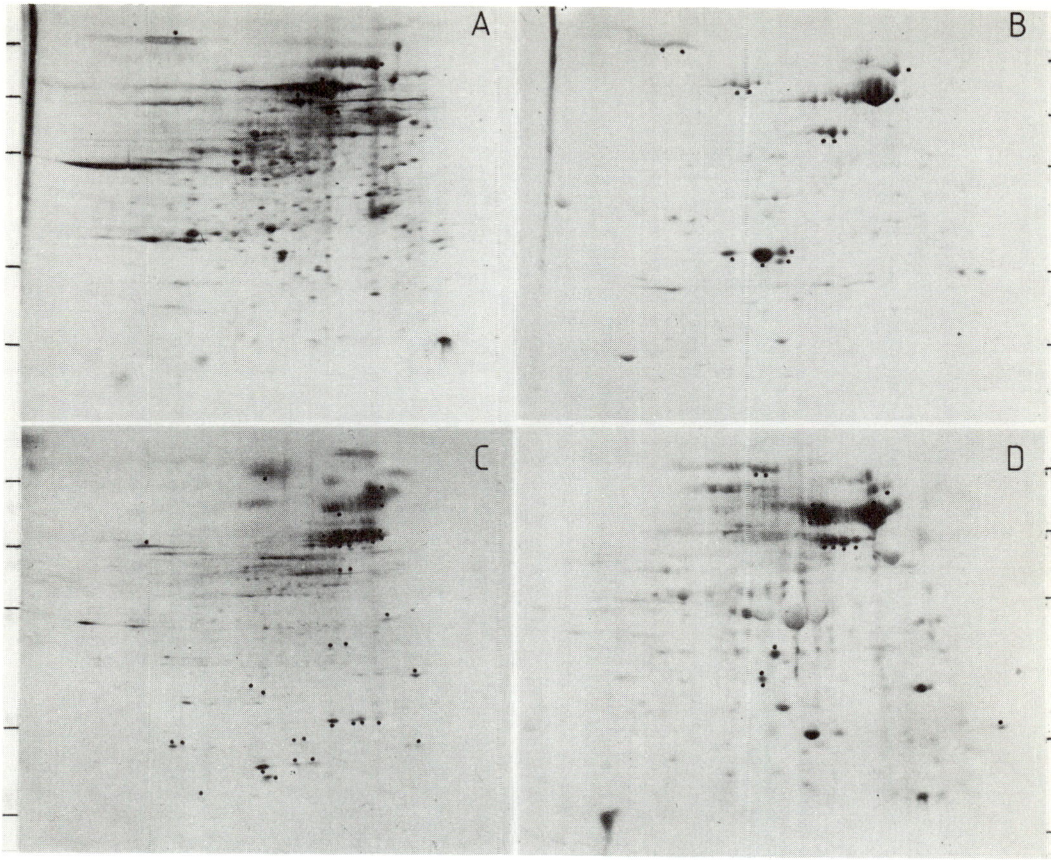

Fig. 2. Heat shock proteins in algae and fungi.
A, *Euglena gracilis* (I, 37°C); **B**, *Claviceps purpurea* (II, 37°C); **C**, *Chlorella pyrenoidosa* (I, 37°C); **D**, *Aspergillus nidulans* (III, 41°C). For explanations see legend to Fig. 3

HSPs with multiple but distinct isoforms the large HSPs of plants form more or less elaborate streaks when separated by isoelectric focusing. The reason for this is unclear. The dominant HSP70 complex contains several subspecies (Wu et al., 1988a). Details are best seen in Fig. 3A and E showing the tomato and parsley HSPs. There are two HSP74 with pI 5.0 and 5.25-5.35, the main spot of HSP70 (pI 5.25-5.5) and HSP68 (pI 5.5-5.7). Probably with exception of the HSP74 (pI 5.0), all are constitutive proteins, whose synthesis increases under hs conditions. Because a detailed sequence analysis of the plant HSP70 family with identification of individual proteins is lacking, their mutual relation is unclear. Antibodies to HSP70 do not cross-react with HSP68 and vice versa (see Sect. B.4.3.3.).

Minor HSPs marked by open arrow heads, are in the range of 53-60 kDa. They are especially prominent in the set of fungal HSPs (Fig. 2B and D), in tomato ovary, petals and roots (Fig. 4) and in soybean roots (Fig. 5A). It remains to be shown, whether they belong to the newly defined HSP60 family of proteins found in mitochondria and chloroplasts (see Sect. B.1.2.4.).

At least three gene subfamilies with one or several members contribute to the multiplicity of small HSPs of plants (see Sect. B.1.2.5.). Characteristically, the pattern of newly synthesized proteins varies with the plant investigated (Figs. 3 to 5), and there are even small differences, if different parts

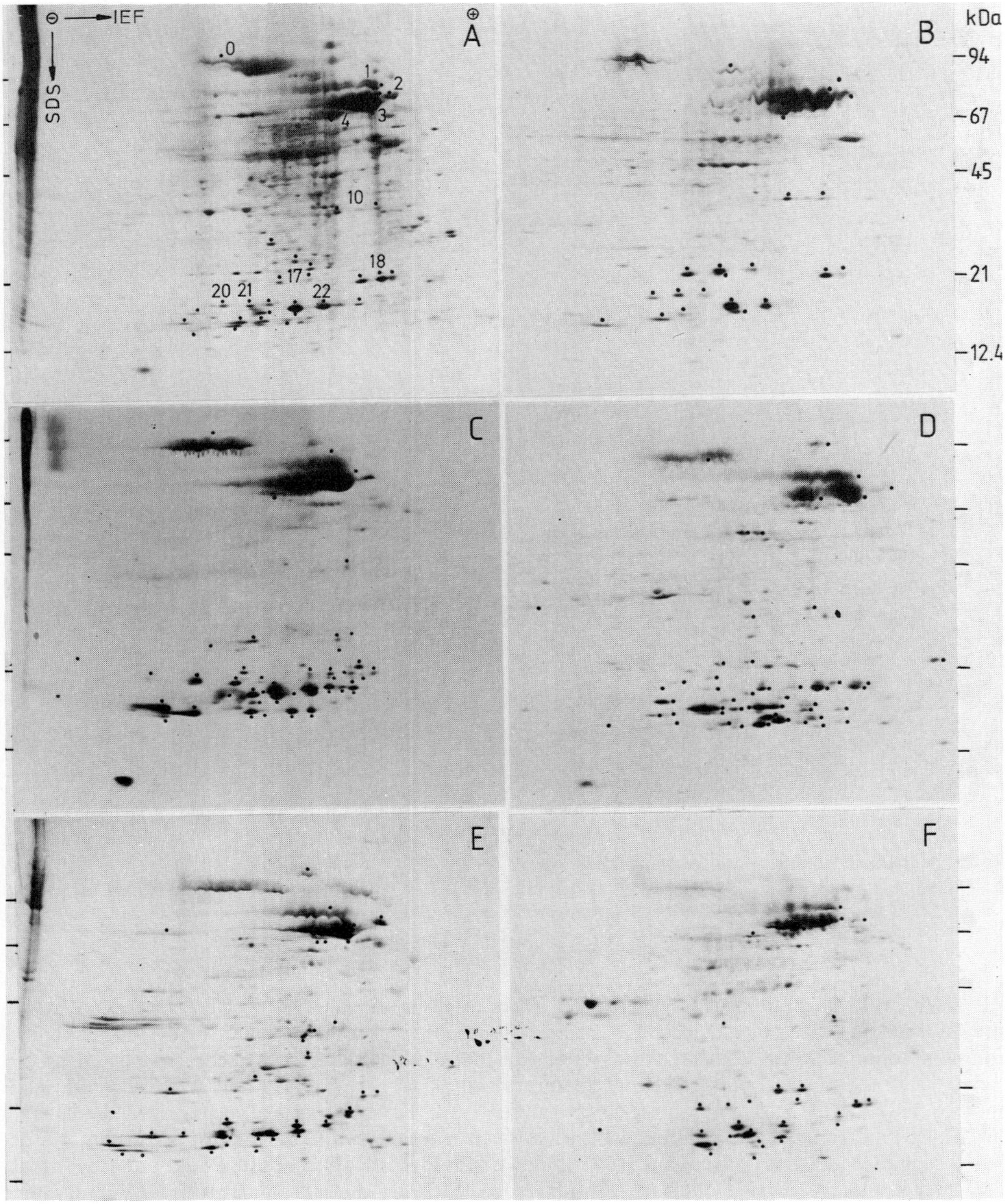

Fig. 3. Heat shock proteins synthesized in plant cell suspension cultures.
A, *Lycopersicon peruvianum* (hs regime III, 40.5 °C); **B**, *Lycopersicon esculentum* (III, 40.5 °C); **C**, *Digitalis lanata* (II, 41 °C); **D**, *Catharanthus roseus* (I, 40 °C); **E**, *Petroselinum crispum* (III, 40 °C); **F**, *Peganum harmala* (I, 38 °C). At the indicated heat shock temperature proteins were labeled for two hours with ^{35}S-methionine and separated by 2D-electrophoresis. Positions of molecular weight marker proteins are indicated at the right margin. Mr values are given in A. HSPs are marked by dots. Arrow heads in Figs. 4, A, C, D and 5A point to HSPs of

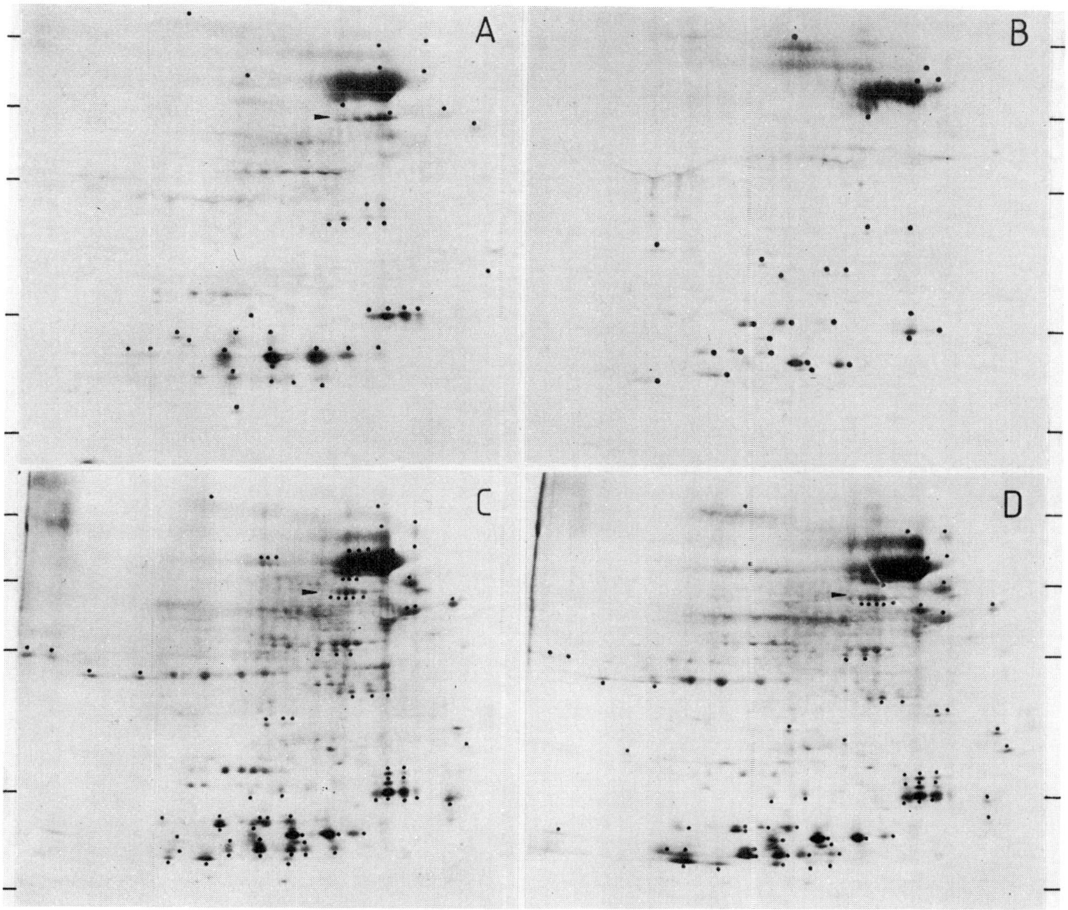

Fig. 4. Heat shock proteins in different tissues of tomato *(Lycopersicon esculentum)*.
A, seedling roots; **B**, leaves; **C**, sepals; **D**, ovaries. In all cases hs regime II with labeling at 40.5 °C was used. For explanations see legend to Fig. 3

of a plant are compared (Fig. 4). In most cases the dominant protein is in the range of Mr 16.5 kDa at pI 5.9 (HSP17). Together with other small HSPs it exists in 15S protein complexes (10 nm pre-hs granules), which under hs conditions form larger aggregates of 40-80 nm (hs granules, see Sect. B.3.3.).

The protein numbered 18 in Fig. 3 A (HSP21) deserves special notice. By immunological methods, it can be demonstrated that it corresponds to the chloroplastic HSP21-25 detected in hs leaves of

---

55-60 kDa (HSP60?). For orientation prominent proteins of the tomato HSP pattern are identified using our numbering published previously (Nover and Scharf 1984):

No	0	1	2	3	4	10	18	22
HSP	95	80	74	70	68	36	21	17
pI	6.5	5.4-6.2	5.0	5.25-5.5	5.5-5.7	5.3, 5.5	5.1, 5.3, 5.4	5.4, 5.6, 5.9

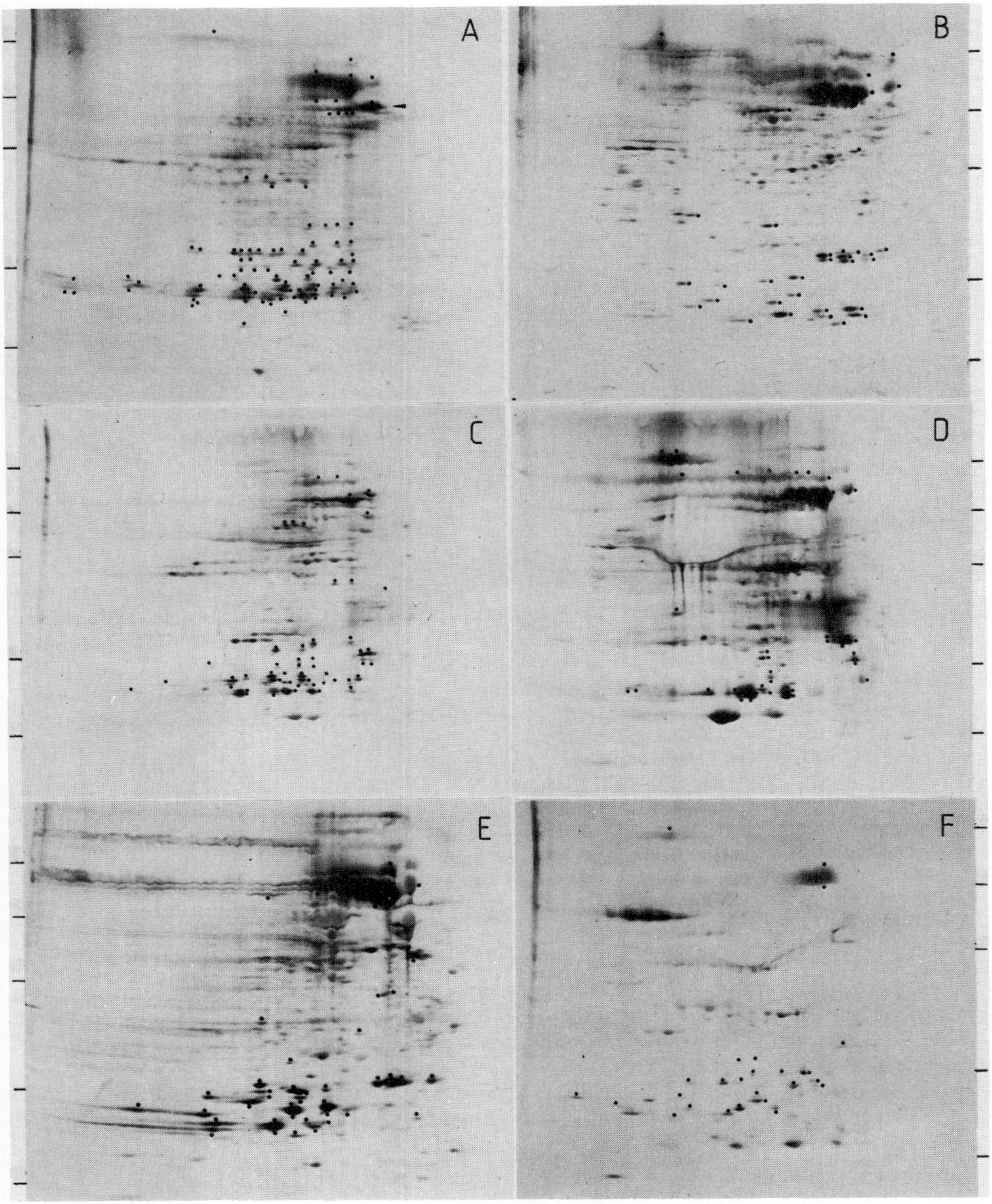

Fig. 5. Heat shock proteins in different plants.
A, *Glycine max* seedling roots (hs regime II, 40.5 °C); **B**, *Zea mays* coleoptiles (II, 40.5 °C); **C**, *Nicotiana rustica* leaves (III, 40 °C); **D**, *Hordeum vulgare* leaves (I, 38 °C); **E**, *Pisum sativum* etiolated leaves (III, 40 °C); **F**, *Vicia faba* cotyledons (II, 40.5 °C). For explanations see legend to Fig. 3

pea, *Arabidopsis*, soybean and others. As summarized in Sect. B.1.2.5. these proteins are synthesized by cytoplasmic ribosomes as precursor molecules with a 5 kDa basic transit peptide removed during entry into the chloroplast compartment. Despite the lack of chloroplasts in tomato suspension cultures and roots, HSP21 is a prominent and stabile protein accumulated in the processed form, whereas in green tissues (Fig. 4C, D), the main series of 2-4 isoforms with 21 kDa is accompanied by corresponding series with slightly higher Mr. It is a matter of speculation that proplastids and mitochondria may be the sites of storage and processing of the HSP21 in suspension cultures and roots (see Sect. B.4.3.4.). In support of the existence of a pre-HSP21 we never found significant amounts of HSP21 after *in vitro* translation of total poly(A)$^+$mRNA in a wheat germ system. Instead, there is a prominent more basic protein of about 25 kDa detectable in 2D-patterns of *in vitro* translated RNA from heat-shocked cells (open arrow heads in Fig. 17D, E).

B.1.2. Multiplicity of hs and related genes—HSP families

B.1.2.1. Heat shock-dependent vs. hs-independent regulation

With the availability of specific antibodies and gene probes it became apparent that HSPs belong to families of related proteins. So far 5 families can be defined. The HSP90 family (Mr 80-95 kDa), the HSP70 family (Mr 63-78 kDa), the organellar HSP60 family (Mr 53-62), the HSP20 family (Mr 14-30 kDa) and the ubiquitin family (Mr 8.5 kDa). Sequencing of an increasing number of cDNA and genomic clones confirmed the structural conservation of proteins belonging to a given family and revealed differences between proteins from bacteria, vertebrates and plants (Figs. 6-13). Moreover, with the exception of bacteria, frequently several genes with regulatory independence code for similar proteins within one HSP family. Though the molecular details remain to be elaborated, most HSPs and/or their related proteins evidently exert vital functions also under normothermic conditions. The apparent need for a hs-independent, constitutive or developmentally controlled expression of the corresponding genes within one family is met by different mechanisms, which are not mutually exclusive. Circumstantial evidence suggest that this is also true for plants, at least for the HSP70 and ubiquitin families. But because analyses are very fragmentary and our knowledge on plant promoter elements and their putative binding proteins in general is very restricted, the following examples can only be derived from data obtained for the hs gene families of yeast and animals. It can be anticipated that bascically similar solutions for the differential regulation of hs and related genes will also be found in plant systems.

1) Complex promoters with independent regulatory elements confer hs-inducibility and other modes of transcription control to the same gene (see Fig. 16 nos. 1, 2, 5). Most remarkable in this respect is the human hsp70 gene (Fig. 16, no. 1), analyzed by Morimoto and coworkers (Wu et al., 1987a; Watowich and Morimoto, 1988). The gene is expressed constitutively, under cell cycle control, after adenovirus infection and in a hs-dependent mode. The promoter region contains 4 independent regulatory elements within 100 bp 5' upstream of the TATA box. They are recognized by 4 independent binding proteins. Other examples are the *Drosophila* small hs genes (Fig. 16, no. 5) induced not only by hs but also during development by three types of developmental signals. One signal is the steroid hormone ecdysterone, whose increase at mid-embryonic stages and at the transition from larval to pupal stages is connected with expression of the small hs genes (Ayme and Tissieres, 1985 Mason et al., 1984; Pauli and Tonka, 1987).
2) A rare case may be represented by the only member of the *Drosophila* HSP90 family, i.e. HSP83 (Fig. 16, no. 3), which is coded by a single intron-containing gene at locus 63BC (Blackman and Meselson, 1986). Its constitutive and hs-induced expression is evidently regulated by the same overlapping trimer of the hs consensus element, which functions as a high affinity trap for small amounts of active hs transcription factor sufficient to maintain the hs mode of transcription activation even at 25 °C (Wu, 1984).
3) Two types of genes are responsible for the hs-induced and the hs-independent modes of expression, respectively. Though details of the promoter structures are not yet available, organization

```
 0 Consensus aa   MW      MXXXXXXXXX XXXXXXXXXX XXXXXXXXXX XXXXXXXXXX XXXXXXXXXX XXXXXXXXXX XXXXXXXXXX XXXETFAFQA EIXQLMSLII NTFYSNKEIF
 1 Hs-HSP90  724  83,303  .--------- ---------- ---------- ---------- ---------- ---------- --PEEVHHGE EEV....... ..A....... ..........
 2 Mm-HSP90  724  84,000  .--------- ---------- ---------- ---------- ---------- ---------- --PEEVHHGE EEV....... ..A....... ..........
 3 Gd-HSP90  728  84,112  .--------- ---------- ---------- ---------- ---------- --------PE AVQTQDQPME EEV....... ..A....... ..........
 4 Mm-GRP94  802  92,475  .RVLWVLGLC CVLLTFGFVR ADDEVDVDGT VEEDLGKSRE GSRTDDEVVQ REEEAIQLDG LNASQIRELR EKS.K..... .vNRm.K... .sL.K.....
 5 Gd-GRP94  795  91,555  .KSAWALALA CTLLLAASVT AEE-VDVDAT VEEDLGKSRE GSRTDDEVVQ REEEAIQLDG LNASQIKEIR EKS.K..... .vNRm.K... .sL.K.....
 6 Dm-HSP83  717  81,858  .--------- ---------- ---------- ---------- ---------- ---------- ---------P EEA....... ..A....... ..........
 7 Zm-HSP82  715  81,929  .--------- ---------- ---------- ---------- ---------- ---------- ASADVHMAGG AET....... ..N..l.... ..........
 8 Bo-HSP80  699  80,056  .--------- ---------- ---------- ---------- ---------- ---------- ---------- ADA....... ..N..l.... ..........
 9 Sc-HSP90  709  81,419  .--------- ---------- ---------- ---------- ---------- ---------- ---------- -AS...E... ..T....... ..V.......
10 Tc-p85    704  80,500  .--------- ---------- ---------- ---------- ---------- ---------- ---------- --T....... ..N....... ..........
11 Ec-HtpG   624  71,492  .--------- ---------- ---------- ---------- ---------- ---------- ---------- KGQ..RG..S .vK..lH.m. HsL.......

 0 LRELISNASD ALDKIRYESL TDPSKLDXXX ELXIRIIPDK XXRTLTIVDT GIGMTKADLI NNLGTIAKSG TKAFMXXXXE ALXAGAXXDI SMIGQFGVGF YSAYLVADKV XVISKHNDDE
 1 .......... .......... .......SGK ..K.D...NP QE....l... .......... .......... .....----. ..Q...--.. .......... .......e.. V..R......
 2 .......... .......... .......SGK ..K.D...NP QE....l... .......... .......... .....----. ..Q...--.. .......... .......e.. V..t......
 3 ......S... .......... .......SGK d.K.Nl..N. HD........ .........v .......... .....----. ..Q...--.. .......... .......e.. T..t......
 4 .......... ......LI.. ..ENA.AGNE ..Tvk.KC.. EKNL.HvT.. .v...rEe.v K......... .SE.lNKMT. .QED.QS-TS El........ ...f...... I.T....N.T
 5 .......... ......LI.. ..ENA.AGNE ..Tvk.KC.. EKNM.HvT.. ......Ee.. K......... .SE.lNKMT. mQDDSQS-TS El........ ...f....r. I.T....N.T
 6 .......... .......... .......SGK ..Y.kl..N. TAG....i.. ......S..v .......... .....----. ..Q...--.. .......... .......... T.T..N....
 7 .......... .......f.. ..K....AQP ..F..lv... ASk..s.i.s .v....S..v .......r.. ..E..----. ..A...T-.v .......... ........r. M.Tt......
 8 ......S... .......f.. ..K....GQP ..F.H..... ANN....i.s .........v .......r.. ..E..----. ..A...--.v .......... .......... I.Tt......
 9 .......... .......K.. s..KQ.etEP d.F...T.K. EQkV.E.R.s .......e.. .......... .....----. ..S...--.v .......... ..lf....r. Q....S....
10 ......S... .C.....Q.. .NQAV.GDES H.R..vv... ANk...vE.. ........e.v .......r.. .....----. ..E..G--.m .......... ........r. T.v..N....
11 .......... .a..l.fRA. sN.DLYeGDG ..Rv.vSF.. DK.T...S.N .v...rDev. Dh........ ..S.l----. S.GSDQAK.S Ql........ ...fi..... T.RtrAAGeK

 0 QYIXXXWESX AGGSFTVXAD XXGEXLGRGT KIXLHLKEDQ TEYLEERRVK ELVKKHSQFI GYPIXLWVEK EREKEVSDDE EEEEKXEEKE EEXXEXEEKE PKIEDVGSDE EDDKEXXEKK
 1 ..a---...S .......R.. H-..Pi.M.. .vI....... .......... .v........ ....T.yl.. .....i.... A....G-... ..DKdD...- .......... ...-SGKd..
 2 ..a---...S .......RX. H-..Pi.... .vI....... .......... .v.....X.. ....T.yl.. ?....i.... A....G-... .?DK.?...- .......... ...-SGKd..
 3 ..a---...S .......Rl. N-..P..... .vI....... ........i. .i........ ....R.f... ..d......d .A...E.... .KEEKT.d.- .E........ .ee.KDGd..
 4 .H.---...D -SNE.s.I.. PR.NT..... T.T.V...eA sd...LDTi. N..r.Y.... Nf..Yv.SS. TETV.EPLe. d.AA.-.... .----Sdd-. AAv.e----- ----.EE...
 5 .H.---...D -SNE.s.ID. PR.NT..... T.T.V...eA sd...LDT.. N....Y.... Nf..Yv.SS. TETV.EPVe. ..A-.-.... .----TddN. AAv..----- ----.EE...
 6 ..v---...S .......R.. NS-.P..... ..V.Yi.... .d....Ski. .i.N...... ....K.L... .......... Add..K-.Gd .KKEMETd-. .......... dA..KDKd-A
 7 ..v---...Q .......TH. TT..Q..... ..T.F..d.. L.......l. d......E.. S...Y..T.. TT...i.... ...dNK--.. ..-------- GDv.e.---d de..dTKd.S
 8 ..v---...Q .......TR. TS..S..... .mV....... L.......l. d......E.. S...S..i.. TI...i.... d.....---.d ..-------- G.v.e.---d .e-..KE...
 9 ...---...N .......Tl. EVN.Ri.... IlR.F..d.. L.....k.i. .vi.r..E.v a...Q.V.T. .V....PIP. ..KKDE.K.d ..KKdEdd.K ..l.e.---- --.-.EE...
10 A.T---...S ...t...TPT PDCdLK-... r.V....... Q.......l. d.i....E.. ..D.E.M... AT....t.ed .d.-----AA ATKN.EG.-. ..v.e.K--- -..A.EG...
11 PENGVF...A gE.Ey..-.. ITK.D--... E.T...r.GE D.f.ddW..r SiiS.Y.DH. aL.v------ ---------- -.I..R...d ---------- --G.T.I.-- ----------

 0 KKTKKVKEXX XDXEELNKTK PIWTRNPDDI TQEEYGEFYK SLSNDWEDHL AVKHFSVEGQ LEFRALLFVP RRAPFDLFEN XXXKKKNNIK LYVRRVFIMD XCEDLIPEYL NFVKGVVDSE
 1 .....i..KY I.Q....... .......... .......... ..t....... .......... ........i. .......... K--....... .......... S.de...... ..ir......
 2 .....i..KY I.Q....... .......... .......... ..t....... .......... ........i. .......... K--....... .......... S.de...... ..ir......
 3 ..K.-i..KY I.E....... .......... .n........ ..t....... .......... .......... .......... R--....... .......... N..e...... ..mr......
 4 P.....EKTV W.W.Lm.DI. ...Q.PSKeV EEd..KA... .F.KeSd.Pm .YI..ta..E vT.kSi.... TS..RG..dE YGS..SDY.. .........T. DFH.mm.K.. .........d
 5 P.....EKTV W.W.Lm.DI. ...Q.PSKeV EEd..KA... tF.KeHd.Pm .YI..ta..E vT.kSi.... NS..RG..dE YGS..SDF.. .........T. DFH.mm.K.. .........d
 6 ..K.Ti..KY TeD....... .......... s......... ..t....... .......... ........i. ..T....... Q--..r.... .......... N......... ..m.......
 7 ..K.....VS HeWVQi..Q. ...L.K.ee. .Rd..aS... ..t....... .......... ...k.i.... ........dT --R..L.... .......... N..e....w. G........d
 8 ..KI.E--VS HeWdLv..Q. ...M.K.ee. NK...aA... .......e.. .......... ...k.v.... k.......dT --K..P.... .......... N...i..... G....i....
 9 P.......EV QeI....... .l.....S.. ......NA... .i......P. Y......... ......i..i. k........S K--....... .........T. EA.......w. S.........
10 ........VT QeFVVQ..H. .l...D.K.v .K...aA... Ai.....eP. ST........ ......i... k.....m..P S--..r.... .......... N....C..w. A..r......
11 ---------- --W.Ki..AQ Al....KSe. .D...K.... HiAH.fN.P. TWS.NR...K Q.yTS..yi. SQ..W.mW-. R--DH.HGl. ...Q...... DA.QFm.N.. R..r.li..S
```

```
 0  DLPLNISREM LQQXKILKVI RKNLVKKXLE LFXEIAEDXK ENYKKFYEXF SKNLKLGIHE DSXNRXRLAK LLRYXXXHSS XSGDEMTSLX DYVSRMKEXQ KXIYYITGES KKXVENSPFV
 1  .......... ...S...... ...i...C.. ..S.l...-. ........A. .......... ..T..R..SE ....---.t. Q........S e.......T. .S........ .EQ.A..A..
 2  .......... ...S...... ...i...C.. ..S.l...-. ........A. .......... ..T..R..SE ....---.t. Q........S e.......T. .S........ .EQ.A..A..
 3  .......... ...S...... .......C.. ..T.l...-. ........Q. ...i...... ..Q..Kk.SE ....---Yt. A.....V..K ..Ct....N. .Hv......t .DQ.A..A..
 4  .....v...T ...H.l.... ..K..r.T.d mIKK..de-. Y.-DT.wKE. GT.i...vI. .HS..T.... ...f---Q.. HHSTdi...D Q..E....K. DK..fmA.S. r.Ea.S....
 5  .....v...T ...H.l.... ..K..r.T.d mIKK...e-. Y.-DT.wKE. GT.v...vI. .HS..T.... ...f---Q.. HHESNl...D Q..E....K. DK..fmA.A. r.Ea.S....
 6  .......... ...N.v.... .......Tm. .IE.lT..-. .......dQ. .......v.. ..N..Ak..D F..f---.t. A...dFC..A .......dN. .Hv.f..... .DQ.S..A..
 7  .........T ...N...... .......Ci. m.F....N-. dD.A...dA. ...i...... ..Q..Ak..D ....---..t K....T...K ...t....G. .D........ r.A......l
 8  .........T ...N...... .......Cm. ..F....N-. .D.N....A. .......... ..Q..Tki.E ....---..t K....l...K ...t....G. ED.f...... ..A......l
 9  .....l.... ...N..m... ...i...Li. a.N.....-S .qfE...SA. ...i...v.. .tQ..AA... ....---N.t K.V..l...T ...t..P.H. .N........ L.A..K...l
10  .........N ...N...... ...i...A.. ..E....N-. .D......Q. G..v...... ..A..Kk.m. ...f---... E..ed..t.K ...t....G. .C...v..d. ..Kl.T...i
11  .....v...i ..DSTvTrNl .NA.T.rV.Q mLEKl.K.DA .K.QT.wQQ. GLV..E.PA. .FA.QEAi.. ...fAST.tD S.AQT-V..E ........G. EK......d. YAAaKS..Hl

 0  ERLRKRGFEV LYMTEPIDEY CIQQLKEFDG KXLVSVTKEG LELXXXEDEE EKKKREEXKA KFEXLCKLMK DILXXDKVEK VVVSNRLVDS PCCIVTSQYG WSANMERIMK AQALXXXRDX
 1  ..v....... v......... .v........ .S........ ...--P.... ....M..S.. ...N...... e..D-K.... .Ti.....S. .......T.. .t........ ....---..N
 2  ..v....... v......... .v........ .S........ ...--P.... ....M..S.. ...N...... e..D-K.... .Ti.....S. .......T.. .t........ ....---..N
 3  .....H.L.. i..I...... .v......e. .T........ ...--P.... ....Q..K.. ...N...i.. ...E-K.... ........T. ........T. .t........ ....---..N
 4  ...L.k.y.. i.l...v... ...A.P.... .RFQN.A... vKF--D.S.K T.ES..ATEK E..P.LNW.. .KaLK..i.. a...q..Te. ..Al.A.... ..G....... ...YQTGk.I
 5  ...L.k.y.. i.l...v... ...A.P.... .RFQN.A... vKF--E.S.K S.ES..ALEK E..P.LNW.. .KaLK..i.. a.l.q..TQ. ..Al.A.... ..G....... ...YQTGk.I
 6  ..vkA..... v......... V..H...yK. .Q........ ...--P...S .......D.. ...S...... S..D-N.... .......... ........f. .......... ....---..T
 7  ...k.k.y.. .f.VdA.... AvG....y.. .K...a.... .K.DDEd... A..r...R.K r..E...vi. ...G .r.... ....D.i... ...l..GE.. .t........ ....---..-
 8  .k.k.k.y.. ...VdA.... A.G.....e. .K...a.... .K.EES...- K...-..L.E ...G...vi. .v.G-..... .i..D.v... ...l..GE.. .t........ ....---..-
 9  dA..AkN... .fl.d..... AFT.....e. .T..Di..d- F..--E.Td. ..AE..KEIK Ey.P.T.al. e..G-.Q... ....Yk.l.A .A-.R.G.f. .......... ....---..-
10  .Qa.r..... .f........ Vm..v.d.eD .KFaCl.... vHF--E.T.. ...Q...E.T Ay.R...a.. .v.G-..... ....E..aT. ..Il...Ef. ...H..Q..r N...---..-
11  .L...k.I.. .LlsdR...W MmnY.T.... .PFQ..s.VD ESGEKLA..V dESAk.AE.. LTPFiDr-v. Al.G-er.KD .RltH..T.t .AIvS.DADE M.TQ.AKLFA .AgQ---kVP

 0  STMXGYMXXK KTXEINPDHP IVEELRKRAE ADKNXDKAVX KDLVVLLFET ALLSSGFSLE DPQTFXXRIX RMIKLGLXID EDEXXXEEXE XXXXDXXXXL EGDEDXSXME EVD*
 1  ...-...MA. .HL....... ...T..Qk.. ....-....- .......... .......... ....HSN..Y .......G.. ...VAA..PN AAVP.EIPP. .....A.R.. ...*
 2  ...-...MA. .HL....... ...T..Qk.. ....-....- .......... .......... ....HSN..Y .......G.. ...VTA..PS AAVP.EIPP. .....A.R.. ...*
 3  ...-...AA. .HL......S .i.T..Qk.. ....-..S.- ....i..y.. .......... ....HAN..Y .......G.. ..dTAA..AS PAVTeEMPP. ...d.T.R.. ...*
 4  ..NY-.aSQ. ..F....R.. liRdmLr.iK E.ED-..T.- M..a.v.... .T.R..yL.P .TKAyGD..E ..lr.S.N.. PeAQVE..P. EEPE.TSEDa .DS.QDEGE. MDAGTEEEEE
 5  ..NY-.aSQ. ..F....R.. liKdmLr.vK ENED-..T.- S..a.v.... .T.R..yM.P .TKEyGD..E ..lr.S.N.. L.AKVE..P. E-PE.AAEEa .Q..eEVDAd AE.SE-----
 6  A..-...AG. .QL....... ...T..Qk.d ....-....- ....i..... S........d S..VHAS..Y .......G.. ...PMTTDDA QSAG.APSLv .DT..A.H.. ...*
 7  .s.Sa..SS. ..M.....NG .m........ ..r.-..S.- ....l..... ...t.....d ..nT.AA..H ..l....N.. ..AAAD.dAd MPAL.EGAAE .-----.K.. ...*
 8  .s.G...SS. ..M....eNA .md....... ....-..S.- ....l..... ...t.....d E.nT.GS..H ..l....S.. d.dAVEAdA. MPPLeDDADa ..----.K.. ...*
 9  .s.SS..SS. ..F..S.KS. .iK..k..vd EGGAQ..T.- ...TK..y.. ...t.....d e.Ts.AS..N .L.S...N.. ...ETETAP. ASTAAPVEEv PA.---tE.. ...*
10  .s.Sa..MS. ..M.....A.. ..K..kr.v. ..E.-....- ....Y...d. ...t...t.d ..TsyAE..H .......Sl. dedNGN..A. PAAAVPAEPv A.T---.S.. N..*
11  EV-------. YIF.l....V l.----...- ..TE-.E.KF SEW.E..LdQ ...AERGt.. ..nL.---.R ..NQ.LvS*

 4  ETEKESTE-K DEL*
 5  -TQKESTDVK DEL*
```

Fig. 6. Sequence comparison of members of the HSP90 family.
The amino acid sequences of eleven members of the HSP90 family are indicated by the one letter code. For comparison a hypothetical consensus sequence was derived (0), and all others are referred to this. Identical residues are marked by dots, lacking residues by hyphens and homologous substitutions within groups of similar amino acid residues by small case letters, with A = I = L = M = V, F = Y = W, S = T, R = K, D = E and N = Q. X denotes positions with lacking homology between the eleven sequences. For further explanations see text. Sequences are: **1**, human HSP90 (Rebbe et al., 1987); **2**, mouse HSP90 (Moore et al., 1987); **3**, chicken HSP90 (Binart et al., personal commun. by E. Baulieu); **4**, mouse GRP94 (Mazzarella and Green, 1987); **5**, chicken GRP94 (Kulomaa et al., 1986); **6**, *Drosophila* HSP83 (Blackman and Meselson, 1986); **7**, maize HSP80 (R. Sinibaldi and Brunke, pers. commun.); **8**, cabbage HSP80 (R. Sinibaldi and Brunke, pers. commun.); **9**, yeast HSP90 (Farrelly and Finkelstein, 1984); **10**, *Trypanosoma cruzi* surface glycoprotein p85 (Dragon et al., 1987); **11**, *E. coli* HtpG (Bardwell and Craig, 1987)

of the hsp70 gene families in *Drosophila* and yeast can illustrate this point. In *Drosophila melanogaster*, 5 hs-inducible hsp70 genes at loci 87A7 and 87Cl plus one hsp68 gene at locus 95D are opposed to one major and two minor constitutively expressed hsc70 genes. The major hsc70 gene at locus 88E codes for a protein (HSC70) whose abundance even under hs conditions always markedly exceeds that of the HSP70 (Palter et al., 1986).

In yeast, eight members of the hsp70 gene family were characterized by promoter fusion and gene deletion analyses. Two proteins of group A (SSA1 and SSA2) are constitutively expressed and hs-induced, whereas the other two members of group A (SSA3 and SSA4) are only detected after hs. In contrast, the constitutive, high expression of the two group B hsc70 genes (SSB1 and SSB2) is severely reduced under hs conditions. The yeast HSP70 family is completed by two further genes (SSC1 and SSD1). The former codes for a mitochondrial hs-induced protein, whereas the product of the latter is unknown (Craig, 1985; Craig et al., 1987; Werner-Washburne et al., 1987).

B.1.2.2. The HSP90 family

A survey of members of this family is given in Fig. 6. Generally 1-2 genes are found in each type of organism. Sequencing of two plant hsp80 genes (nos. 7, 8) allows a clear assignment to this family (Sinibaldi and Brunke, unpublished). Also a member of this family is a special glucose deficiency-induced protein (GRP94) found in vertebrates. Together with its counterpart from the HSP70 family (GRP78), it represents a type of abundant ER proteins, whose expression is not induced by hs but under conditions of disturbed proteinogen processing in the ER/Golgi system. Both GRPs are characterized by an N-terminal extension (leader sequence) and by the C-terminal-KDEL tetrapeptide responsible for their retention in the lumen of the ER (Fig. 6 nos. 4, 5, Fig. 8 nos. 6, 7, see Munro and Pelham, 1986, 1987).

The basic structure of the HSP90 proteins derived from the consensus sequence (Fig. 7A) shows a mosaic pattern of regions with very high to no conservation. Remarkable are two hydrophilic regions with high density of charged amino acid residues (arginine and lysine, but mainly glutamic

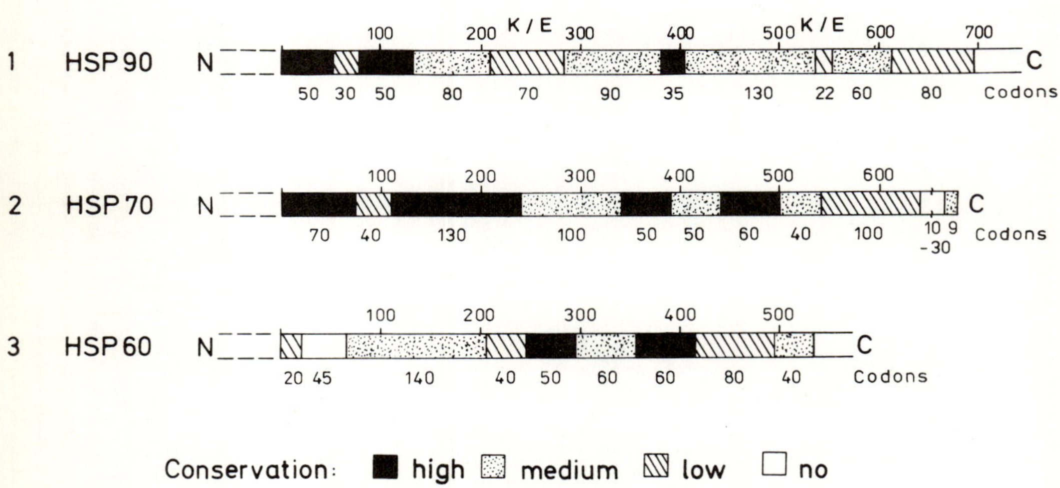

Fig. 7. Extent of sequence conservation between members of the HSP90, HSP70 and HSP60 families.
The proteins are depicted by block diagrams derived from data in Figs. 6, 8 and 9. The large HSPs are composed of a mosaic of regions with high, medium, low or no conservation between individual members of the same family (see symbols given below)

and aspartic acid residues (K/E)). The larger one in the N-terminal half is characterized by more or less extensive deletions, and it is totally lacking in the *E. coli* HtpG protein. This part of the HSP90 is flanked by two highly conserved proline residues (dots in Fig. 6), which evidently mark border regions of an acidic helix capable to shield the DNA binding site of steroid hormone receptors (BAULIEU and CATELLI, 1989).
With the exeption of the GRP94, members of the HSP90 family are abundant cytoplasmic phosphoproteins (NOVER and SCHARF, 1984, see summary by NOVER, 1990) which, in vertebrate cells were found in protein complexes with the inactive steroid hormone receptor (BAULIEU, 1987; BAULIEU and CATELLI, 1989), with actin (NISHIDA et al., 1986) and with certain onc-gene products (BRUGGE et al., 1983). However, at present there is no evidence for a stress-related function in these complexes (see Table 2).

B.1.2.3. The HSP70 family

Compared to the proteins of the HSP90, HSP60 and HSP20 families the extent of sequence conservation is much higher between the 19 members of the HSP70 family compiled in Fig. 8. Considerable deviations from the consensus sequence are only observed for the two ER-localized GRP78 of vertebrate origin (nos. 6, 7) and for the two bacterial DnaK proteins (nos. 18, 19). Generally, the C-terminal part of the HSP70 family proteins is less conserved. But with the exception of the GRPs and the tomato HSC70 all representatives from eukaryotes contain characteristic repetitive motifs of -MGG(G)P- and end in the nonapeptide -SGPTIEEVD.
Sequencing of plant HSP70 genomic clones from maize (ROCHESTER et al., 1986), *Petunia* (WINTER et al., 1988) and *Arabidopsis* (WU et al., 1988a) in each case revealed one intron of 300–700 nucleotides inserted into codons 71–73. At least 5 different members of the plant HSP70 complex can be identified by two dimensional electrophoresis (see Fig. 3 and WU et al., 1988a). But only extensive characterization of corresponding gene clones and of their regulatory behaviour can clarify the actual complexity of the plant HSP70 family. This also concerns the question of the existence of a plant 'GRP78' or of organelle-specific members. In vertebrates, fungi, bacteria and plants some members of the HSP70 family are phosphorylated and/or methylated (see NOVER and SCHARF, 1984; and summary of data in NOVER, 1990).
The high degree of conservation and the abundance of the HSP70 proteins under control and hs conditions facilitates studies on their intracellular localization and possible function.
The data so far accumulated mostly concern non-plant systems (for summary see NOVER, 1990). They are briefly summarized, because they characterize general properties of HSP70 -type proteins and of their interaction with heterologous proteins, which may be relevant to understand their function also in plants (see Table 2):

1) Clathrin uncoating ATPase: By binding to coated vesicles the prominent constitutive member in vertebrates (HSC71–73) acquires ATPase activity and helps to disaggregate the clathrin cages of coated vesicles (CHAPPELL et al., 1986; UNGEWICKELL, 1985). Recently coated vesicles were also identified in plants (ROBINSON, personal commun.).
2) Interfacing proteins of cytoskeletal systems: In vertebrate cells the mutual interconnection as well as stress-dependent rapid reorganization of the cytoskeletal systems and their contact to the plasma membrane involve members of the HSP70 family (HUGHES and AUGUST, 1982; NAPOLITANO et al., 1987; PALTER et al., 1986; WIEGANT et al., 1987a, b).
3) Reversible nuclear/nucleolar binding during hs: In different types of organisms, including plants, hs-induced and constitutive members are highly concentrated in the nuclear compartment of stressed cells. Two different phenomena are related to this: (i) nucleolar binding of HSP70 correlates with protection of the transcription and processing machinery for ribosome biosynthesis (see Sect. B.3.2.) and (ii) HSP70 may help to dissociate aggregates of non-chromosomal nuclear proteins tightly bound to the nuclear matrix (CLARKE et al., 1988; LEWIS and PELHAM, 1985; LITTLEWOOD et al., 1987; ROTI ROTI et al., 1986; STÜRZBECHER et al., 1987).
   It is well known that nuclear translocation depends on recognition of the corresponding proteins

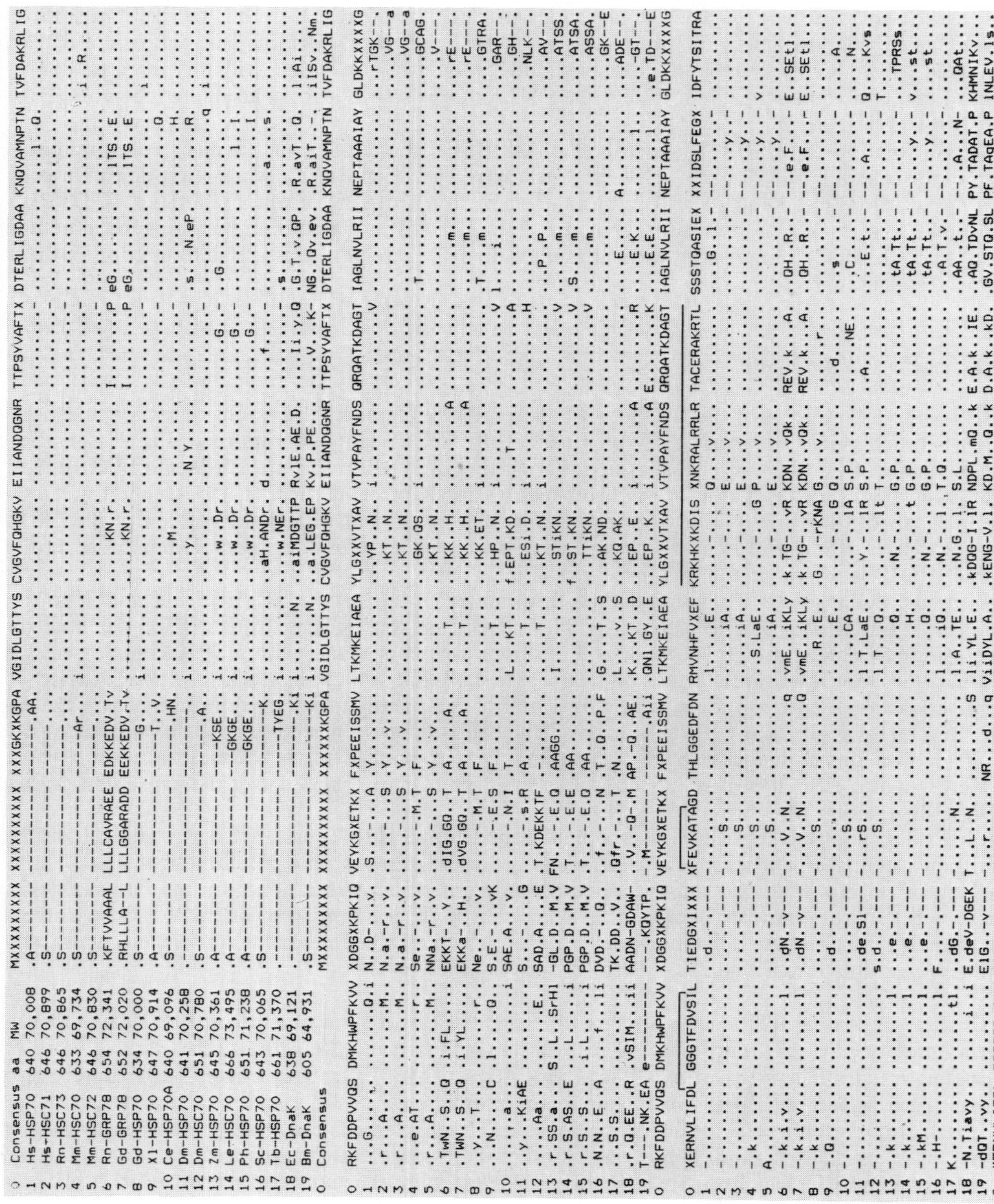

Fig. 8. Sequence comparison between members of the HSP70 family.
For details and explanations see legend to Fig. 6 and text. Sequences are: **1**, human HSP70 (Hunt and Morimoto, 1985); **2**, human HSC71 (Dworniczak and Mirault, 1987); **3**, rat HSC73 (Sorger and Pelham, 1987a; O'Malley et al., 1985); **4**, mouse HSC70 (Zakeri et al., 1988); **5**, mouse HSC72 (Giebel et al., 1988); **6**, rat GRP78 (Chang et al., 1987; Munro and Pelham, 1986); **7**, chicken GRP78 (Stoeckle et al., 1988); **8**, chicken HSP70 (Morimoto et al., 1986); **9**, *Xenopus* HSP70 (Bienz, 1984); **10**, *Caenorhabditis elegans* (nematode) HSP70

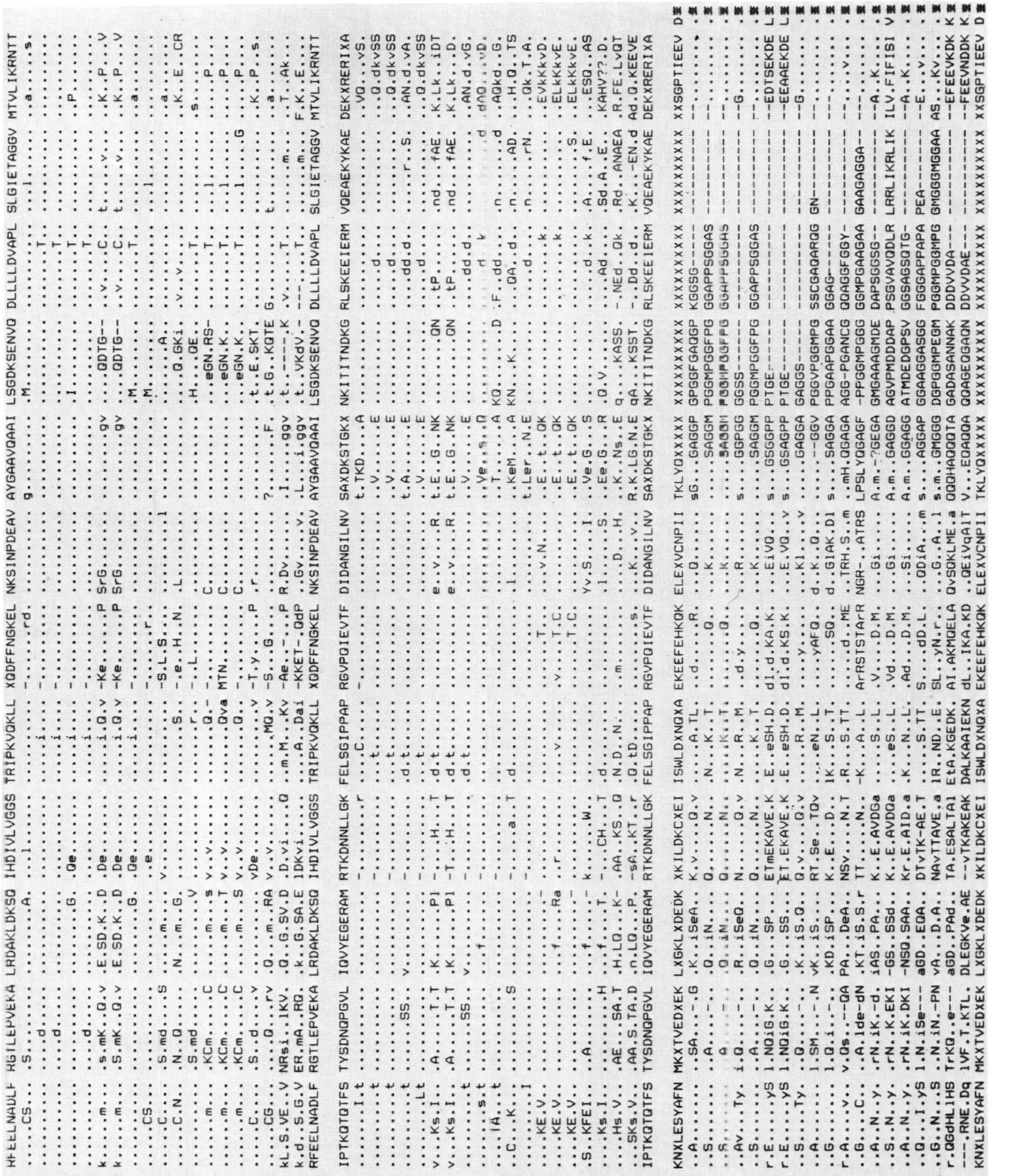

(SNUTCH et al., 1988); **11**, *Drosophila* HSP70 (INGOLIA et al., 1980; KARCH et al., 1981; TÖRÖK and KARCH, 1980); **12**, *Drosophila* HSC70 (A. SLATER and E. A. CRAIG, pers. commun.); **13**, maize HSP70 (ROCHESTER et al., 1986; **14**, *Lycopersicon esculentum* (tomato) HSC70 (J. WINTER, pers. commun.); **15**, *Petunia hybrida* HSP70 (WINTER et al., 1988); **16**, yeast HSP70 (A. SLATER and E. A. CRAIG, pers. commun.); **17**, *Trypanosoma brucei* HSP70 (GLASS et al., 1986); **18**, *Escherichia coli* HSP70 = DnaK (BARDWELL and CRAIG, 1984); **19**, *Bacillus megaterium* HSP70 = DnaK (SUSSMAN and SETLOW, 1987)

Table 2. Protein or RNP aggregates associated with heat shock or related proteins
The far spread association of HSPs with heterologous proteins may indicate a general function for stabilization of abnormal or immature proteins, for protein assembly or intracellular protein transport (for reviews see DESHAIES et al., 1988b, PELHAM, 1986). But in many cases these interactions are not necessarily stress-related.

Compartment	Process (organism[a])	Stress-related	HSP associated or involved	Remarks	References
Plasma membrane	– Formation of intramembrane protein aggregates (V)	yes	?	–	RICE et al., 1987
	– Uncoating of clathrin cages (V)	no	HSC 72/73	Dependent on ATP cleavage	CHAPPELL et al., 1986; UNGEWICKELL, 1985
Cytoskeleton	– Collapse of intermediate filament system and reorganization of microfilaments and microtubules (I, V)	yes/no	HSP90, HSP70, HSC70, ubiquitin	Essential for stress survival, role of HSC70 as interfacing protein	HUGHES & AUGUST, 1982; MURTI et al., 1988; NAPOLITANO et al., 1987; NISHIDA et al., 1986; WIEGANT et al., 1987a, b
ER	– Retention of defective proteins, stabilization of immature subunits of protein complexes in the ER lumen (V)	yes/no	GRP78, GRP94	Dissociation ATP dependent	DORNER et al., 1988; HENDERSHOT et al., 1988; KASSENBROCK et al., 1988; ROTHMAN & SCHMID, 1986; SHARMA et al., 1985
	– Protein transport through ER membranes (Y, V)	no	HSP70, HSC70	ATP required, HSP70 as unfoldase	CHIRICO et al., 1988; DESHAIES et al., 1988a, b; ZIMMERMANN et al., 1988
Nucleus/nucleolus	– Non-histone proteins tightly bound to nuclear matrix, including cellular and viral onc-gene products (V, I, Y)	yes	HSP70/HSC70	Binding of proteins of the HSP70 family released in presence of ATP, shuttle function of HSP70?	BERRIOS & FISHER, 1988; CLARKE et al., 1988; LEWIS & PELHAM, 1985; LITTLEWOOD et al., 1987; MCCONELL et al., 1987; PELHAM, 1984; ROTI ROTI et al., 1986; STÜRZBECHER et al., 1987; VELAZQUEZ & LINDQUIST, 1984
	– Formation of dense nucleoplasmic and nucleolar RNP aggregates (perichromatin granules) (P, V, I)	yes	HSP70	Represent unprocessed precursors of mRNP and rRNP, maturation and removal in recovery period depends on HSP70 (?)	HEINE et al., 1971; NEUMANN et al., 1987; NOVER et al., 1986; YOST & LINDQUIST, 1986; SIMARD & BERNHARD, 1967

Compartment	Process (organism[a])	Stress-related	HSP associated or involved	Remarks	References
Chloroplasts	– Assembly of ribulosebisphosphate carboxylase from small and large subunits (P)	no	HSP60	Function as molecular chaperonin?	Hemmingsen et al., 1988; Roy & Cannon, 1988
	– Aggregation of thylakoid membrane proteins (P)	yes	HSPs20–21	Protection of thylakoid membranes?	Glaczinski & Kloppstech, 1988; Schuster et al., 1988
Mitochondria	– Uptake and cleavage of preproteins (Y)	no	HSP70		Chirico et al., 1988; Deshaies et al., 1988a, b; Murakami et al., 1988
	– Assembly of matrix proteins (Y)	no	HSP60	Molecular chaperonin?	Cheng et al., 1989; Reading et al., 1989
Cytoplasm	– Aggregation of small HSP to heat shock granules, mRNA storage (P)[b]	yes	HSP70 and small HSP	Release of mRNA may require HSP70 and ATP	Nover et al., 1983, 1989
	– Stabilization and shielding of DNA binding site of inactive steroid or dioxin receptor (V)	no	HSP90	HSP90 detached by binding of ligand (hormone, dioxin)	Denis et al., 1988; Mendel et al., 1986; Perdew, 1988; Sabbah et al., 1987; Sanchez et al., 1987; Schuh et al., 1985; Ziemiecki et al., 1986
	– Processing of $pp60^{v\text{-}src}$ and related viral oncgene products (V)	no	HSP90	Transient maturation complex	Brugge et al., 1983; Schuh et al., 1985; Ziemiecki et al., 1986
Bacteria *(E. coli)*	– Accumulation of abnormal proteins by partial denaturation or aggregation cause an increased demand on the proteolytic system	yes	Lon (protease), DnaK (= HSP70), GroE (= HSP60), GrpE	Restoration of proteolytic homeostasis	Straus et al., 1988
	– DNA synthesis of λ-phage, processing of preprimosomal complex		DnaK, GrpE, DnaJ	Release of P-protein requires DnaK and ATP	Georgopoulos et al., 1989; Johnson et al., 1989; Liberek et al., 1988

Compartment	Process (organnism[a])	Stress-related	HSP associated or inolved	Remarks	References
	– Phage head assembly	yes/no[c]	GroE	Morphopoietic complex	GEORGOPOULOS et al., 1989
	– Complexing with RNA polymerase	no	DnaK	Function unclear	SKELLY et al., 1988
	– Assembly of ordered membrane complexes of sn-glycerol-3-phosphate acyltransferase	no	DnaK, GroEL	–	WILKISON & BELL, 1988

[a] Organisms are abbreviated: V, vertebrates; P, plants; I, insects *(Drosophila);* Y, yeast

[b] Similar cytoplasmic aggregates of small HSPs were also detected in *Drosophila* and vertebrate cells (see Sect. B.3.3.)

[c] Phage infection and multiplication certainly generates a situation of stress in the bacterial cell. But there is evidence that all HSPs involved also have vital functions in the normal cell.

by the nuclear pore complex. In many cases investigated in animal cells the signal is a continuous stretch of 5–6 lysine or arginine residues (KLEINSCHMIDT and SEITER, 1988; LANFORD et al., 1986). The only sequence element of this kind in the HSP70 molecule starts at about aa 250 (Fig. 8, bars). Remarkably, the following 25 highly conserved amino acid residues of all eukaryotic HSP70 exhibit striking homology to several nuclear DNA binding proteins. These are transcription factors of mammals and nematodes characterized by the POU homeo box (BODNER et al., 1988; HERR et al., 1988; see example of the human Oct1 protein), the yeast GCN4 transcriptional activator protein and the c-jun oncogene product (VOGT et al., 1987):

HSP70 cons.:	KRKHKK-DISXN-KRALRRLRTACERAKRT	
human Oct1:	RRRKKRTSIETNIRVALEKSF--LENQKPT	(405)
yeast GCN4:	KRARNTEAARRSRARKLQRMKQLEDKVEEL	(260)
avian c-jun:	KRMRNRIAASKSRKRKLERIARLEEKVKTL	(252)

Referred to the HSP70 consensus sequence identical or homologous amino acid residues are underlined. There are 15 out of 28 for Oct1, 14 out of 30 for GCN4 and 17 out of 30 residues for c-jun. The significance of this homology, extending beyond the putative nuclear pore recognition site, remains to be established. But it is worth noticing that sequence conservation in this region is much less for the four non-nuclear members of the HSP70 family included in Fig. 8. These are the vertebrate GRP78 (nos. 6, 7) and the bacterial DnaK proteins (nos. 18, 19).

4) Assembly of protein complexes: GRP78 (BiP) in the rER of vertebrate cells forms tight complexes with immature or abnormal proteins. It may participate in the proper assembly of multimeric proteins from the precursor subunits, e.g. of immunoglobulins in B-lymphocytes (DORNER et al., 1988; KASSENBROCK et al., 1988; MUNRO and PELHAM, 1986, 1987; SHARMA et al., 1985).
5) Protein sorting and temperature sensitive growth: In yeast, deletion of the four hsp70 genes of the hs-inducible SSA subfamily led to cells unable to grow at higher temperatures and defective in protein sorting to mitochondria, vacuoles and the extracellular milieu. In contrast, deletion of the two constitutively expressed SSB genes resulted in cold sensitive strains (CHIRICO et al., 1988; CRAIG et al., 1987; DESHAIES et al., 1988a, b; WERNER-WASHBURNE et al., 1987). A role for protein import into dog pancreas microsomes was also reported for rat HSC70 (ZIMMERMANN et al., 1988).
6) Proteolytic activity: An autoproteolytic activity of HSP70-type proteins in *Drosophila* and mammalian cells was reported by MITCHELL et al. (1985) and UNGEWICKELL (1985).
7) Autorepression of HSP synthesis: HSP synthesis is evidently self-limiting. Though proof for a di-

rect role in autorepression is lacking, there is evidence for *Drosophila* (Di Domenico et al., 1982a, 1982b), yeast (Craig et al., 1989) and vertebrates (Mizzen and Welch, 1988) that the HSP70 level may serve as indicator of the stress state. But in *Escherichia coli* autorepression clearly involves HSP70 (DnaK), which affects the level of the hs transcription factor (sigma[32]; Tilly et al., 1983; Grossman et al., 1987; Paek and Walker, 1987).

8) Bacterial DNA replication: The DnaK protein and other HSPs of *E. coli* were characterized as essential component of the phage DNA replication complex. But at elevated temperatures, it is also required for DNA and RNA synthesis in general (Dodson et al., 1986; Georgopoulos et al., 1989; Paek and Walker, 1987; Zylicz et al., 1983).

A key for an understanding of most of the stress and non-stress related functions of proteins of the HSP70 family, compiled under items 1 to 5 (see also Table 2), is their capability for ATP binding, which can be used for affinity purification (Welch and Feramisco, 1985; Beaulieu and Tanguay, 1988). The binding site is in the N-terminal part (Chappell et al., 1986; Munro and Pelham, 1984). It is tempting to speculate that a highly conserved region starting at about aa residue 200 may be relevant in this respect (see brackets in Fig. 8). There is a definite homology with the ATP binding sites of members of the protein kinase family (Hannink and Donoghue, 1985):

```
                   L    F    Y              V   LK
PK, consensus       GXGX GXV  -7to15aa-VA  KX
                   I    Y    W              I   IR

HSP70, consensus  LGGGTFDVSI-7aa----FEVKATA
```

ATPase activity of HSP70 depends on the association with other proteins, and HSP70 release from stress-induced nuclear binding or its dissociation from other protein complexes can be stimulated *in vitro* by addition of ATP (Clarke et al., 1988; Collier and Schlesinger, 1986; Kassenbrock et al., 1988; Lewis and Pelham, 1985; Rothman and Schmid, 1986). Thus stabilization of unassembled protein precursors, of unprocessed preribosomes, or of partially denatured nuclear and cytoplasmic protein complexes as well as their release, proper processing or degradation may all depend on the same ATP consuming activity of HSP70 (Pelham, 1984, 1986). However, many of these functions are not necessarily stress-related (Table 2), and at this state of our knowledge it is certainly premature to refer to HSP70 proteins as “unfoldase”.

B.1.2.4. The HSP60 family

Proteins of this type are constitutive components of mitochondria and chloroplasts. They were originally identified as a hs-induced mitochondrial protein of *Tetrahymena thermophila* (McMullin and Hallberg, 1987). But the chloroplast homolog was well known before as the Rubisco subunit binding protein (Barraclough and Ellis, 1980; Musgrove and Ellis, 1986). It is presumably implicated in the post-translational assembly of the ribulose-1,5-bisphosphate carboxylase from preformed large and small subunits (Musgrove and Ellis, 1986; Roy and Cannon, 1988). Eukaryotic HSP60 proteins are homologous to the major *E. coli* HSP, the GroEL protein (Fig. 9; McMullin and Hallberg, 1987, 1988; Hemmingsen et al., 1988). This homology is stressed by the finding that GroEL itself can also promote assembly of Rubisco subunits in *E. coli* transformed with the *Anacystis nidulans* Rubisco genes (Goloubinoff et al., 1989). The role of HSP60 as ‘molecular chaperonin’ (Hemmingsen et al., 1988) is supported by analyses of a ts-mutant in yeast. At the non-permissive temperature (37 °C), the altered HSP60 complex is unable to maintain the assembly of intramitochondrial oligomeric protein complexes, but the import of precursor proteins through the mitochondrial membrane is not affected (Cheng et al., 1989; Reading et al., 1989).

Similar to the bacterial GroEL, the eukaryotic HSP60 form oligomeric complexes with ATPase activity and 12 nm hollow core morphology (McMullin and Hallberg, 1988; Hemmingsen et al., 1988). In contrast to the *Tetrahymena* HSP60, there is no evidence for an increased synthesis of the chloroplast HSP60 in heat-shocked plant cells. But so far unidentified plant HSPs in this size range can be detected in several objects (see arrow heads in Figs. 4, 5A). They were also found in chloroplast fractions of *Vigna sinensis* and *Sorghum vulgare* (Krishnasamy et al., 1988) and in mito-

```
0 CONSENSUS AA  MW     MXXXXXXXXX XXXXXXXXXX XXXXXXAKXI KFDXDARXXL XRGVNXLADA VKVTLGPKGR NVVLEKSWGA
1 Hs-HSP60  573 61,049 .LRLPTVFRQ MRPVSRVLAP HLTRAY..Dv ..GA...ALm LQ..DL.... .A..m..... T.ii.Q...S
2 Ta-HSP60  543 57,516 G--------- ---------- ----AD..E. A..QKS.AA. QA..EK..N. .G.....r.. ....dE-y.N
3 Sc-HSP60  572 60,830 .LRSSVVRSR ATLRPLLRRA YSS---H.El ..GVeG.AS. Lk..ET..e. .Aa....... ..li.QPf.P
4 Mt-GroEL  540 56,751 .--------- ---------- ------..T. Ay.Ee..RG. E..l.A.... .......... ......K...
5 Ml-GroEL  540 56,695 .--------- ---------- ------..T. Ay.Ee..RG. E..l.S.... .......... ......K...
6 Ec-GroEL  547 57,140 .--------- ---------- -----A..Dv ..GN...VKm L....V.... .......... ....d..f..
7 Cb-htpB   552 58,249 .--------- ---------- -----A..Vl ..SHevLHAm S...EV..N. .......... ....d..f..

0 PTITNDGVSI AKEIELEDKY ENIGAXLVKE VAXKTNDXAG DGTTTATVLA QAIIKEGLRX VAAGANPVXL KRGIEKAVXA VIEELXKXAK
1 .Kv.K...tv ..S.d.K... K....K..Qd ..NN..eE.. .......... RS.A...FEK iSK.....Ei r..vML..D. ..A..K.QS.
2 .KvV....t. .rA...ANPM ..a..A.ir. ..S....S.. ......C... RE...L.iLS .TS.....S. .k..d.T.QG l....ErK.r
3 .K..K...TV ..S.V.K..f ..m..K.lQ. ..S...eA.. ....s....G R..FT.SvkN ....c..mD. r..SQV..EK ...F.SANK.
4 .......... ........P. .K...E.... ..K..D.V.. .......... ..lR.....N .......lG. ........EK .T.T.L.G..
5 .......... ........P. .K...E.... ..K..D.V.. .......... ..lv.....N .......lG. ........DK .T.T.L.D..
6 ....K....v .r.......f ..m..Qm... ..S.A..A.. ..NN...... ....T...KA ....m..mD. ....d...T. av...KALSV
7 ....K....v .........f ..m..Qm... ..Sr.S.D.. .......... ...lVd.ikA .i..m..mD. ....d...T. avA..K.IS.

0 PVXTXEXIAQ VATISAXGDX SIGXLIAEAM DKVGXEGVIT VEDGXTFXXX LELXEGMXFD RGYISPYFVT XXEXXXAELE XPYILLXDKK
1 ..T.P.E... ......N..K E..Ni.Sd.. K...RK.... .K..K.LNDE ..iI...K.. .........iN TSKGQKC.FQ DA.v..Se..
2 ..KGSGD.KA ..s...GN.E L..Am..d.i ....Pd..ls i.SSSs.ETT vdvE...EI. ......Q... NL.KSIv.F. NARvliT.Q.
3 EiT.S.E... ......N..S Hv.K.l.S.. e...K..... iRe.R.LEDE ..vT...R.. ..f.....i. DPKSSKv.F. K.Ll..Se..
4 E.E.K.Q..A T.A...-..Q ...D...... ....N..... ..eSN..GLQ ...T...R.. k....G.... DP.RQE.V.. D.....VSS.
5 E.E.K.Q..A T.A...-..Q ...D...... ....N..... ..eSN..GLQ ...T...R.. k....G.... DA.RQE.V.. E.....VSS.
6 .CSDSKA... .G....TS.E tv.K...... ....K..... ....TGLQDE .dvV...Q.. ...l....iN KP.TGAv... S.f...A...
7 .CKDQKA... .G....NS.K ...Di..... e...K..... ....SGLENA ..vV...Q.. ...l....iN NQQNMS.... N.f...V...

0 ISSIKDILPL LEXVXXAGKP LLIIAEDVEG EALATLVVNK IRGXVKVVAV KAPGFGDRRK AMLQDMAILT GGXVISEEXL GLXLENAXLX
1 ....QS.v.a ..IaNAHR.. .v......d. ...S...l.r lkVGlQ.... ........N.. NQ.K....a. ..A.FG..G. T.N..DvQPH
2 .t...e.i.. ..QTTQLRC. .F.v...iT. .......... l..IiN.a.i ...S..e... .v...i..v. .AEYlAKd-. ..Lv...TvD
3 ....Q....a ..ISNQSRr. ........d. ....ACil.. l..Q...c.. ........N.. NTiG.i.v.. ..T.Ft..-. D.KP.qCTiE
4 v.tv..l... ..K.IG.... .......... ...S...... ...TF.S... .......... .......... ..Q.....-v ..T....D.S
5 v.tv..l... ..K.IQ...S .......... ...S...... ...TF.S... .......... .......... .AQ.....-v ..T...TD.S
6 ..N.rem..v ..A.AK.... .......... ......a..T ...I...a.. .......... .....i.T.. ..T.....-i .mE..K.T.E
7 ..N.reli.. ..N.AKS.r. ..v....i.. .........N ...V...a.. .......... .....i.v.. ..K.....-v ..S..A.S.D
```

```
0 XLGKARKVVV TKDDTTIVEG AGDXDAIXGR VAQIRXEIEX SDSDXYDREK LQERLAKLAG GVAVIKVGAA TEVELKERKX RIEDALXATR
1 D...vGE.i. ....AMllK. K..KAQ.EK. iQE.IEQldV tT.e-.ek.. .N......SD ....l...GT sd..vN.k.D .vT...N...
2 Q..T...iTi HQTT..liAD .ASK.E.QA. ...lkK.lSE t..I-..S.. .A..i...S. .........T ..T..Ed.QL .....KN..F
3 N..SCDSiT. ..e..v.lN. S.PKe..QE. iE..kGS.dI tTtNS.ek.. ........S. .....r..G. s...vG.k.D .Yd...N...
4 L......... ...e...... ...T...A.. ......Q..N ....-..... .......... ......a... .........H .....vRNAk
5 L........m ...e...... ...T...A.. ......T..N ....-..... .......... ......a... .........H .....vRNAk
6 D..Q.kr..i N..T....id. v.eEA..Q.. .....QQ..E AT..-..... ....v..... .......... ....m..k.A .v....H...
7 D..S.kr... .......id. S..AGD.KN. .E...K..EN .S..-..k.. .......... .......... ....m..k.A .v....H...

0 AAVEEGIVAG GGVALLXXXX XAPALDXLXX XNXDXXVGAX IIKRALEAPL XQIAXNXGVE PXVVAEKVXX XXXXXXXXGY NAXTGEYEDL
1 ........l. ..C...R--- Ci....S.TP A.E.QKi.iE ....T.Ki.A MT..K.A... GSliv..iMQ S---SSEV.. D.MA.dfVNm
2 ..i.....P. ..a.YVHLST Yv..iKE-TI EDH.ERl..D ..Qk..Q..A SL..N.A... GE..i..iKE S---EWEM.. ..M.DK..N.
3 .......lP. ..T..vK--- ASRv..EvVV D.F.QKl.vD ..r..iTR.a K..iE.A.E. GS.iiG.lID EYGDDFAK.. D.SKS..T.m
4 .......... ...T..Q--- A..T..E.K- LEG.EAT..N .v.V...... K...F.S.l. .G......RN L---PAGH.L ..Q..V....
5 .......... ...T..Q--- A.....K.KL TGDeAT-..N .v.V...... K...F.S.m. .G......RN L---SVGH.L ..A.......
6 ......v... .....iR--- V.SK.AD.RG Q.E.QN..iK vaL..m.... R..vL.C.E. .S...NT.KG G---DGNY.. ..A.E..GNm
7 ......v.P. .....iRVLK S---..SvEV E.E.QR..vE .ar..mAYP. S..vK.T..Q AA...d..LN H--KDVNY.. ..A....G.m

0 LXXGVIDPXK VTRSALQNAA SVAGLMLTTE AVVXDKPKXX XAXAXGAXXM GGMGGGMGGM GGMX*
1 vEK.i...T. .V.t..LD.. G..S.lT.A. V..TeI..EE KDPGM..--. .......GG- ...F*
2 iES.....A. ...C...... ..S.mv...Q .i.Ve...PK PKV.EP.EGQ l--------- --SV*
3 .AT.i...F. .V..G.VD.S G..S.la... vaiV.A.EPP A.A--..--- ...P...P.. P..M*
4 .AA..a..V. .......... .i...F.... ...A...EKE K.SvP.GGD. ...------- --DF*
5 .KA..a..V. .......... .i...FT.-. ...A...EKT A.P.SDPT-- .....-.--- --DF*
6 iDM.il..T. .......Y.. ......i... Cm.T.L..ND -.ADL..AG- -....-.... ...M*
7 iEM.il..T. ...t...... .i....i... Cm.TeA..KK EESmP.GGD. .....-.... ...M*
```

Fig. 9. Sequence comparison of 7 members of the HSP60 family.
For explanations and details of presentation see legend to Fig. 6 and text. Sequences are from the following sources: **1**, human mitochondrial HSP60 (Jindal et al., 1989); **2**, wheat chloroplast HSP60 = ribulose-bisphosphate carboxylase subunit binding protein (Hemmingsen et al., 1988); **3**, yeast mitochondrial HSP60 (Reading et al., 1989); **4–7**, bacterial GroEL proteins from *Mycobacterium tuberculosis* (**4**, Shinnick et al., 1987), from *M. leprae* (**5**, Mehra et al., 1986), from *E. coli* (**6**, Hemmingsen et al., 1988) and from *Coxiella burnetii* (**7**, Vodkin and Williams, 1988). The N-terminal extensions of 1 and 3 presumably represent the tags for transition into mitochondria

chondria of maize and cabbage (NEBIOLO and WHITE, 1985; SINIBALDI and TURPEN, 1985). In both cases there is evidence that the proteins are synthesized within the organelles (see Sect. B.5.2.). This contrasts to the results of HEMMINGSEN and ELLIS (1986) on the two related subunits with Mr 61 and 60 kDa of the Rubisco subunit binding proteins, which are nuclear encoded, synthesized on cytoplasmic ribosomes and imported into chloroplasts. Thus, it remains to be shown, whether both types of plant HSPs with Mr 52-61 kDa belong to the HSP60 (GroEL) family. So far, this was only verified by sequencing of the 61 kDa α-subunits of the wheat and castor bean Rubisco binding protein (HEMMINGSEN et al., 1988; see Fig. 9). At least in the case of the human and yeast mitochondrial HSP60, there is clear evidence for the synthesis of a cytoplasmic precursors with N-terminal extensions of 22-26 aa residues (see Fig. 9).

B.1.2.5. The HSP20 family

Unlike the other families, the HSP20 family is much more variable with respect to size, number of representatives and primary sequence. There is a single low molecular weight HSP in yeast (HSP26, PETKO and LINDQUIST, 1986), chicken (HSP24, COLLIER and SCHLESINGER, 1986) and mammals (HSP27, GAESTEL et al., 1989; HICKEY et al., 1986). In *Drosophila melanogaster* genes of the 4 prominent small HSPs (HSPs 22, 23, 26 and 27) are clustered together with three additional hs-inducible genes in a single chromatin domain 67B (AYME and TISSIERES, 1985; PAULI and TONKA, 1987). Four closely related hsp16 genes were characterized in the nematode *Caenorhabditis elegans* (JONES et al., 1986; RUSSNAK and CANDIDO, 1985), whereas 8-10 small HSPs were detected in the slime mould *Dictyostelium* (Mr 30 kDa, LOOMIS and WHEELER, 1980), in the unicellular ciliate *Tetrahymena* (AMARAL et al., 1988; GUTTMAN et al., 1980), or in *Xenopus laevis* cells (DARASCH et al., 1988).

```
1 CONSENSUS    AA  MW      MSLIPSFFGG RRSNVFDPFS LDVWDPFKDF XFPNTXSSAS
2 Gm-HSP17.9D 159 17,862 .DFRVMGLES PLFHTLQHMM DMSE.GAG.N KTH.A-PtW.
3 Gm-HSP18.5  161 18,486 .....N.... ..N....... .......L.. P....L....
4 Gm-HSP17.5E 154 17,535 .....G.... .......... ..m....... HV.-.-..v.
5 Gm-HSP17.5M 153 17,546 ......I... .......... .......... H..-.--.l.
6 Gm-HSP17.6L 154 17,572 ......I... P......... ..m....... HV.-.-..v.
7 Gm-HSP17.3B 153 17,347 .......... ...S...... .......... P..---..l.
8 Ps-HSP17    158 18,100 .......S.. .......... .....L.... P.S.sSL...

1  FXXXXAENSA FVSTRVDWKE TPEAHVFKAD IPGLKKEEVK VQIEDDRVLQ ISGERNVEKE
2  y---VRdAK. MaA.Pa.V.. y.NSy..Ei. m....SGdi. ..v...Nl.L .c...LRdE.
3  .PEFSR.... .......... .......... .......... ......k... ..........
4  -----..... .......... .......... .......... .......... ..........
5  -----..... ..N....... .......E.. .......... .......... ......l...
6  -----..... ..N....... .Q...L.... .......... .......... ..........
7  -----..... .......... .......... .......... lE.Q.G.... ..........
8  .P---R..P. .......... .......... l......... .Ev....... .....S....

1  DKNDTWHRVE RSSGKFMRRF RLPENAKVEQ VKASMENGVL TVTVPKEEVK KPDVKAIEIS G*
2  KEGAKyL.m. .RV..L..k. V.....NTdA iS.VCQD... s...Q.LPPP E.KKPRTIQV KVA*
3  .......... .......... .......... .......... .......... .......... .*
4  .......... .....T.... ........NE .......... .......... .......... .*
5  .......... ....N..... .......... .......... .......... .......... .*
6  .........d .......... .......... ...c...... ...i...... .S...P.... .*
7  .......... .....Lv... ........d. .......... ........i. .......d.. .*
8  ....E..... .....l.... .......mdK .......... ........i. .Ae..S.... .*
```

Fig. 10. Sequence conservation of HSP17 and related plant heat shock proteins.
For explanation of the presentation see legend to Fig. 6. Sequences of soybean **(Gm)** and pea **(Ps)** HSPs are from the following sources: HSP17.9 D and HSP18.5 (RASCHKE et al., 1988); HSP17.5E (CZARNECKA et al., 1985); HSP17.5M and 17.5L (NAGAO et al., 1985); HSP17.3B (SCHÖFFL et al., 1984a) and pea HSP17 (E. VIERLING, personal commun.)

Contrasting to all other organisms including green algae and fungi (Fig. 2), the HSP20 family of higher plants is more complex with 15-20 major and minor proteins (Figs. 3-5). As a result of the extensive studies of the soybean genes by J. KEY anf F. SCHÖFFL and coworkers, there are at least four subfamilies. They were originally identified by hybrid release translation using cDNA clones (SCHÖFFL and KEY, 1982, 1983) and later on confirmed by sequencing (CZARNECKA et al., 1985, 1988; NAGAO et al., 1985; RASCHKE et al., 1988; SCHÖFFL et al., 1984a; VIERLING et al., 1988; see also summaries by NAGAO et al., 1986 and SCHÖFFL et al., 1986). According to these results the plant HSP20 multigene family is composed of two closely related subfamilies I and VI (SCHÖFFL and KEY, 1982, 1983) each coding for 9-12 HSPs in the 17-18 kDa region. A third subfamily (IV) comprises one or two genes for 20-23 kDa proteins, which are localized in chloroplasts (VIERLING et al., 1988, 1989; see Sect. B.5.2.). Recently a fourth type of small HSP with Mr 26 kDa was identified from soybean (CZARNECKA et al., 1988; for sequence see Fig. 11). This first intron-containing small HSP gene of plants is induced by heavy metals, osmotic stress, abscisic acid, 2,4-D and by hs. The protein probably represents a special type of general stress protein. But the existence of similar proteins in other plants remains to be shown.

Compared to the other four HSP families, sequence conservation between members of the HSP20 family is low. The data are summarized in Figs. 10 to 13. The high homology observed between representatives of the HSP17 subfamilies I and VI (Fig. 10) may be misleading, because so far only

```
A:Cr-HSP20 (157 aa; MW 16,702)
     MALSNYVFGN SAADPFFTEM DRAVNRMINN ALGVAPTSAG KAGHTHAPMD IIESPTAFEL
     HADAPGMGPD DVKVELQEGV LMVTGERKLS HTTKEAGGKV WRSERTAYSF SRAFSLPENA
     NPDGITAAMD KGVLVVTVPK REPPAKPEPK RIAVTGA

B:Ps-HSP21 (232 aa; MW 26,176)
     MAQSVSLSTI ASPILSQKPG SSVKSTPPCM ASFPLRRQLP RLGLRNVRAQ AGGDGDNKDN
     SVEVHRVNKD DQGTAVERKP RRSSIDISPF GLLDPWSPMR SMRQMLDTMD RIFEDAITIP
     GRNIGGGEIR VPWEIKDEEH EIRMRFDMPG VSKEDVKVSV EDDVLVIKSD HREENGGEDC
     WSRKSYSCYD TRLKLPDNCE KEKVKAELKD GVLYITIPKT KIERTVIDVQ IQ

C:Nc-HSP30 (228 aa; MW 25,262)
     MALFPRGFYG SYGSDPSFTN LFRLLDDFDT YTREVQGSAP ETGSRRHTQP TRTFSPKFDV
     RETEQTYELH GELPGIDRDN VQIEFTDPQT IVIRGRVERN YTAGTPPAQV AGVLTEKGEP
     HSPAAHHATV EDDVDEDNRS VATTATGANN QNNQQVAQRA SAPTTEEKPK APAEKYWVSE
     RSIGEFSRTF NFPGRVDQNA VSASLNNGIL TITVPKAKKH ETIRIAIN

D:Gm-HSP26A (225 aa; MW 25,946)
     MAATQEDVKL LGIVGSPFVC RVQIALKLKG VEYKFLEENL GNKSDLLLKY NPVHKKVPVF
     VHNEQPIAES LVIVETIDET WKNNPILPSD PYQRALARFW SKFIDDKIVG AVSKSVFTVD
     EKEREKNVEE TYEALQFLEN ELKDKKFFGG EEFGLVDIAA VFIAFWIPIF QEIAGLQLFT
     SEKFPILYKW SQEFLNHPFV HEVLPPRDPL FAYFKARYES LSASK

E:Ml-HSP18 (148 aa; MW 16,705)
     MLMRTDPFRE LDRFAEQVLG TSARPAVMPM DAWREGEEFV VEFDLPGIKA DSLDIDIERN
     VVTVRAERPG VDPDREMLAA ERPRGVFNRQ LVLGENLDTE RILASYQEGV LKLSIPVAER
     AKPRKISVDR GNNGHQTINK TAHEIIDA

F:Hv-UBIQUITIN (76 aa; MW 8,564)
     MQIFVKTLTG KTITLEVESS DTIDNVKAKI QDKEGIPPDQ QRLIFAGKQL EDGRTLADYN
     IQKESTLHLV LRLRGG
```

Fig. 11. Sequences of small heat shock proteins.
Only sequences of those small HSPs are included that do not belong to the HSP17 subfamily (see Fig. 10). Data are from the following sources. **A**, *Chlamydomonas reinhardii* chloroplast HSP20 (GRIMM et al., 1989); **B**, pea chloroplast HSP21 (VIERLING et al., 1988); **C**, *Neurospora crassa* mitochondrial HSP30 (N. PLESOVSKY-VIG and BRAMBL, pers. commun.); **D**, soybean stress protein HSP26A (CZARNECKA et al., 1988); E, *Mycobacterium leprae* P18 (NERLAND et al., 1988); **F**, barley ubiquitin, which differs only in three positions form the human ubiquitin (GAUSING and BARKARDOTTIR, 1986; GRAHAM et al., 1989)

genes from two closely related species were sequenced (soybean, pea). At least, immunological data demonstrate that cross-reactivity between HSP17 proteins from monocots and dicots is low (see Sect. B.4.3.3.).

Sequences of three organellar proteins of the HSP20 family are included in Fig. 11. These are the pea HSP21, the *Chlamydomonas* HSP22 and a HSP30 of *Neurospora crassa* mitochondria. Only for the pea HSP21 gene, sequences for an N-terminal transit peptide could be defined. The evident lack of similar sequence elements in the other two organellar proteins is unexplained.

Search for common sequence elements between different types of HSP20 proteins leads to results presented in Figs. 12 and 13. A mosaic of various homology regions is recognized, which are found

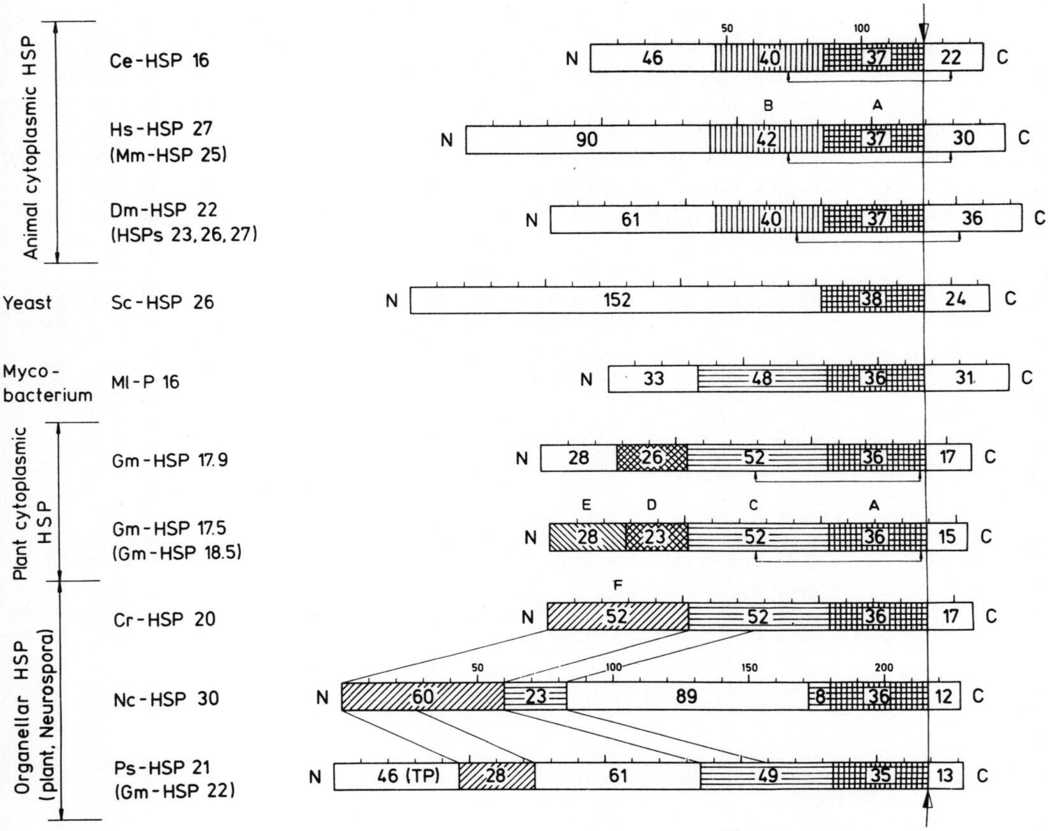

Fig. 12. Homology elements of small heat shock proteins from animals, microorganisms and plants. The C-terminal border of the only sequence element of 35–38 amino acid residues conserved in all small HSPs (segment A) was used for alignment. Other homology elements are marked by the type of shading. Element B and C form the N-terminal part of the homology region in animal and plant genes respectively (for details see Fig. 13). Other sequence elements are specific for the small cytoplasmic HSPs of plants (D, E) or for the organellar HSPs of *Chlamydomonas, Neurospora* and plants (segment F). Gene-specific sequences at the N- and C-termini are indicated by open boxes. The same is true for inserts of 93 and 71 amino acid residues in the mitochondrial HSP30 of *Neurospora* and the pea chloroplast HSP21 respectively. TP, transit peptide element of the pea pre-HSP21. Brackets indicate protein regions coding for the consecutive domains of predominantly hydrophilic, hydrophobic and hydrophilic aa residues respectively (see summaries by Lindquist and Craig, 1988; Nover, 1990, Schöffl et al., 1986). Sequences of members of the HSP20 family are from the following sources: Ce-HSP16 (Jones et al., 1986; Russnak and Candido, 1985), Hs-HSP27 (Hickey et al., 1986), Mm-HSP25 (Gaestel et al., 1989), Dm-HSPs 27, 26, 23, 22 (Southgate et al., 1983), Ml-HSP18 (Nerland et al., 1988), Sc-HSP26 (Susek and Lindquist, pers. commun.); Gm-HSPs18.5 and 17.9. (see legend to Fig. 10); Nc-HSP30 (Plesovsky-Vig and Brambl, personal commun.); Cr-HSP22 (Grimm et al., 1989); Ps-HSP21 (Vierling et al., 1988)

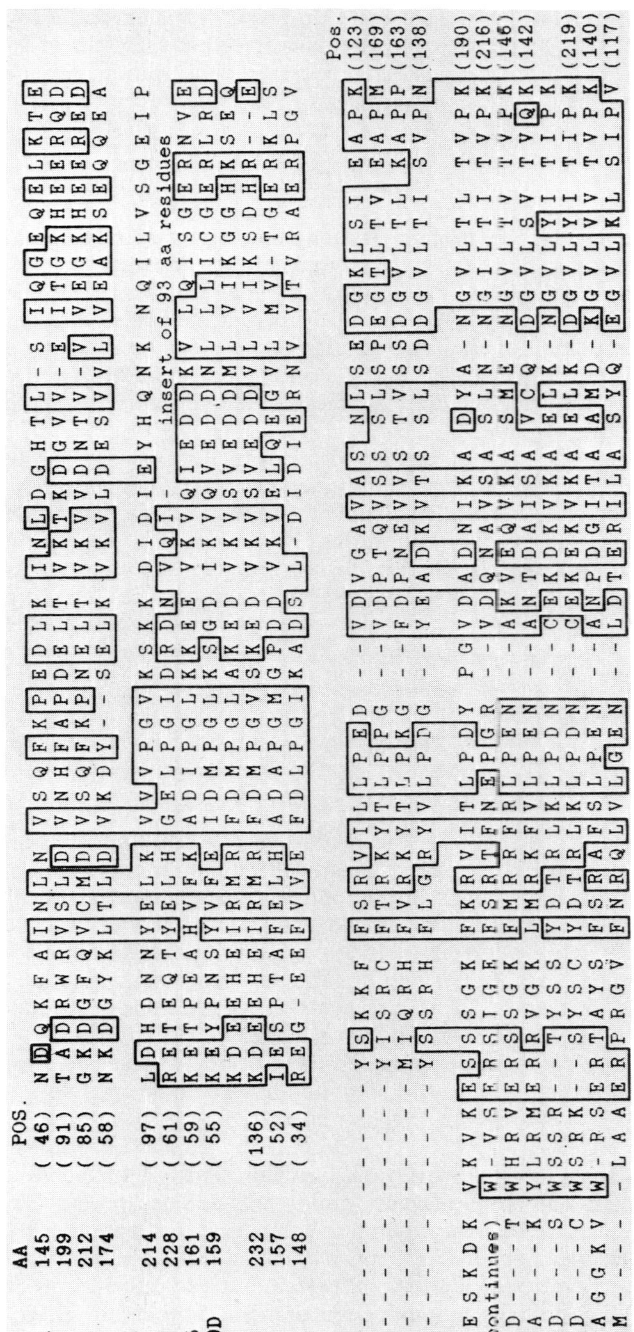

Fig. 13. C-terminal homology of HSP20 proteins.
Sequences compared represent the homology boxes. B/A and C/A marked in Fig. 12. In some cases only selected representatives of a group of closely related proteins were included. Exact positions of the homology regions within the indicated protein are given at the beginning and end of the sequences. Identical or homologous amino acid residues are boxed. In the first part of the homology region (boxes C and B respectively of Fig. 12), the group of animal HSP20 (1–4) differs in several positions from that of microorganisms, plants and the bacterial HSP18 (5–12). This part is separated from the general homology box A by a variable linker of 2–15 amino acid residues. References are given in legend to Fig. 12

in all HSP20 proteins (region A and the two related regions B and C in Fig. 12) or in characteristic subgroups (regions D, E, F). They are flanked or interrupted by non-conserved parts. Details of sequences especially in the short region A (Fig. 13) show a remarkable degree of conservation between bacterial, plant and animal small HSPs in their C-terminal parts.

B.1.2.6. The HSP8.5 (ubiquitin) family

Attachment of ubiquitin via an isopeptide bond is the primary reaction of protein catabolism in eukaryotes. Ubiquitin is a highly conserved, low molecular weight protein with 75–76 amino acid residues (Fig. 11; MW 8,500, see reviews by Rechsteiner, 1987, Schlesinger and Bond, 1987). The hs response, evoked by hyperthermic conditions or by many chemical stressors, coincides with imbalances in proteolysis.—Transient accumulation of abnormal or defective proteins is evidently a key event in the signal transformation process probably connected with the activation of hs genes (see summary by Nover, 1990, Chapter 1). Heat shock-induced synthesis of ubiquitin may serve the restoration of the homeostasis of protein degradation.

Characterization of the ubiquitin multigene families of chicken (Bond and Schlesinger, 1986; Bond et al., 1988), yeast (Finley et al., 1987; Özkaynak et al., 1987; Tanaka et al., 1988) and *Drosophila* (Lee et al., 1988) led to the discovery of hs-inducible polyubiquitin genes, whose translation products are processed to the monomeric proteins. But ubiquitin as HSP8.5 was also reported in *Xenopus laevis* (Krone and Heikkila, 1987), *Trypanosoma cruzi* (Swindle et al., 1988) and plants (B. Atkinson, personal commun.). In fact, ubiquitin may have a vital function for stress tolerance and recovery. Deletion of the yeast polyubiquitin gene leads to stress-hypersensitive strains unable to survive hs or starvation conditions (Finley et al., 1987).

B.2. Control of gene expression

B.2.1. Chemical stressors inducing HSP synthesis

At the beginning of the molecular analysis of the heat shock response was the fortuitous observation of Ritossa (1962) that puffing, i.e. the gene activity pattern of polytene chromosomes of *Drosophila* larvae, drastically changed after a short heat shock. But he found also that almost identical effects were brought about by treatment with chemical stressors, like 2,4-dinitrophenol or salicylic acid. Meanwhile, the lists of biologically active compounds inducing HSP synthesis comprises more than 100 entries including respiratory drugs, heavy metals, sulfhydryl reagents, amino acid analogues, ionophores, inhibitors of gene expression, lipotropic agents, steroid hormones, wounding, anoxia, nutrient starvation, etc. (for a recent summary see Nover; 1990).

Though details are far from clear for most of them, it seems appropriate to classify these chemical stressors with respect to their mode of action into hs-like and non-hs-like inducers (Nover, 1990):

1. Heat shock-like inducers act on the hs-specific signal transformation chain. This is probably true for many respiratory drugs, heavy metals, sulfhydryl reagents, amino acid analogs, puromycin, ethanol, anoxia and recovery and others. Characteristically, these compounds or treatments induce the whole set of HSPs in a broad range of organisms. Central to the hs-specific signal transformation is evidently the transient accumulation of abnormal proteins caused either by hs or drug-induced denaturation of preexisting proteins by production of 'xeno-proteins' in genetically engineered cells or by translation of nonfunctional proteins/protein fragments in the presence of amino acid analogs or puromycin. In several cases the identity of the inducing mechanism for hs and chemical stressors was directly verified by defining the promoter elements involved (see Sect. B.2.4.) or by demonstrating the activation of the hs transcription factor. This is true for abnormal proteins, arsenite, cadmium ions, canavanine, azetidine carboxylic acid, 2,4-dinitrophenol, puromycin and salicylate tested in bacteria, *Xenopus* oocytes, *Drosophila* or

vertebrate cells (ANANTHAN et al., 1986; GOFF and GOLDBERG, 1985; KAY et al., 1986; MIRAULT et al., 1982; MORIMOTO et al., 1989; MOSSER et al., 1988; ZIMARINO and WU, 1987).

2. In contrast to the former group, the action of non-hs-like inducers is more restricted to few organisms or even to certain tissues within an organism only. Depending on the promoter structure and competence state, only some representatives of the total set of HSP are affected. Typical examples of this type of inducers are the steroid hormones ecdysone or teratogens in *Drosophila* (AYME and TISSIERES, 1985. BOURNIAS-VARDIABASIS and BUZIN, 1986; MASON et al., 1984), cell cycle factors, infection with DNA tumor viruses or wounding in vertebrates (IMPERIALE et al., 1984; MILARSKI and MORIMOTO, 1986; SIMON et al., 1987; WU et al., 1987a) and cell cycle arrest or nutrient starvation in eukaryotic microorganisms (see Sect. C.8.). In some cases promoter elements were defined that are clearly different from the heat shock element. Promoters of some hs genes with several regulatory elements are complexly responding to different binding proteins (for examples see Sect. B.1.2.1. and the summary by NOVER, 1990).

The situation for plants is more complex and not extensively studied. On the one hand, well characterized stress situations inducing HSPs in cells of microorganisms and animals (wounding, nutrient, starvation, anoxia) are characterized by different sets of stress proteins in plants (WSPs, SSPs, ASPs, see Sects. C.5., C.8. and C.9.). On the other hand, because of the high general toxicity of chemical stressors, uptake of amino acid precursors and their incorporation into proteins are drastically reduced. Meaningful experiments require a preinduction protocol, i.e. analysis of the synthesis of HSPs in the recovery period following the removal of the stressor. On the basis of the coordinate induction of HSP70 and HSP17, the following hs-like inducers were found effective in tomato cell cultures: arsenite, Cd^{2+}, Hg^{2+} and amino acid analogs like azetidine carboxylic acid and p-fluorophenylalanine. This is documented in Fig. 14 by Northern blots showing increased levels of the two mRNAs. EDELMAN et al. (1988) investigated arsenite and Cd^{2+} effects on soybean seedlings. Using Northern blotting with 8 hs-specific cDNA clones and *in vitro* translation, they detected massive induction of mRNAs coding for HSP80, 75, 70, 26/27, 21–24, 17 and 15/16. The same two inducers were also applied to sunflower tumor tissue transformed with a soybean hspl7 gene (GURLEY et al., 1986). Finally, WINTER et al. (1988) found that arsenite, Cd^{2+}, Cu^{2+} and Zn^{2+} strongly stimulate hsp70 mRNA synthesis in *Petunia hybrida* leaves. Other stress conditions (osmotic stress by PEG, salt stress, anaerobiosis, inhibitors of respiration) have no detectable effects

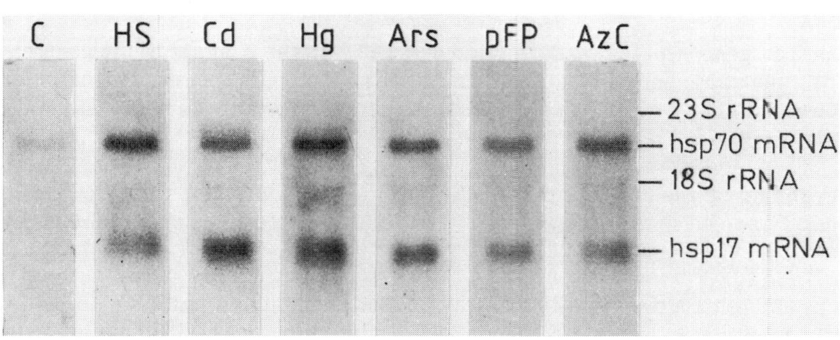

Fig. 14. Chemical stressors induce increased levels of hsp70 and hsp17 mRNAs in tomato cell cultures. Total RNAs from stressed tomato cell cultures were analyzed by Northern hybridization using labeled Drosophila hsp70 and soybean hsp17 clones (see legend to Fig. 15). Size of the mRNAs is 2.2 and 0.9 kb respectively (see markers of the two ribosomal RNAs on the margin). The following samples were used for preparation of the RNAs: **C**, control; **HS**, 2 hrs hs at 38.5 °C; **Cd**, 2 hrs 1 mM $CdSO_4$ + 2 hrs recovery; **Hg**, 2 hrs 70 µM $HgCl_2$ + 3½ hrs recovery; **Ars**, 2 hrs 70 µM sodium arsenite + 2 hrs recovery; **pFP**, 2 hrs 1 mM p-fluorophenylalanine + 2 hrs recovery; **AzC**, 2 hrs 1 mM azetidine carboxylic acid + 3½ hrs recovery

on HSP synthesis or hs mRNA levels in soybean (Czarnecka et al., 1984) or tomato cell cultures (Scharf, unpubl.).

A remarkable exception to this may be the increased synthesis of a general stress protein (HSP26) observed in all cases. The gene is constitutively expressed and intron-containing in soybeans and shows hs and hs-independent regulation (Czarnecka et al., 1984, 1988). The existence of similar genes in other plants is unclear.

Increased levels of HSP70 mRNA only were detected in corn mesocotyls under conditions of water stress, wounding stress or after addition of abscisic acid (Heikkila et al., 1984). If this can be reproduced and extended to other plants, this report describes for the first time non-hs-like induction of HSP70 also in plants. Evidence for a developmental control of hsp70 genes in plants is accumulating from *in situ* hybridization data (E. Vierling, J. Winter, unpublished).

Similar to the situation in brewer's yeast and animal cells, toxic levels of transition series metal ions do not only induce HSP synthesis but also formation of chelating compounds (see Sect. C.4.). In the former, these are low molecular weight sulfur-containing proteins (metallothioneins, see reviews by Brady, 1982; Hamer et al., 1985; Karin, 1985), whereas in plants and fission yeast non-proteinaceous peptide derivatives of glutathione (phytochelatins) are involved (Grill et al., 1985, 1987, 1988; Jackson et al., 1987; Steffens et al., 1986).

B.2.2. Coordinate regulation of HSP synthesis

Frequently, heat shock proteins are considered as a group of coordinately regulated proteins. This is true only to a certain extent. From careful studies of time and temperature effects on HSP synthesis in tomato cell cultures it is apparent that the pattern changes during the hs response. There is the bulk of 'normal' HSPs (HSPs80, 70, 17 in Fig. 3 A), whose synthesis rapidly increases to maximum values after about 1-2 hours of the hs period and then declines to control levels within 12 hours. In addition, there are 'transient HSPs' with effective labeling only at the beginning of the hs treatment (HSPs95 and 22 in Fig. 3 A), and there are 'delayed HSPs' (HSPs18/19) whose synthesis is optimum only after a long-term hs of 12 hrs (Nover and Scharf, 1984). Though not so detailed in the analytical procedure, similar results were reported for seedlings of maize (Cooper and Ho, 1983) barley and wheat (Necchi et al., 1987) as well as in regenerating tobacco mesophyll protoplasts (Meyer and Chartier, 1983). But the plant HSP patterns can also vary with the tissue or organ investigated and genetic variability within a plant species (Cooper et al., 1984; Necchi et al., 1987; Pitto et al., 1983; Yacoob and Filion, 1986; Zivy, 1987, for similar results with animal systems see summary by Nover, 1990).

The accumulating data on hs gene structure and function as well as on the changes of the gene expression apparatus during the stress period (see Sects. B.2.4., B.2.6.) provide a molecular basis for an understanding of the individuality of HSPs with respect to their regulatory behaviour. The main points will be briefly summarized:

1) Though basically dependent on the presence of the HSE, details of the response of hs promoters, i.e. the extent, duration and threshold temperature of transcription activation is determined by the exact HSE combination and presence of other promoter elements, e.g. for hs-independent activities (see Sect. B.1.2.1.).
2) The interference of hs and other stress conditions with effective splicing of pre-mRNP limits the expression of intron-containing hs genes (see Sect. B.2.5.). Optimum synthesis of these HSPs may be restricted to moderate stress conditions or to the recovery period.
3) Stability and translational efficiency of individual hs mRNA species may vary considerably, especially if stress and recovery conditions are compared. Though corresponding information for plant systems is lacking at present, there is some evidence on this point for animals (Di Domenico et al., 1982b; Dura, 1981; Theodorakis and Morimoto, 1987).
4) In *Drosophila*, yeast, and vertebrate cells, there are results indicating an autorepression of HSP synthesis, which mainly concerns the HSP70 level, whereas other HSPs are not or less affected (see Sect. B.1.2.3.).

5) The stability of individual HSPs certainly depends on their intracellular localization and probably on structural binding during the stress period. Generally, they are long-lived proteins accumulating to visible amounts (see Fig. 30). But there are also exceptions, e.g. the transiently synthesized HSPs22 and 95 of plants.

B.2.3. Reprogramming of transcription

Heat shock induced synthesis of new stress proteins needs transcription of new mRNAs. In fact, light-microscopic observations on the differential activation of hs genes and repression of control genes using polytene chromosomes of *Drosophila* larvae marked the beginning of the molecular biological era of the hs response (RITOSSA, 1962). But this exceptional method to document reprogramming of transcription is restricted to few experimental objects only. Several other techniques are at hand and applied to plant systems:

1) Induced HSP synthesis is inhibited by inhibitors of transcription, e.g. actinomycin D (NOVER and SCHARF, 1984).
2) Using isolated nuclei from heat-shocked soybean seedlings and *in vitro* transcription assay (nuclear run-off assay) SCHÖFFL et al. (1987b) documented the marked increase of hs gene-specific transcripts and the corresponding decline of other transcripts, e.g. of auxin-induced genes or of rRNA genes.
3) The highly selective and efficient accumulation of new hs mRNAs to about 20,000 copies per soybean cell after 2 hrs at 40 °C considerably facilitated the cloning of the small hs genes of plants (SCHÖFFL and KEY, 1982, 1983, see Sect. B.1.2.). Increasing levels of hs mRNAs can be directly monitored either by their *in vitro* translation in reticulocyte lysates or wheat germ extracts followed by analysis of the labeled products (see Fig. 17; BASZCZYNSKI et al., 1983; BELANGER et al., 1986; EDELMAN et al., 1988; KEY et al., 1981; KLOPPSTECH et al., 1985, 1987; NOVER et al., 1989; REES et al., 1987; VIERLING et al., 1986, 1989; WU et al., 1988a) or by blotting techniques (Northern or dot/slot blots) using corresponding clones derived from hs genes (CZARNECKA et al., 1984; EDELMAN et. al., 1988; KIMPEL and KEY, 1985b; MASCARENHAS and ALTSCHULER, 1985; NAGAO et al., 1986; ROCHESTER et al., 1986; SCHÖFFL and KEY, 1982; VIERLING and KEY, 1985; WINTER et al., 1988). Northern analyses of HSP70 and HSP17 mRNA levels after hs induction of 8 different plants are illustrated in Fig. 15. The relatively weak signal obtained with barley and maize RNA and the hsp17-specific probe from soybean may result from reduced cross-hybridization (see Sect. B.4.3.3.).
4) The development of new techniques for *in situ* hybridization allows the detection of hs mRNAs in single cells and analyses of their distribution within tissues or organs. This was pioneered in animal systems (SPRANG and BROWN, 1987), but was recently applied to plants as well (E. VIERLING, J. WINTER, personal commun.). The method found broad application for induced mRNAs of the plant defense-related complex (see Sect. C.10. and review by SCHEEL and HAHLBROCK, 1989).

B.2.4. Promoter structure and function

Sequence comparison of several *Drosophila* heat shock promoters and functional analysis of the hsp70 promoter by PELHAM (1982) led to the definition of the 14 bp palindromic heat shock promoter element (HSE):
5'-CtaGAAnnTTCtaG-3'
Nucleotides indicated by large case letters are highly conserved, those indicated by small case letters are less conserved. The central two nucleotides (n) are variable.
A synthetic oligonucleotide of this type, introduced 5'upstream of the TATA-box of the Herpes simplex virus thymidine kinase gene, was found sufficient to confer hs-inducibility to this heterologous gene, if transferred into monkey COS cells or *Xenopus* oocytes (PELHAM and BIENZ, 1982). Se-

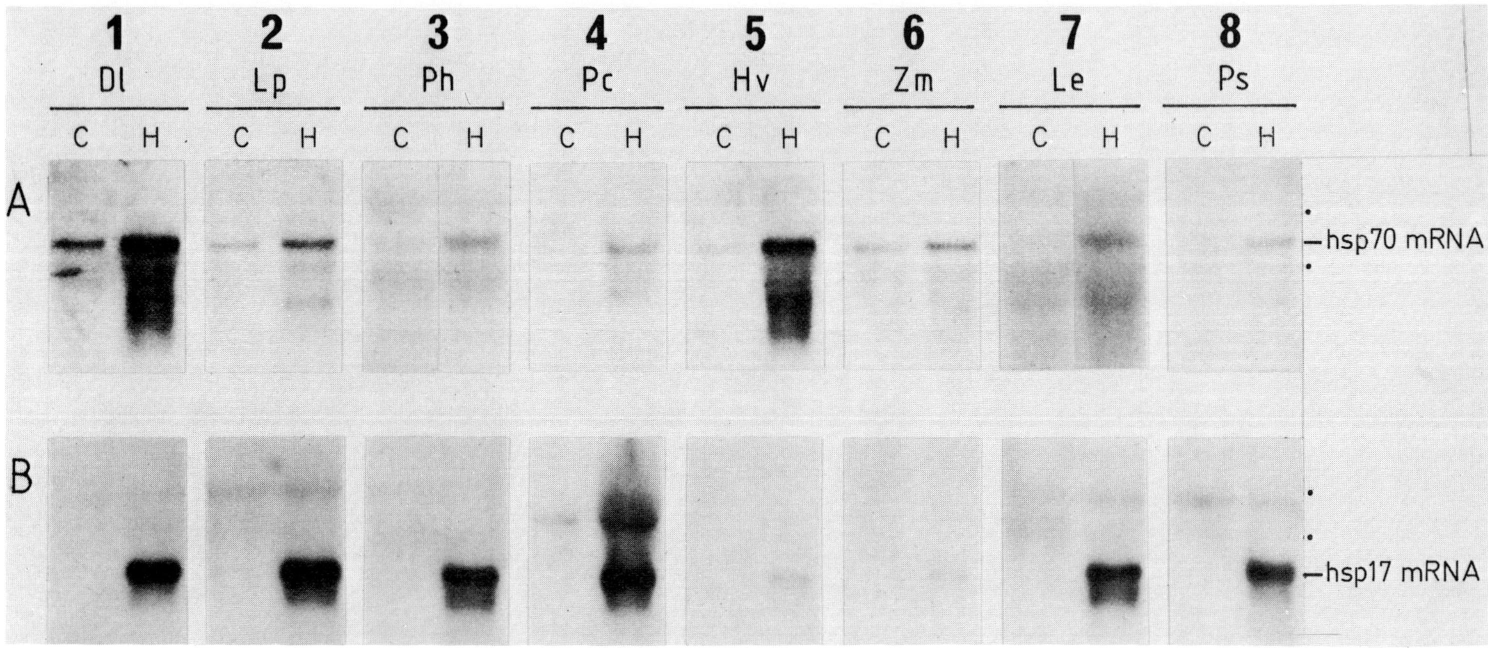

Fig. 15 Northern analysis of hsp70 and hsp17 mRNA levels in different plants after heat shock induction.
Total RNA was isolated from the indicated plant material and separated on formaldehyde/agarose gels. All materials were preinduced, followed by a two hours heat shock at 40 °C (H). C, RNA from control samples. Probes for hybridization are a Dm-hsp70 clone (SalI fragment of 56H8) and a Gm-hsp17 clone (Pst fragment of pFS 1968). Samples **1–4** were obtained from cell cultures of *Digitalis lanata* **(1)**, *Lycopersicon peruvianum* **(2)**, *Peganum harmala* **(3)**, *Petroselinum crispum* **(4)**, samples **5,7,8** are from leaves of *Hordeum vulgare* **(5)**, *Lycopersicon esculentum* **(7)** and *Pisum sativum* **(8)**, sample **6** is from *Zea mays* coleoptiles. Dots at the margin mark positions of 23S and 18S rRNAs

quencing of many other hs genes throughout the eukaryotic kingdom in all cases revealed similar HSEs in the promoter regions (see summaries by Nover, 1987, 1990). Selected examples from animals and the known hs promoter sequences of plants are summarized in Fig. 16.

In most cases several HSEs are found in variable positions 5'upstream of the TATA-box. But only thorough functional analyses with deletions and site-specific mutations as well with protein footprint studies can unequivocally demostrate their significance for hs-induced transcription activation. The peculiarities of the HSE monomer, with a purin-rich 5' half, a pyrimidine-rich 3' half and a direct 4 bp repeat -CtaG- at both ends are the basis for different types of repetitive structure formed by overlapping HSE oligomers (see promoters no. 2–4, 10–16 in Fig. 16). This leads to more general definitions of the HSE (Nover, 1987, 1990; Xiao and Lis, 1988) as alternating pentanucleotide patterns of purins and pyrimidines: (nGAAn) (nTTCn) (nGAAn) (nTTCn). The purine and pyrimidine motifs form opposite sites of the DNA α-helix. This may be essential for proper recognition by the hs transcription factor (HSTF). Generalization of the HSE structure in this way leads to a reinterpretation of the yeast hsp90 promoter, for which no Pelham-type of HSE could be defined (Farrelly and Finkelstein, 1984). In fact, there ist a long stretch of regularly spaced -GAA- and -TTC- between bp -216 and -145. They form four coherent pyrimidine and purine sites of the DNA (underlined). But the interspaced pentanucleotide on the opposite DNA site is lacking in each case:

(-216)GAA3nTTC7nTTC3nTTC3nGAA6nGAA3nTTC6nTTC2nGAA7nGAA(-146)

A functional analysis must show whether this 60 nucleotide stretch really harbors the hsp90 recognition sequence of the recently isolated yeast HSTF (see below).

Interaction between several HSEs is required for optimum function. Most extensively studied in this respect is the Dm-hsp70 promoter (Fig. 16, no. 4, Amin et al., 1987, 1988; Xiao and Lis, 1988). The overlapping HSE dimer at bp -170 is dispensible, but cooperativity between the two TATA proximal HSEs centered at bp -77 and -53 is important not only for high promoter activity *in vivo* (Amin et al., 1987, 1988; Xiao and Lis, 1988) but also for HSTF binding (Shuey and Parker, 1986). According to these investigations the following general structure of the essential promoter parts must be given:

-(-87)TTC2nGtt2nTTC2nGAA10nGAA2nTTC2nGgA13nTATAAAT(-27)-

B A

In both cases at least 3 pentanucleotide motifs are present. If site B is totally deleted only 2 % of the hs promoter activity is maintained. *In vivo*, a similar effect may be observed under moderate hs conditions (33–35 °C), because, at low concentrations of active HSTF, only site A is occupied. Cooperativity between both sites and full promoter activity requires higher HSTF concentrations, i.e. temperatures > 35 °C. Details of this interaction were studied by *in vitro* DNP formation using partially purified HSTF (Shuey and Parker, 1986). For optimum interaction the distance and relative positions of A and B are crucial. One turn of the DNA helix (10 nucleotides) but also 2–4 turns are tolerated, whereas increase of the distance by two nucleotides or reduction by 5 has deleterious effects (Amin et al., 1987; Xiao and Lis, 1988).

The extensive characterization and availability of hs promoters led to their broad application in genetic engineering as vector parts conferring selective inducibility and high efficiency of transcription to the adjacent reporter gene (see summaries by Nover, 1987, 1990). Tightly related to this are investigations on the HSE binding protein, the hs transcription factor (HSTF). By a four step procedure including affinity chromatography, active, apparently homogenous HSTF preparations were obtained from *Drosophila* cells (M_r 110 kDa, Wu et al., 1987b) and yeast (M_r 130 kDa; Sorger and Pelham, 1987b). After screening of phage expression libraries with HSTF antibodies, the yeast HSTF gene was isolated and sequenced (Sorger and Pelham, 1988; Wiederrecht et al., 1988). It codes for a protein of 833 amino acid residues (MW 93,218 dalton), which is essential for growth of yeast cells at any temperature between 15 and 35 °C. Analyses of deletion proteins, screened with synthetic HSE oligonucleotides (Southwestern-Screening), defined a stretch of 118 amino acid resi-

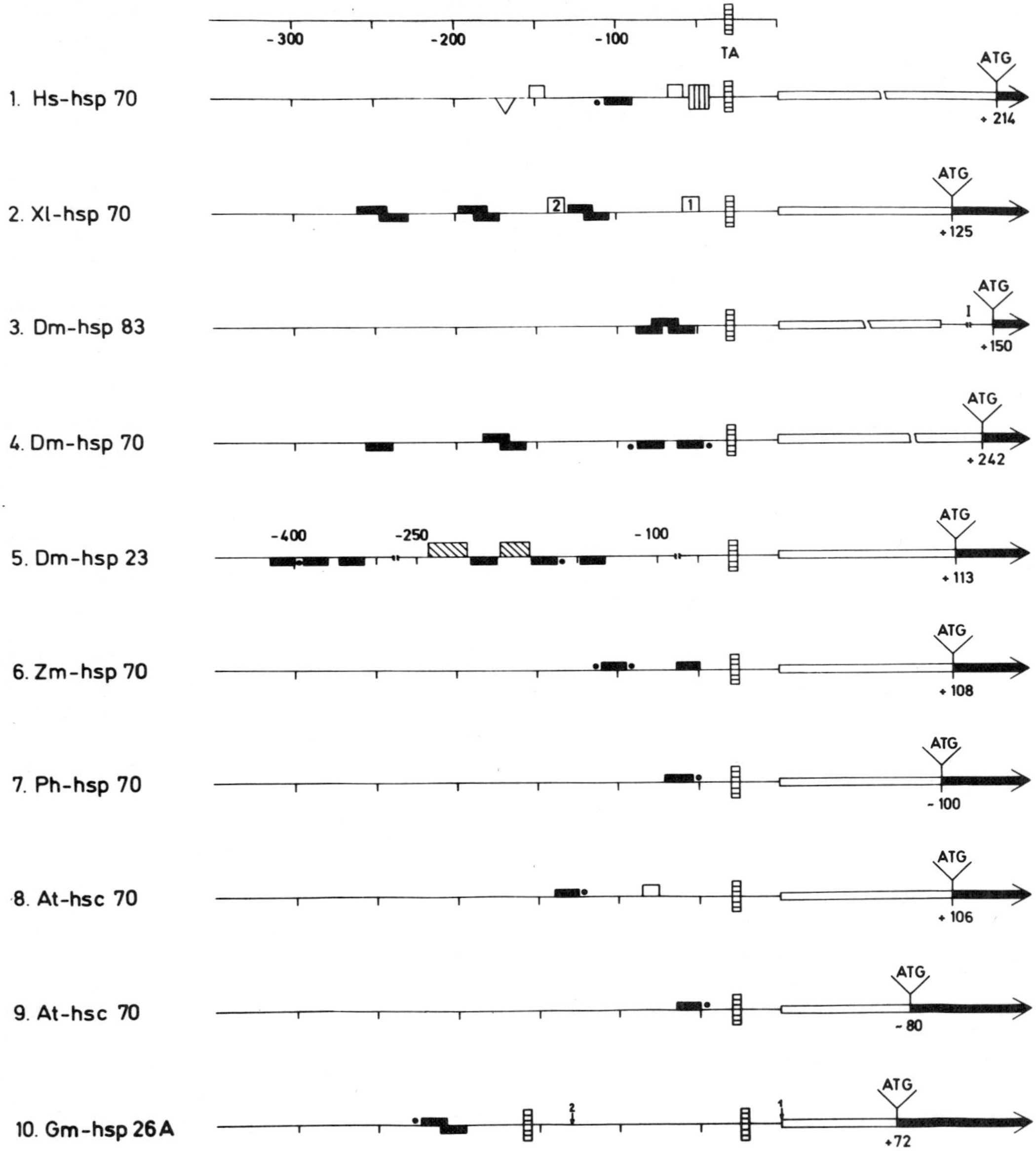

Fig. 16. Basic promoter structures of heat shock genes (from **Nover 1990, modified**).
Symbols used to identify characteristic promoter elements are explained at the bottom. Black bars represent the 14 nucleotide palindromic heat shock element as defined by Pelham (1982). In frame extensions are denoted by dots (see text for explanation). The transcription units are indicated by the thick arrows on the right site with the ATG marking the start of the open reading frame. **1–5**, animal hs genes promoters mentioned in the text: **1**, human hsp70 (Hunt and Morimoto, 1985; Wu et al., 1987a); **2**, ***Xenopus*** **hsp70** (**Bienz**, 1986; Bienz and Pelham, 1986); **3**, *Drosophila*, hsp83 (Hacket and Lis, 1983; **Blackman and Meselson, 1986**); **4**, *Drosophila* hsp70 (Bienz and Pelham, 1982; Dudler and Travers, 1984); **5**, *Drosophila* **hsp23** (**Ayme** et al., 1985; Pauli et al., 1986); **6–24**, plant hs promoters: **6**, maize hsp70 (Rochester et al., 1986), **7**, ***Petunia*** hsp70 (**Winter** et al., 1988), **8,9**, *Arabidopsis* hsc 70 genes 1 and 2 (Wu et al., 1988a); **10–16** soybean hs genes (for references see legend to Figs. 10 and 11)

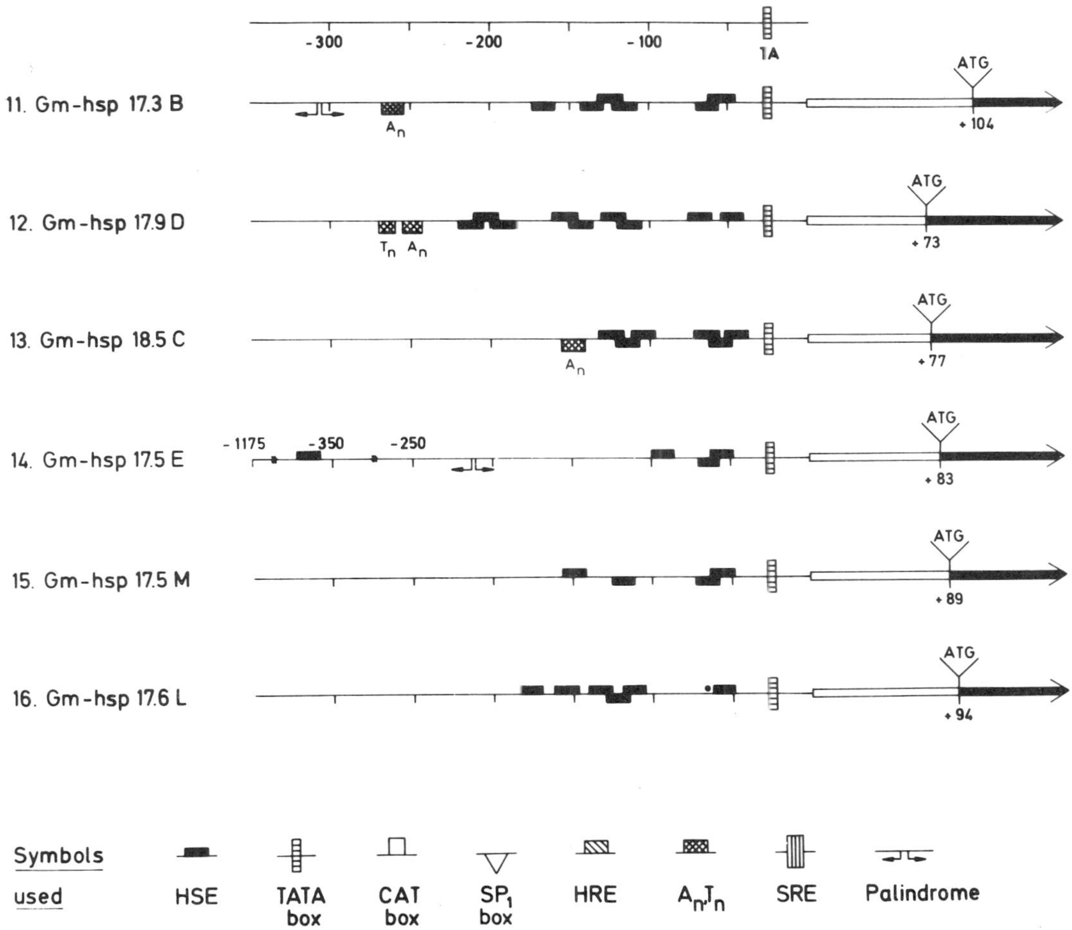

dues in the N-terminal half which is responsible for HSE binding (WIEDERRECHT et al., 1988). Heat shock activation is connected with phosphorylation of the yeast HSTF (SORGER and PELHAM, 1988; SORGER et al., 1987).

Promoter studies with plant hs genes are much more restricted. Sequence comparison frequently reveals the existence of overlapping HSE oligomers or, similar to the *Drosophila* hsp70 promoter at least 1 1/2 HSE. Deletion analyses used Ti plasmids and stable transformations with *Agrobacterium tumefaciens* infection of tobacco or sunflower leaf discs (BAUMANN et al., 1987; GURLEY et al., 1986; SCHÖFFL and BAUMANN, 1985). Especially the investigations by BAUMANN et al. (1987) and SCHÖFFL et al. (1989) on the Gm-hsp17. 3 promoter (Fig. 16, no. 11) revealed the unusual complexitiy of the plant hs promoters. Deletion of all sequences upstream of the HSEs to bp -139 reduced the hs activity by about 85 %. Though the multiple HSEs between bp -142 and -147 alone confer hs-inducibility, they need interaction with a quantitative enhancer element positioned 5' upstream of this. Search for sequence elements with putative regulatory functions in this region only revealed an imperfect palindrome centered at bp -305 and a homopurine stretch (A_{14}) extending from bp -255 to -268. It will be important to test whether, in contrast to the results with animal hs genes, this separation of essential promoter functions into a neutral quantitative element (bp -336 to -139) and the qualitative element for hs inducibility (bp -142 to -47) is a general trait of the plant hs promoters. A possible role of the uninterrupted sequence with 14 adenine residues ('simple sequence') can be seen in connection with nuclear scaffold attachment sites necessary for fixation of active chromatin domains (GASSER and LAEMMLI, 1987). At any rate, insertion of bp -76 to -41 or a corresponding synthetic oligonucleotide, containing the overlapping HSE dimer of the hsp17.3 promoter, into appropriate test promoters gave high levels of hs-induced transcription. Test promoters were the light-inducible ribulose-bisphosphate carboxylase small subunit promoter (STRITTMATTER and CHUA, 1987) and a truncated cauliflower mosaic virus 35S promoter (SCHÖFFL et al., 1989). In the combination with the light-inducible promoter, hs-induction is restricted to leaves but is not observed in roots of transgenic tobacco. Evidently a dominant, tissue-specific silencer element in the rbcs promoter is detected after combination with the neutral hs enhancer (STRITTMATTER and CHUA, 1987).

Considering the high conservation of HSEs in the eukaryotic kingdom, it was not totally surprising that gene constructs containing the *Drosophila* hsp70 promoter fused to the neomycin phosphotransferase gene (npt II) were also expressed in a hs-dependent manner in tobacco cells (SPENA and SCHELL, 1987; SPENA et al., 1985). In this case a Dm-hsp70 promoter/leader fragment from bp -258 to +199 was used. High levels of npt mRNA were detected after 45 min. hs at 40 °C. But translation of NPT requires 2 hrs of recovery. Evidently, the *Drosophila* leader sequence of the mRNA, which confers selective translation to the hsp70 mRNA in heat-shocked *Drosophila* cells, does not function in plant cells (see Sect. B.2.6.2.). Heat shock induction of the hybrid gene was detected in all tissues of transgenic tobacco plants except in pollen (SPENA et al, 1985). Unfortunately, other gene constructs using parts of the hsp70 promoter were not tested, and results on the overall transcription efficiency are lacking. Thus, the two types of analyses with the homologous Gm-hsp17 (BAUMANN et al., 1987; SCHÖFFL et al., 1989) and the heterologous Dm-hsp70 promoters (SPENA and SCHELL, 1987; SPENA et al., 1985) cannot be compared.

B.2.5. Processing of pre-mRNP

Similar to the block of preribosomal RNP processing in the nucleolus (see Sect. B.3.2.), maturation of pre-mRNP is affected by hs as well. SPINELLI et al. (1982) reported on the accumulation of histone mRNP precursors in sea urchin mesenchyme blastula, and MAYRAND and PEDERSON (1983) observed immediate and drastic changes in the assembly of nuclear pre-mRNP of heat-shocked *Drosophila* and mammalian cells. PEDERSON (1983) hypothesized that the processing pathway of intron-containing pre-mRNP is particularly heat sensitive. In keeping with this, BOND (1988) demonstrated a disruption of the U4/U5/U6 and U2 RNA-containing snRNPs in HeLa cells.

Usually the inhibition of the splicing machinery is noticeable only under severe hs conditions and affects the expression of intron-containing hs-genes as well. This was extensively studied in *Drosophila melanogaster* cell cultures (YOST and LINDQUIST, 1986). Pre-mRNP of the only intron-containing hsp gene of this organism (hsp83) accumulates at 38 °C, i.e., by virtue of its promoter structure (Fig. 16, no. 3) and its intron, the hsp83 gene exhibits optimum expression under moderate stress conditions (33-35 °C), whereas products of the *Drosophila* hsp70 genes become dominant at ≧ 37 °C. Removal of the intron can be maintained even at 38 °C, if the cultures are conditioned by a prior hs treatment (YOST and LINDQUIST, 1986). Similar results, including a protective effect in thermotolerant cells were also reported for mammalian cells (BOND, 1988; KAY et al., 1987).

Particularly interesting is the situation in parasitic protozoa of the *Trypanosoma* species. These organisms contain no internal introns but need trans-splicing of a separately encoded 5'-capped mini exon of 39 nucleotides to the 5' ends or its pre-mRNAs. Transfer of the miniexon is blocked during hs, and the corresponding transcription products accumulate. It is not clear, how the production of hsp70 and hsp83 mRNAs, which also contain the miniexons, proceeds under these conditions (MUHICH and BOOTHROYD, 1988).

No hs effects on intron splicing were observed in plants. But instead. $CdCl_2$ at 10^{-4} to 10^{-3} M causes a partial block of intron removal of the hsp70 pre-mRNP in *Petunia* (WINTER et al., 1988) and of the hsp26 pre-mRNP in soybean (CZARNECKA et al., 1988). Unlike the results in *Drosophila* and mammals, the Cd^{2+} inhibition of splicing could not be released by prior hs conditioning (WINTER et al., 1988).

B.2.6. Translational control

B.2.6.1. Selective synthesis, storage and degradation of mRNAs

The high selectivity of HSP synthesis depends on the massive synthesis of new mRNAs encoded by the hs genes and a number of alterations of the translation apparatus leading to an efficient discrimination of non-hs mRNAs. The effect, initially described for *Drosophila* cell cultures (Krüger and Benecke, 1981; Storti et al, 1980), was also oberserved in many other systems, including plants (Key et al., 1981). By comparing two-dimensional separations of proteins from control (Fig. 17A) and heat-shocked (Fig. 17B) cells labeled *in vivo* or by *in vitro* translation of the respective RNA fractions (Fig. 17C, D), it can be seen that the selectivity of HSP synthesis in heat shocked cells (B) does not correspond to the composition of the mRNA fraction (D). Evidently, most of the 'control' mRNAs, though not translated, are maintained intact and can be reactivated in the recovery period (Nover and Scharf, 1984; Nover et al., 1989).
With respect to the particular hs effects at least four clases of mRNAs can be discerned (Key et al., 1985a; Nover and Scharf, 1984; Nover et al., 1989):

(i) Heat shock mRNAs rapidly accumulate during hs and preferentially occupy the polysomal apparaturs (Fig. 17E). In animal cells, there is evidence that hs mRNAs are rather stable during the stress, but rapidly decay in the recovery period (Di Domenico et al., 1982b; Theodorakis and Morimoto, 1987).

(ii) Control mRNAs probably coding for 'house keeping proteins' are not translated. In plants, their protection from degradation and rapid reactivation in the recovery period is tightly associated with formation of the cytoplasmic hs granules (Nover et al., 1989, see Sect. B.3.3.).

(iii) For unknown reasons, translational discrimination does not affect all control mRNAs. Examples are mRNAs coding for the unidentified tomato proteins in boxes b and e (Fig. 17B), for many ribosomal proteins (see Sect. B.3.2.), for some soybean storage proteins (Altschuler and Mascarenhas, 1982; Mascarenhas and Altschuler, 1985) and for three prominent tobacco mosaic virus proteins expressed in infected tobacco leaves (Dawson and Boyd, 1987).

(iv) Some mRNAs are degraded. Examples documented so far concern mRNAs coding for specialized proteins. Thus, in barley aleurone layers, the α-amylase mRNA, induced by giberellic acid, rapidly decays together with an almost complete destruction of the rER (Belanger et al., 1986) The authors speculate that the latter may be the cause for the instability of the α-amylase mRNA. In soybean, an auxin-induced mRNA, defined by the clone pJCW1, is degraded during hs (Walker and Key, 1982; Key et al., 1985a). A more complex example was described by Vierling and Key (1985) analyzing green cell suspension cultures of soybean. The cytoplasmic mRNA for the small subunit of ribulosebisphosphate carboxylase is degraded during hs (38-40 °C), whereas the large subunit mRNA in the chloroplasts is stable (see Sect. B.5.1.).

Analyses in other organisms are as fragmentary as in plants but point to an equally complex situation. In yeast and similar primitive eukaryotes the specificity of gene expression under hs conditions is dominated by the rapid decay of most control mRNAs (Gwynne and Brandhorst, 1982; Kim and Warner, 1983; Lindquist, 1981). By using different cDNA clones Hallberg et al. (1984) characterized different classes of mRNAs in *Tetrahymena thermophila* that are largely identical with those defined for plants. In *Drosophila*, noncoordinate expression of histone genes is primarily due to the differential behaviour of the corresponding mRNAs (Farrell-Towt and Sanders, 1984; Spadoro et al., 1986). Under hs conditions, H2B mRNA is effectively transcribed and translated whereas H2A, H3 and H4 mRNAs, though preserved in active form, are barely translated. Finally, H1 mRNA is degraded. Selective destabilization of mRNAs was also reported for estrogen-induced vitellogenin mRNA in *Xenopus* hepatocytes (Wolffe et al., 1984), for mouse mammary tumor RNA in mammary carcinoma cells (Ralhan and Johnson, 1986) and for tubulin mRNAs in *Tetrahymena* (Coias et al., 1988). In animal cells the half-life of tubulin mRNAs is strongly reduced by association of the polysome-bound nascent tubulin chain with free tubulin subunits (see review by Cleveland, 1988). It is tempting to speculate that the hs-induced rearrangements of the cytoskeletal systems may cause a similar effect also in *Tetrahymena* and possibly other organisms.

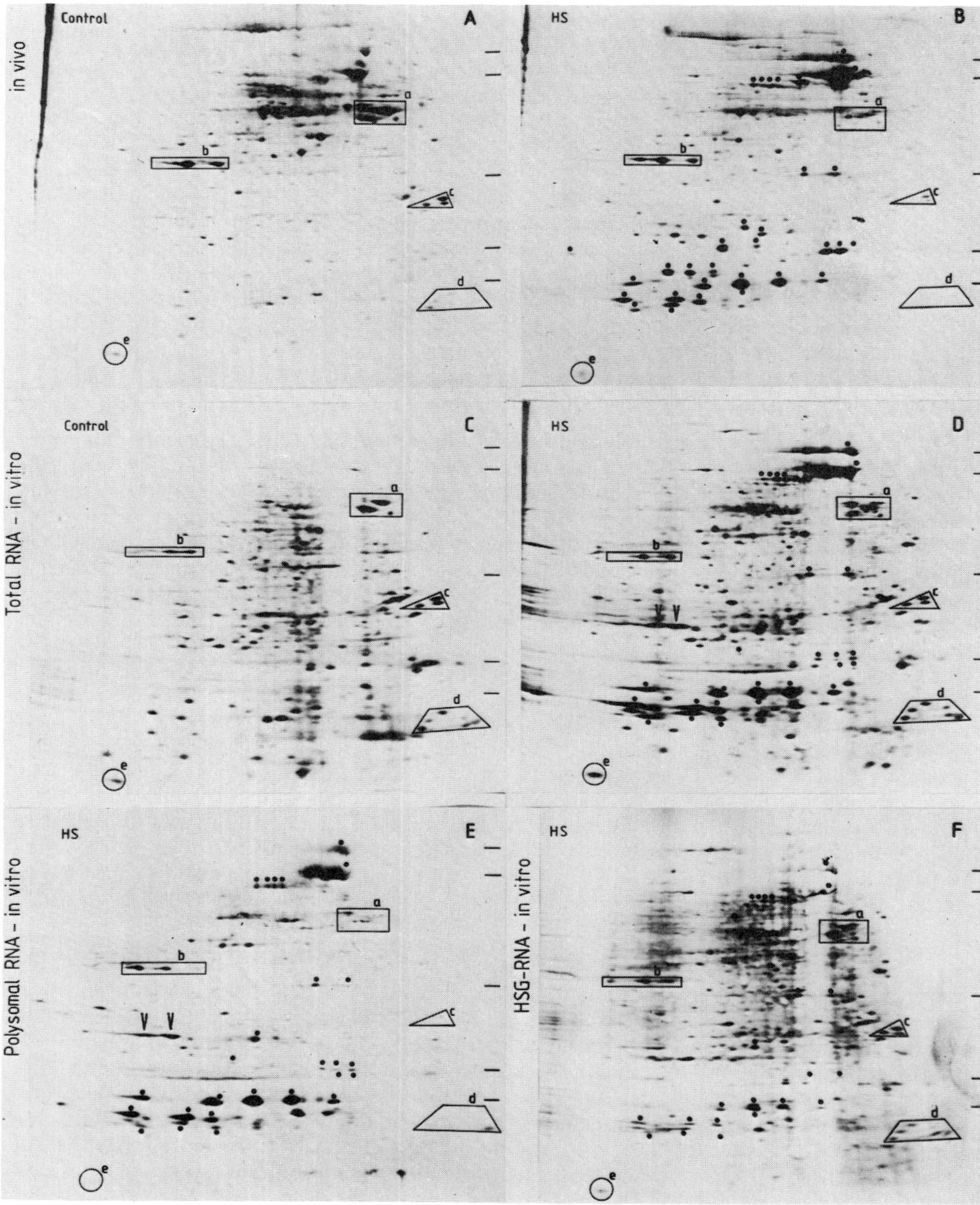

Fig. 17. *In vivo* and *in vitro* translation patterns of RNAs from control and heat-shocked tomato suspension culture cells (from Nover et al., 1989, with permission).
2D-autoradiographs of ^{35}S-methionine labeled proteins are shown. **A, B**, proteins labeled in whole cells at 25 °C (Control) and 39 °C (HS) respectively. **C–F**, *in vitro* translation patterns obtained with a mRNA-dependent wheat germ system. The following RNA sources were used: C, total RNA from control cells; **D**, total RNA from heat shocked cells; **E**, polysomal RNA from heat shocked cells; **F**, RNA from purified heat-shock granules. Note the differ-

B.2.6.2. Mechanisms involved in translational control

The remarkable specificity of hs protein synthesis with respect to the discrimination of control mRNAs led to intensive investigations on possible mechanisms involved. Though a coherent picture is still not available, several hs-induced changes of the ribosomal apparatus have been observed and will be briefly summarized:

1) *In vitro* translation systems obtained from heat shocked cells of *Drosophila* (KRÜGER and BENECKE, 1981; STORTI et al., 1980), mammals (DE BENEDETTI and BAGLIONI, 1986a, 1986b; PANNIERS et al., 1985) and plants (KLOPPSTECH and OHAD, 1986; KLOPPSTECH et al., 1987) retained their potential for selective translation. The molecular mechanisms were not studied in great detail. But modifications of soluble factors participating in initiation complex formation are probably important (de BENEDETTI and BAGLIONI, 1986a; DUNCAN and HERSHEY, 1984; MAROTO and SIERRA, 1988; SANDERS et al., 1986).
2) The initial rapid decay of polysomes is evidently due to a transient block of initiation (McCORMICK and PENMAN, 1969). The depletion of polysomes, observed at the ultrastructural level in many different organisms (see Sect. B.3.1.), were analyzed also for soybean (KEY et al., 1981) and tomato cell cultures (NOVER, 1984). It was speculated that the collapse of the intermediate filament cytoskeleton, observed in *Drosophila* and mammalian cells is tightly connected with the clearance of the translation machinery from non-hs mRNAs (FALKNER et al., 1981; HOWE and HERSHEY, 1984). Polysome decay is a general phenomenon of stress-induced reprogramming of translation in plants. Besides hs, this includes starvation (CARDELLI and DIMOND, 1981; WALTER and HAHLBROCK, 1985), anaerobic stress (LIN and KEY, 1967; SACHS et al., 1980), or water deficiency (BARTELS et al., 1988; HSIAO, 1970), or wounding (THEILLET et al., 1982).
3) Heat shock-induced changes also affect ribosome structure. Most prominent is the dephosphorylation of the small subunit protein S6, which is localized in the initiation center (GLOVER, 1982; SCHARF and NOVER, 1982). The reversible dephosphorylation/rephosphorylation of r protein S6 in a hs/recovery regime is illustrated for different plants (Fig. 18). It was observed in all types of organisms, but is not related to the decay of polysomes (see summary by NOVER, 1990). Other modifications of ribosomes of heat shocked cells are the association with a new basic 22 kDa protein and a newly formed scRNA in *Tetrahymena thermophila* (KRAUS et al., 1987; McMULLIN and HALLBERG, 1986) and with a 57 kDa protein in cells of stressed maize leaves, i.e. during hs. pathogen infection, oxidations stress (WU et al., 1988b).
4) Preferential translation of hs mRNA s evidently depends on parts of their untranslated leader sequences preceding the AUG codon. This was studied for the *Drosophila* hsp22 (HULTMARK et al., 1986) and hsp70 genes (McGARRY and LINDQUIST, 1985). Contrasting to the hs transcription element (see Sect. B.2.4.), this translation element acts species-specific If the *Drosophila* promoter/leader sequence was hooked up to an appropriate reporter gene and transferred into plants (SPENA et al., 1985) or *Xenopus* oocytes (LAWSON et al., 1984), high levels of hs-inducible mRNAs were detected. But effective translation was restricted to the recovery period. Interestingly, similar results were also reported for plants (SCHÖFFL et al, 1989). (i) Translation under hs conditions in transgenic tobacco plants requires the leader sequence of the Gm-hsp17.3 gene. (ii) The leader of the hs gene can be replaced in this particular function by the cauliflower mosaic virus leader sequence. Evidently some viral mRNAs are preferentially used by the translation apparatus of hs cells (see also DAWSON and BOYD, 1987).

   Though the species-specificity of this control is unexplained, an essential element for the overall effects may be the lacking secondary structure in the leader region (HOFFMAN and CORCES, 1984; McGARRY and LINDQUIST, 1985). Two arguments support this: (1) Under hypertonic stress the

---

ences of mRNA populations between polysomes (E) and HSG (F). Both add up to the pattern in D (total hs RNA). Control proteins of groups a, c, d are not or barely translated under hs conditions (B), but their mRNAs are preserved intact (D) and mainly sequestered to the HSG fraction (F). Open arrow heads mark the position of the putative HSP21 precursor (see text)

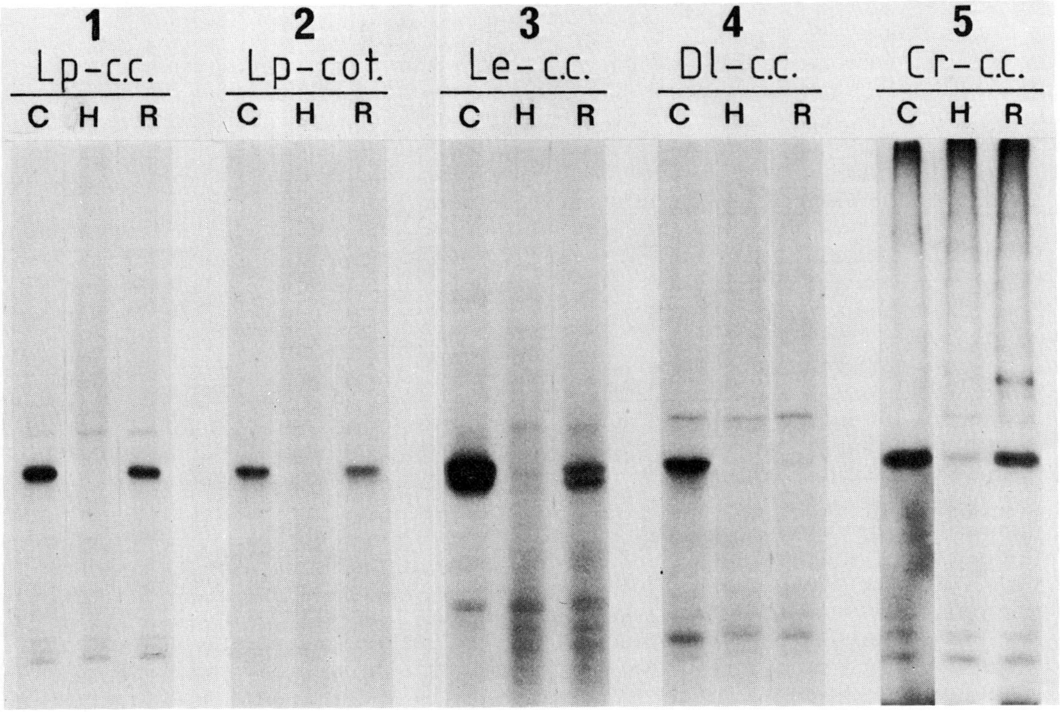

Fig. 18. Reversible dephosphorylation of small subunit ribosomal protein S6.
The plant material was labeled *in vivo* by feeding two hours with ^{32}P-o-phosphate. Ribosomes were isolated, and the ribosomal proteins were separated by SDS-PAA gel electrophoresis (for details see SCHARF and NOVER, 1982). After two hours labeling at 25° (C) cultures were shifted for two hours to 39° (samples **1–4**) and 41 °C (sample **5**) respectively and subsequently allowed to recover at 25 °C for another two hours. The dominant phosphate labelled band in the middle coincides with the small subunit protein S6 of about 30 kDa. The protein is desphosphorylated during hs (H) and rephosphorylated in the recovery period (R). The following plant materials were used: Cell cultures of *Lycopersicon peruvianum* **(1)**, *L. esculentum* **(3)**, *Digitalis lutea* **(4)**, *Catharanthus roseus* **(5)** and cotyledons of *L. peruvianum* **(2)**

translation apparatus of monkey COS cells strongly prefers mRNAs with non-structured leader (KOZAK, 1988) and (2) mRNA competition is a main factor for reduction of normal protein synthesis in heat-shocked *Drosophila* cells (JACKSON, 1986). It can be speculated that the unwinding of the mRNA leader sequence during scanning of the initiating ribosome from the 5'-cap to the AUG codon (see KOZAK, 1988 and review by RHOADS, 1988) is abolished by modification of initiation factors and the striking dephosphorylation of the small ribosomal subunit S6 protein. But there is good evidence that unstructured leader sequences allow an internal, cap-independent start of translation (EDERY et al., 1989).

From all the data, a hypothetic multistep model of essential translation events during hs can be put forward (NOVER, 1990; NOVER et al., 1989): (i) An immediate initiation deficiency and the collapse of the cytoskeleton lead to the destruction of polysomes (0–5 minutes). (ii) Newly formed hs mRNAs with unstructured leader sequence and other 'strong' cell or viral mRNAs are capable for polysome reformation with altered initiation factors and dephosphorylated ribosomes (10–30 minutes). (iii) Non-translated control mRNAs attached to the cytoskeleton are protected from degradation by increasing deposition of HSPs, i.e. formation of heat shock granules in plants beginning 15 minutes after the onset of hs. (iv) In the late phase of the hs or in the recovery period stored mRNAs are reactivated. It is tempting to speculate on an ATP-consuming, active role of the HSP70 for mRNA release from the complex of hs granules and cytoskeletal aggregate (see Sect. B.1.2.3.).

B.3. *Changes of cell ultrastructure and biochemical functions related to this*

B.3.1. *Survey of ultrastructural changes in animal and plant cells*

The dramatic response of cells to heat stress is extensively documented by biochemical and molecular biological results. Contrasting to this, investigations on the cell ultrastructure are limited with respect to the experimental objects, the heat shock regimes and the methods applied. In the physiological range of the hs response, which is characterized by a high metabolic and gene expression activity, the integrity of cellular substructures is largely retained. Yet systematic studies with highly sophisticated methods reveal a considerable number of reversible hs-related changes. It is worthwhile to recall the fact that 125 years ago the German plant physiologist Julius Sachs published experimental studies to define the upper temperature limits of plant survival (Sachs, 1864). He also described changes of cell ultrastructure, e.g. the leakage of anthocyanins by disruption of the tonoplast membrane and the stop of the protoplasmic streaming, which nowadays can be connected with changes of the actin microfilament system.

Depending on the method applied, the following alterations of cell ultrastructure were observed in heat-shocked plant and animal cells:

1. The reprogramming of gene expression at the onset of hs (see Sect. B.2.) leads to a transient, visible loss of polysomes and dictyosomes (Heine et al., 1971; Love et al., 1970; Mansfield and Key, 1988; Risueno et al., 1973; Simard and Bernhard, 1967; Schnepf and Schmitt, 1981). In barley aleurone layers, a complete desctruction of the extensive rER may be related to the decay of α-amylase mRNA (Belanger et al. 1986). It is remarkable that, associated with the hs granules in plant cells, a massive reformation of all substructures actively engaged in protein synthesis and processing (polysomes, rER, dictyosomes) is observed in the recovery period (Nover et al., 1989).
2. The accumulation of a new type of RNP material (perichromatin granules) in nuclei of heat-shocked mammalian cells was originally described by Simard and Bernhard (1967). For the nucleolus these granules can be identified with the unprocessed preribosomal RNP (see below). But the evidence for a block of pre-mRNP splicing under stress conditions (Kay et al., 1987; Winter et al., 1988; Yost and Lindquist, 1986) suggests a similar conclusion also for the non-nucleolar perichromatin granules. In addition, a marked increase of chromatin-associated non-histone protein content in *Drosophila* and vertebrate cells (Littlewood et al., 1987; McConell et al., 1987; Roti Roti et al., 1986; Tomasovic et al., 1978) and a visible accumulation of actin in form of paracrystallin rods were reported for nuclei of mammalian cells and the water mold *Achlya* (Iida et al., 1986; Pekkala et al., 1984; Welch and Suhan, 1985).
3. Occasionally, effects on mitochondrial ultrastructure were detected, e.g. a swelling of mitochondria in mouse neuroblastoma cells or rat fibroblasts (Van Bergen en Henegouwen et al., 1987; Welch and Suhan, 1985) and the formation of crystals or granules in mitochondria of a chrysophycean flagellate (Schnepf and Schmitt, 1981) and plants (Mansfield et al., 1988; Risueno et al, 1973).
4. Using antibodies against cytoskeletal proteins dramatic rearrangements of cellular fine structure could be visualized in insect and vertebrate cells. This concerns the microtubules (tubulin), intermediate filaments (vimentin and related proteins) and the microfilaments (actin). Particularly striking and relevant is the collapse of the intermediate filament system around the nucleus (Borelli et al., 1986; Collier and Schlesinger, 1986, Glass et al., 1985; Falkner et al., 1981; Leicht et al., 1986; Wiegant et al., 1985; Welch and Suhan, 1985). There is good evidence that the cytoskeletal rearrangements are necessary for cell survival (Van Bergen en Henegouwen and Linnemans, 1987; Wiegant et al., 1987a, b) and that members of the HSP70 family play an

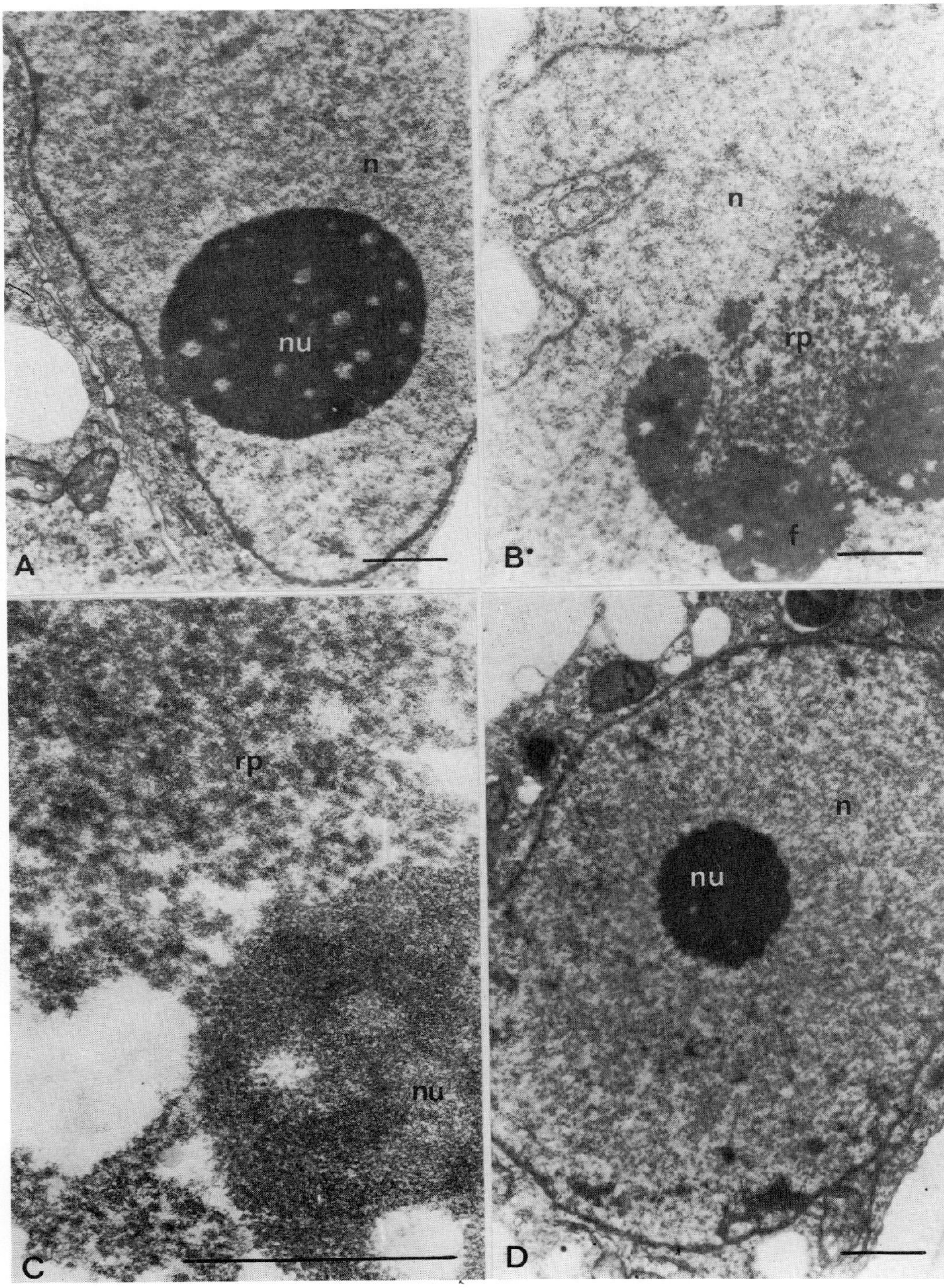
n
nu
A
n
rp
f
B
rp
nu
C
n
nu
D

important role in the restoration of the original state (see Sect. B.4.3.). Disregarding the frequent use of protoplasmic streaming for analysis of hs effects on the function of the plant microfilament system (SACHS, 1864; see summaries by ALEXANDROV, 1977, 1979), direct experimental data on cytoskeletal rearrangements in plant systems are lacking.

5. The changes of the cytoskeleton also affect the tight interaction with membrane constituents and the cell-cell or cell-substrate contacts of animal cells. Cells in culture loose their substrate attachment and round up. A microsorting of membrane components may lead to formation of intramembranous protein aggregates and the extrusion of membrane-surrounded parts of the cortical cytoplasm (blebbing). Under severe stress conditions, formation of large blebs is correlated with the extent of cell killing (BASS et al., 1982; BORELLI et al., 1986; KAPISZEWSKA and HOOPWOOD, 1986; RICE et al., 1987; WIEGANT et al., 1987b)

In the following we will concentrate on two generally observed alterations of cell ultrastructure and their possible functional significance in the frame of the regulatory network operative in the hs response. These are the changes of nucleolar ultrastructure and the accumulation of small HSPs in cytoplasmic aggregates (heat shock granules).

B.3.2. The nucleolus and effects on ribosome biosynthesis

The first step of ribosome biosynthesis is the binding of ribosomal proteins to the nascent pre-rRNA in the nucleolus. Formation of 80 S pre-ribosomes is followed by a complicate multistep maturation pathway by pre-rRNA processing and sequential addition of further r proteins in the nucleus. The final steps of this processing line are accomplished after export of the pre-ribosomal subunits to the cytoplasm. From the point of view of nucleolar morphology, the two basic activities of pre-ribosome assembly and their processing can be assigned to the fibrillar parts containing the rDNP and RNA polymerase I and the granular parts containing the ribosomal processing intermediates (pre-rRNP) (for summaries see BIELKA, 1982; FAKAN and PUVION, 1980; GRIERSON, 1982; HADJIOLOV, 1985; SOMMERVILLE, 1986; SOLLNER-WEBB and TOWER, 1986).

Early investigations with *Drosophila* and mammalian systems demonstrated that ribosome synthesis is a typical housekeeping function of cells, which is immediately blocked by heat shock. Simultaneously, the nucleolar ultrastructure undergoes marked changes characterized by disintegration, loss of the fine granular material (degranulation) and formation of large perichromatin granules (ARRIGO et al., 1980; BELL et al., 1988; ELLGAARD and CLEVER, 1971; HEINE et al., 1971; LOVE et al., 1970; SIMARD and BERNHARD, 1967; SPRADLING et al., 1977; WAROCQUIER and SCHERRER, 1969; see summaries by FAKAN and PUVION, 1980 and HADJIOLOV, 1985). Similar observations, either based on biochemical or on ultrastructure research, were reported for other organisms as well, e.g. for *Neurospora crassa* (TON-THAT and TURIAN, 1984; TON-THAT et al., 1981), *Physarum polycephalum* (CADRIN et al., 1984; MATSUMOTO et al., 1987), *Saccharomyces cerevisiae* (JOHNSTON and SINGER, 1980; KIM and WARNER, 1983; WARNER, 1982), *Tetrahymena* spec. (BYFIELD and SCHERBAUM, 1967; HERMOLIN and ZIMMERMAN, 1976; YUYAMA and ZIMMERMAN, 1972) and plants (DAWSON and GRANTHAM, 1981; FRANSOLET et al., 1979; KHOKHLOVA et al., 1987; MANSFIELD et al., 1988; RISUENO et al., 1973).

In Figs. 19 to 21 these ultrastructural changes of the nucleolus are documented for a great number

---

Fig. 19. Heat-shock induced changes of nucleolar ultrastructure in tomato suspension culture cells.
A, control, **B–D**, hs for 2 hrs at 41 °C (regime III). Depending on the state of ribosome synthesis prior to the hs treatment, the nucleolus appears highly dispersed with accumulation of preribosomal RNP (B, C, rp), or it becomes pyknotic (D). Bars, 1 µm. For electronmicroscopy cells were prefixed with 3 % glutaraldehyde in phosphate buffer, fixed with 1 % OsO_4/Palade buffer and embedded in Araldite. Ultrathin sections were stained with Pb (for details see NEUMANN et al., 1984, 1987). The following abbreviations are used in this and similar pictures: c, cytoplasm; d, dictyosome; er, endoplasmic reticulum; hg, heat shock granules; m, mitochondria; n, nucleus; nu, nucleolus; p, plastids; ps, polysomes; rer, rough endoplasmic reticulum; rp, ribonucleoprotein particles; v, vacuole; w, cell wall

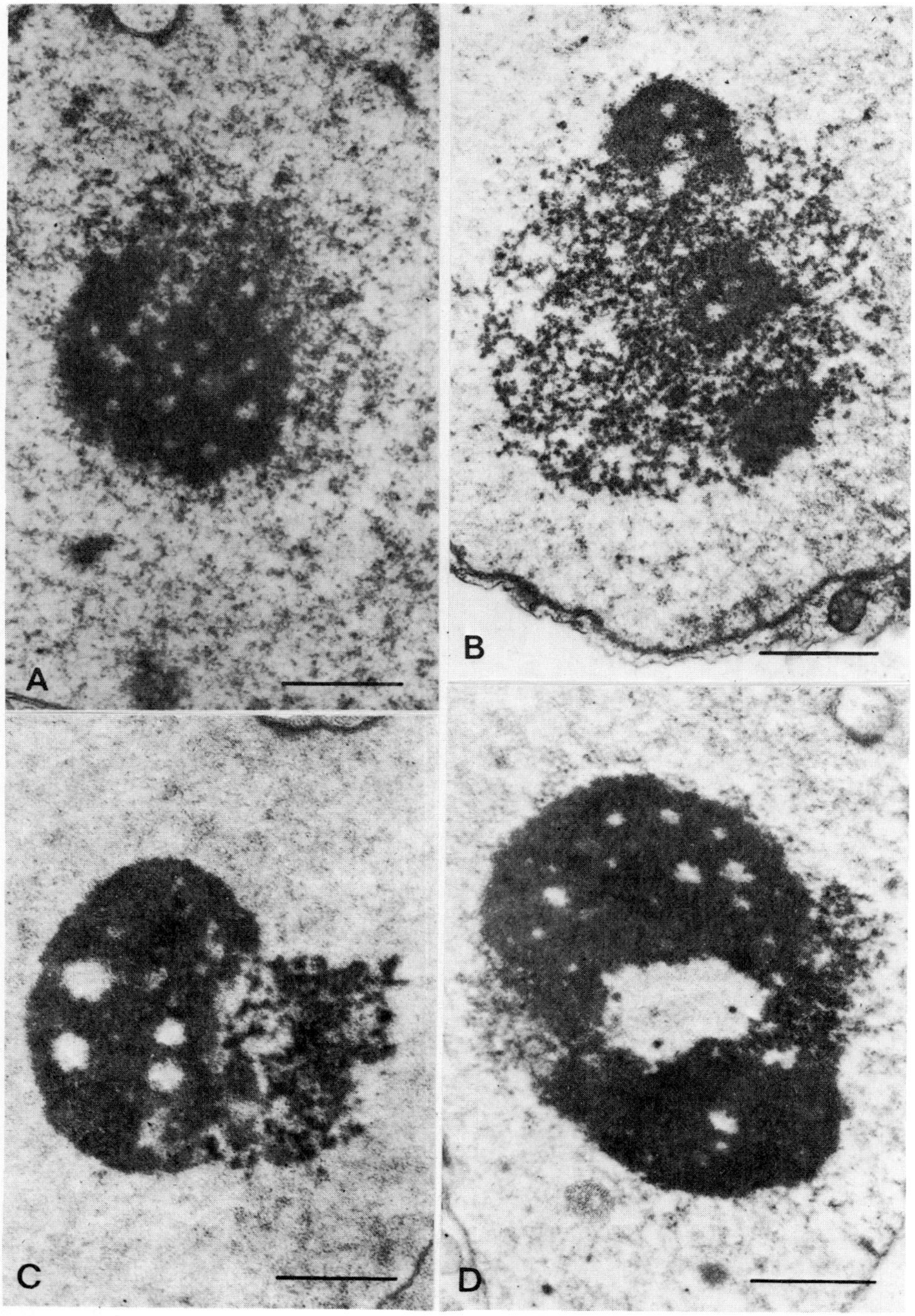

Fig. 20. Nucleoli from heat shocked cell cultures.
Heat shock regime I. For explanations see legend to Fig. 19. **A**, *Catharanthus roseus* (40 °C); **B**, *Peganum harmala* (38 °C); **C**, *Digitalis lanata* (40 °C); **D**, *Petroselinum crispum* (40 °C). Bars, 1 μm

of plants, including suspension cultured cells and different plant organs. Using *L. peruvianum* suspension cultures we studied many aspects of the corresponding changes of ribosome biosynthesis in detail (NEUMANN et al., 1984, 1987; NOVER et al., 1986; SCHARF and NOVER, 1982, 1987). The following results are noteworthy:

(i) With ongoing hs the initial loosening of nucleolar structure and the disappearance of the fine granular material is followed by a total disintegration and spectacular accumulation of 50–80 nm granules. Finally, the fibrillar material with the rRNA genes may form small islets at the periphery of the nucleolar region (Fig. 19B). The extent of nucleolar transformations are evidently dependent on the activitiy of ribosome synthesis, i.e. they are most pronounced in rapidly dividing suspension culture cells. Cells of organized plant tissues are intermediate, whereas in slowly growing or resting cultures nucleoli undergo shrinkage and become pyknotic (Fig. 19D, see NEUMANN et al., 1984).

(ii) With the onset of hs the formation of mature ribosomal subunits stops immediately and completely. No labeling of ribosomal RNA or proteins can be detected in the cytoplasmic fraction (NOVER et al., 1986; SCHARF and NOVER, 1982, 1987).

(iii) In contrast to the lacking labeling of mature ribosomes, ^{32}P-phosphate or ^{35}S-methionine incorporation into ribosomal precursor particles in the nucleolus goes on at about 30–50% of the level found in non-hs cells. Pre-rRNP processing is evidently the most heat-sensitive step of ribosome biosynthesis. Changes of nucleolar ultrastructure are mainly the result of the proceeding accumulation of pre-ribosomal material as coarse granular material. Under appropriate hs conditions freshly synthesized heat shock preribosomes are functionally intact and can be quantitatively processed into mature cytoplasmic subunits, if the cells are allowed to recover. But this integrity is evidently maintained only for a relatively short time. If the stress period is continued, most of the pre-rRNP becomes defective forming 'ribosomal wastage', whose removal with regeneration of the normal nucleolar structure and function also proceeds only possible in the recovery period (NOVER et al., 1986; SCHARF and NOVER, 1987).

(iv) Comparative investigations with preinduced vs. non-preinduced cultures show a pronounced thermotolerance effect at the level of nucleolar function. Maintenance of a relatively high level of activity during hs and the almost immediate recovery after the stress period are only observed in preinduced cultures. This protective effect is correlated with a massive accumulation of HSP70 in the granular material of the nucleolus (NEUMANN et al., 1987; NOVER et al., 1983 and Sect. B.4.3.). In agreement with experiments of PELHAM (1984), using monkey COS cells transformed with a constitutively expressed *Drosophila* hsp70 gene construct, the normalization of nucleolar functions in the recovery period seems to be much enhanced in the presence of excess HSP70.

B.3.3. Cytoplasmic heat shock granules

A second remarkable alteration of cell ultrastructure is the formation of electron-dense cytoplasmic granular complexes (heat shock granules, HSG). They were identified and extensively studied in tomato suspension cultures (NEUMANN et al., 1984, 1987; NOVER and SCHARF, 1984; NOVER et al., 1983, 1989). But probably the earliest electron microscopic documents for the hs-induced formation of 'dense corpuscules' in root tip meristems of onion and corn can be found in papers of RISUENO et al. (1973) and FRANSOLET et al. (1979), respectively.

The protein part of HSGs is mainly composed of a distinct subset of HPSs(see Fig. 30 and NOVER et al., 1983). Their formation depends on hs conditions and is not a consequence of HSP synthesis per se. Under normothermic conditions these HSPs are rather part of much smaller precursor particles of 15S (pre-HSG), which were identified and purified from the postpolysomal supernatant (NOVER et al., 1989). Reversible heat shock-induced aggregation to the HSG complexes involves 5-10% of the total HSP70 and 50–80% of the small HSPs (NEUMANN et al., 1984; NOVER and SCHARF, 1984).

HSG are ubiquitous cytoplasmic structures of heat shocked plant cells. They can be easily detected

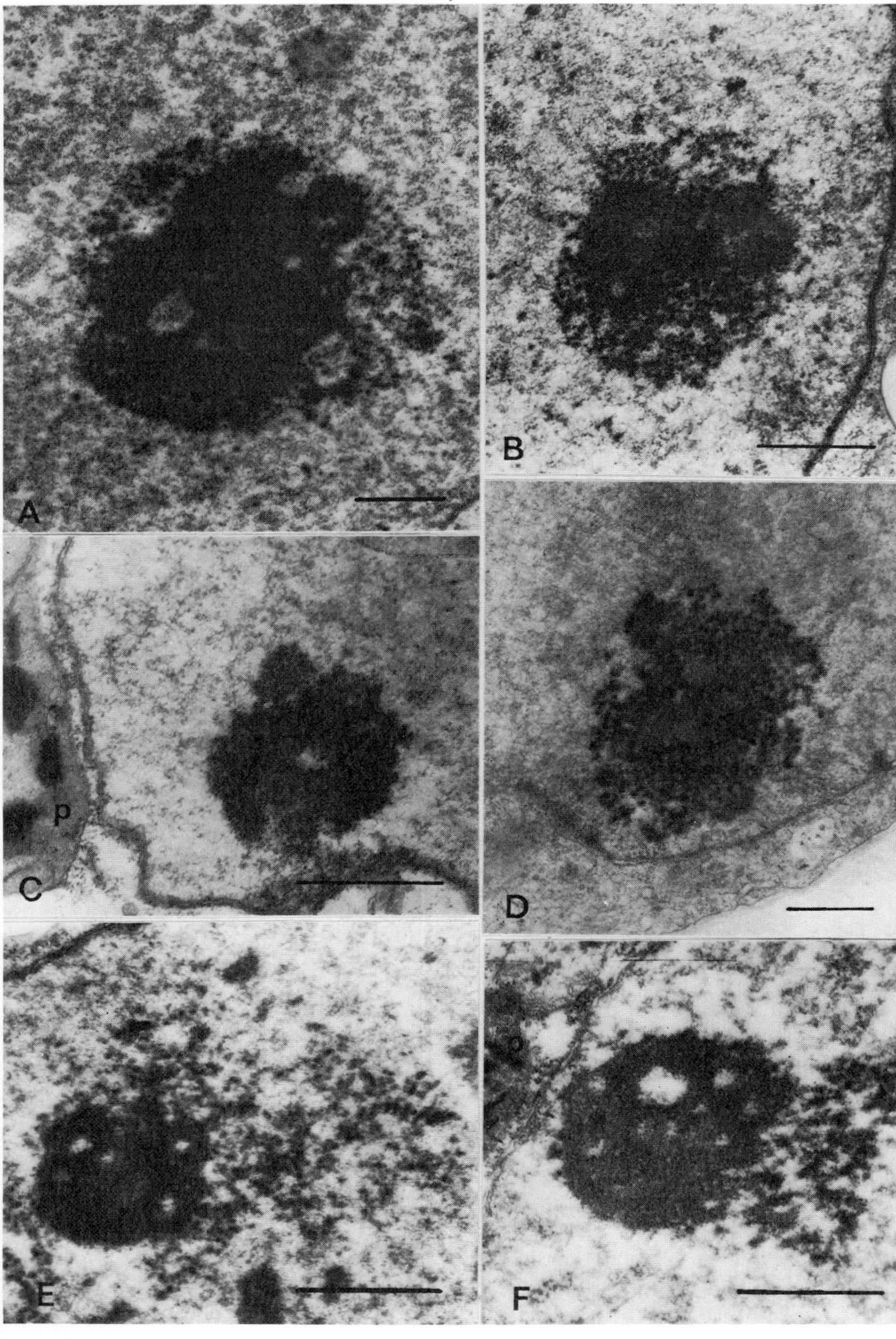

Fig. 21. Nucleoli from heat shocked plant organs.
Heat shock regime I. For explanations see legend to Fig. 19. **A–E**, leaves of **A**, *Hordeum vulgare* (40 °C); **B**, *Nicotiana rustica* (40 °C); *Cucurbita pepo* (40 °C); **D**, *Pisum sativum* (40 °C); **E**, *Catharanthus roseus* (40 °C); **F**, *Lyco-*

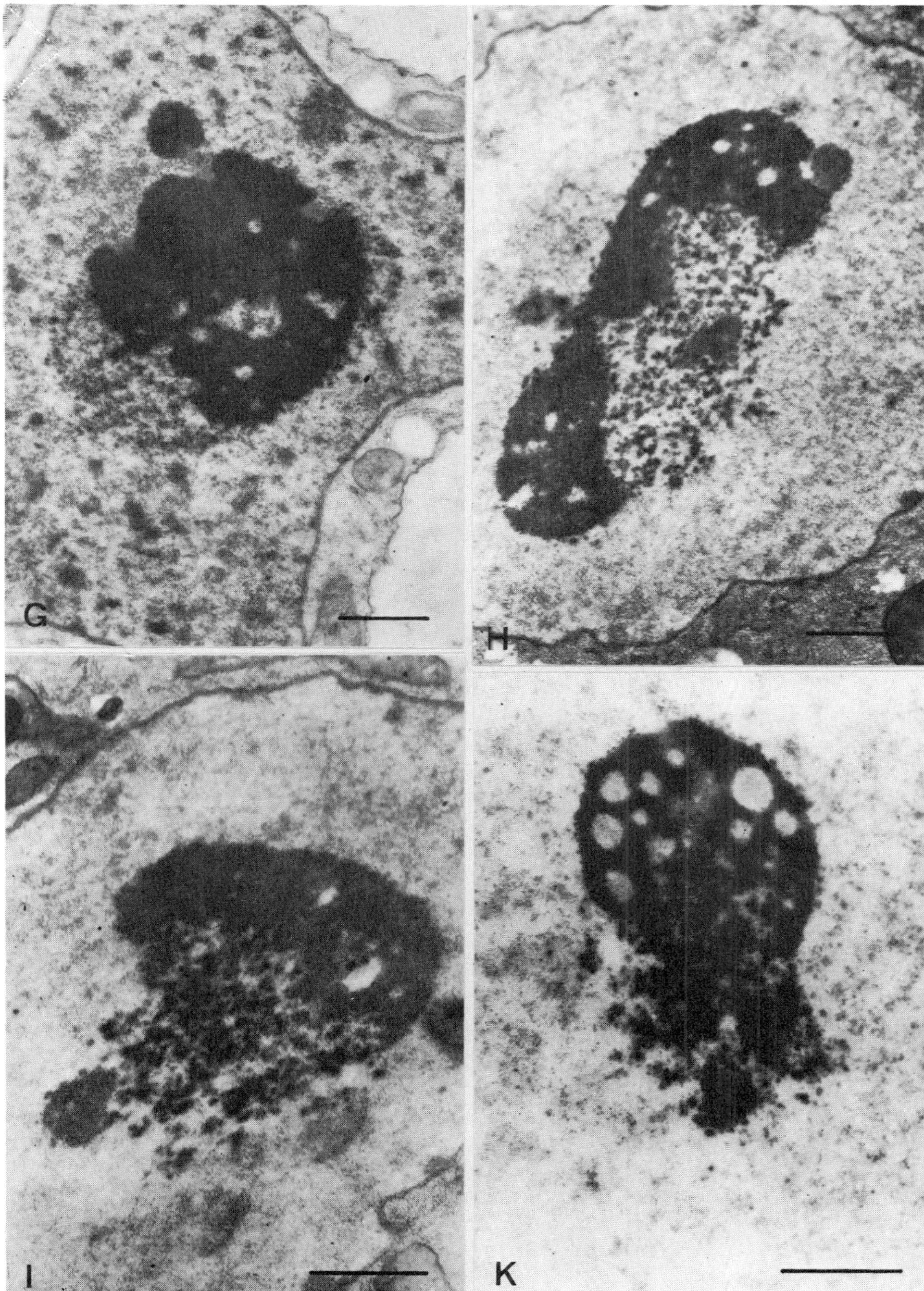

persicon esculentum ovary (41 °C); **G–I** roots of **G**, *Zea mays* (41 °C); **H**, *L. peruvianum* (41 °C); **I**, *L. esculentum* (41 °C) and **K**, *Glycine max* hypocotyl (43 °C). Bars, 1 μm

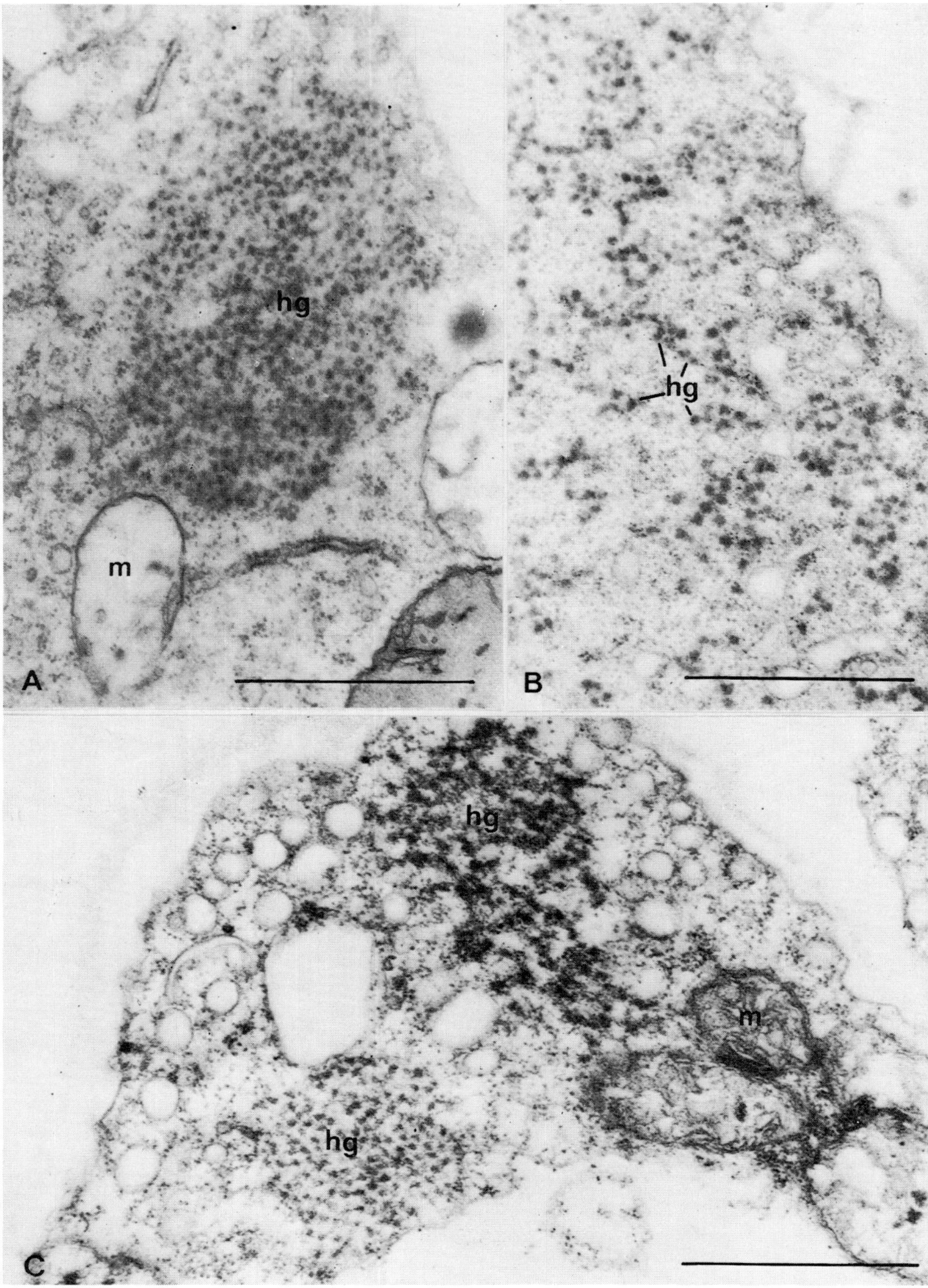
hg
m
A
hg
B
hg
m
hg
C

by their electron dense structure and their size. For unknown reasons, the morphology of the HSG is variable. In tomato cell cultures 40 nm granules form either large complexes (Fig. 22 A) or dispersed networks (Fig. 22 B). In addition, 80 nm granules can be observed (Fig. 22 C).
These figures are complemented by extensive documentations on HSG formation and morphology in cell suspension cultures and plant organs (Figs. 23 to 25). Though Mansfield and Key (1988) confirmed, by biochemical data, the existence of HSG in *Glycine max* seedlings, they were unable to detect them in electronmicrographs. The authors speculate that the HSGs exist in a precursor state and aggregate during the isolation procedure. This is certainly not true since HSG can also be detected in cells of heat-shocked soybean seedlings (Fig. 25 M).
As described earlier (Scharf and Nover, 1982), under appropriate hs conditions more than 90 % of the labeled precursor amino acid is incorporated into HSP. This provides a second simple criterium for HSG identification using electronmicroscopic microautoradiography. The preferential ^{3}H-leucin labeling clearly marks two major cellular sites of HSP accumulation. These are the nucleolus and the HSG (Neumann et al., 1984; Nover et al., 1983). Corresponding results for HSG obtained with 10 different plant objects are illustrated in Figs. 26–28.
HSG formation is most pronounced in metabolically active cells, e.g. in rapidly growing suspension cultures or meristems of whole plants. There is probably a simple correlation with the amount of HSPs accumulated during hs. Due to their sedimentation behaviour (1 hour at 61.500 x g_{av}) and their extraordinary stability in the presence of 500 mM KCl or NaCl, 30 mM EDTA and 1 % Nonidet P 40 or Triton X100 HSGs can be relatively easily separated from most other cell constituents (Nover et al., 1983, 1989), which are destroyed or not sedimentable under these conditions. However, two important problems may arise:

(i) In organized plant tissues mass accumulation of HSGs is usually restricted to few layers of meristematic cells, whereas only few HSG complexes are detected in specialized cells. Thus, the overall yield of HSGs may be very low compared to a cell suspension culture (see the example given by Mansfield and Key (1988) for soybean seedlings).
(ii) If relatively mild buffer conditions are applied to preserve the integrity of cell organelles (nuclei, mitochondria, chloroplasts, polysomes), the heterogeneity in size of HSG and their tight connection with other cellular structures (cytoskeleton?) evidently leads to an irreversible contamination of other organelle fractions. Careful inspection of isolated nuclei and chloroplasts by electron microscopy always reveals the presence of residual HSG. This is also true for polysomes, particularly if they are sedimented together with HSG prior to fractionation on a sucrose gradient (Mansfield and Key, 1988). Thus, the specific subset of HSPs in mitochondria, chloroplasts and nuclei may be obscured by HSG-bound HSP. This explains many artifactual results of cell fractionation procedures published in the literature (Cooper and Ho, 1987; Lin et al., 1984; Mansfield and Key, 1988; Restivo et al., 1986). Only immunological studies combined with electronmicroscopic techniques can help to overcome these difficulties (see Sect. B.4.3.).

Heat shock granules contain RNA synthesized in the period immediately preceding the hs. This was shown by microautoradiography (Neumann et al., 1984) and by banding of fixed HSG in CsCl gradients (Nover et al., 1989). By careful purification of HSG and comparison with polysomal fractions, two distinct mRNA populations could be identified: (i) mRNAs coding mainly for the transiently non-translated control proteins in the HSG and (ii) hs-mRNAs in the polysomal fraction. Thus, probably in connection with cytoskeletal elements, plant HSGs evidently serve as mRNA storage sites (Nover et al., 1989). Of particular interest in this respect are morphological analyses of HSG complexes in the recovery period. The granules become less compact and are surrounded by organelles indicative of a highly active protein synthesis (polysomes, rER, dictyosomes see Fig. 29). Evidently, the rapid reactivation of the HSG-bound mRNAs signalizes an urgent need for short

---

Fig. 22. Morphology of heat shock granules in tomato *(L. peruvianum)* suspension culture cells.
Heat shock regime III, 2 hrs at 41 °C. For explanations see legend to Fig. 19. **A**, HSG cluster with 30-40 nm particles; **B**, dispersed form; **C**, clustered HSG together with highly electron dense 80 nm particles. Bars, 1 µm

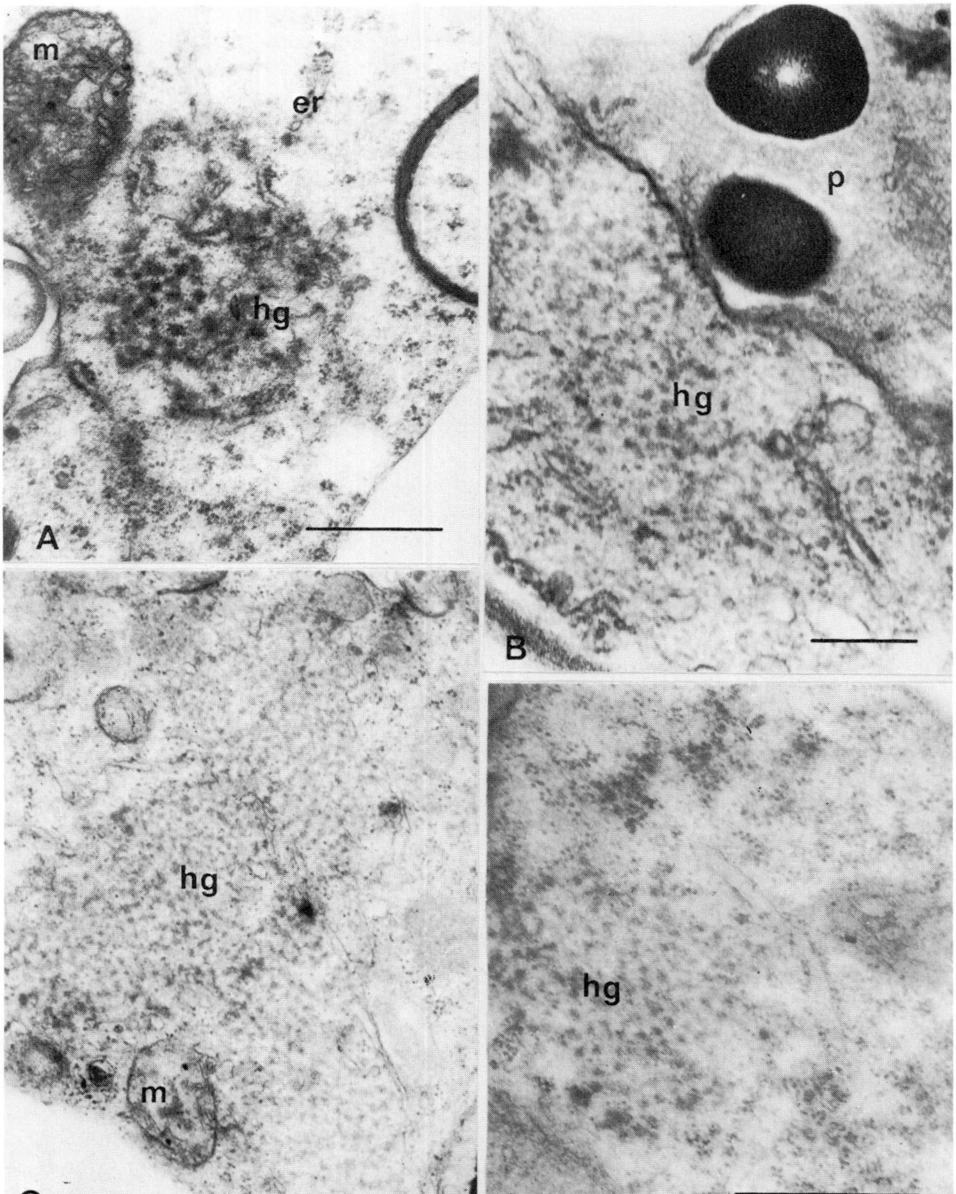

Fig. 23. Heat shock granules in cell cultures.
Heat shock 3 hrs, regime I. For explanations see legend to Fig. 19. **A**, *Catharanthus roseus* (40 °C); **B**, *Digitalis lanata* (40 °C); **C**, *Petroselinum crispum* (40 °C); **D**, *Neobesseya similis* (cactus, 48 °C). Bars, 0.5 μm

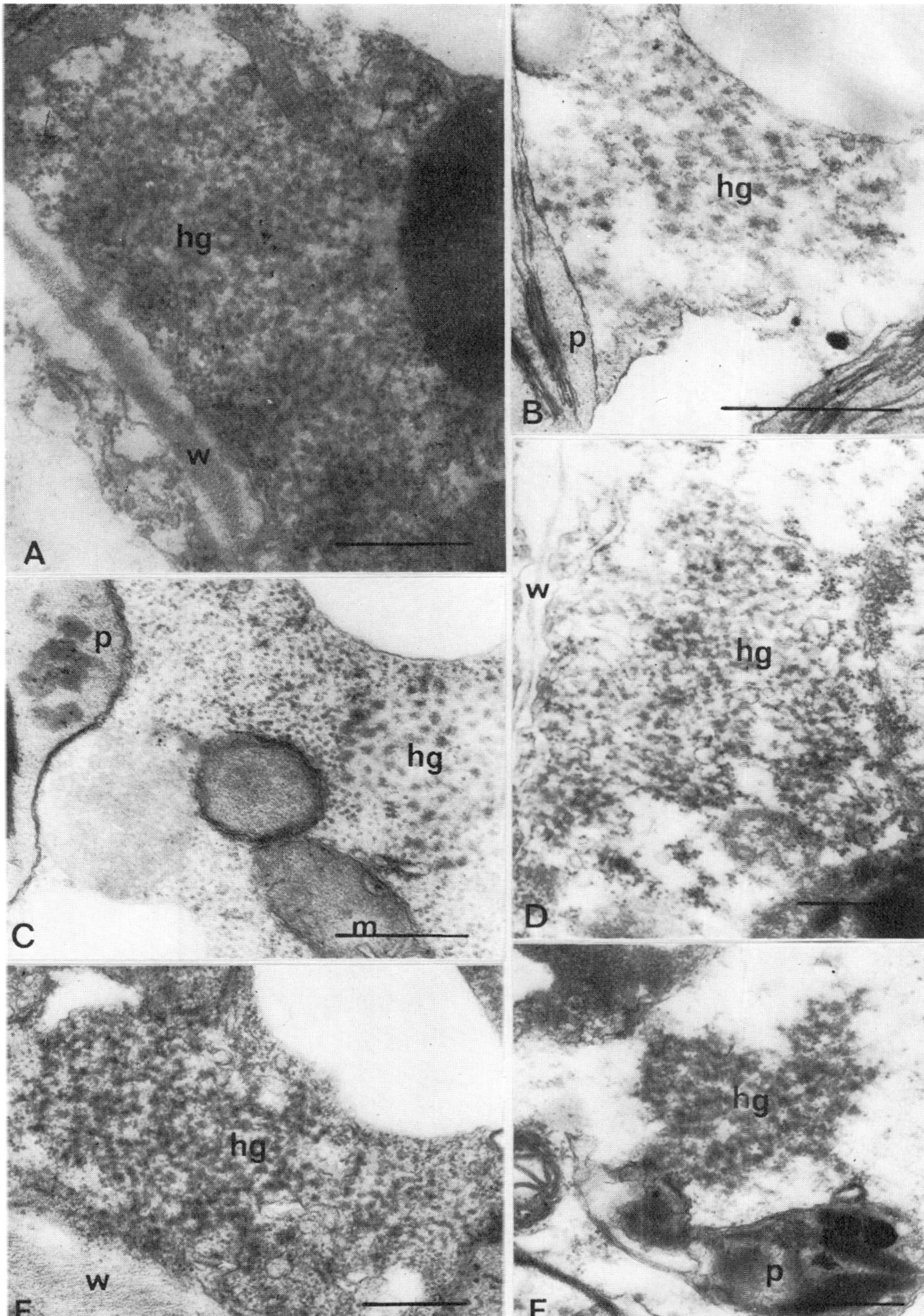

Fig. 24. Heat shock granules in different organs of tomato *(Lycopersicon esculentum)*.
Heat shock for 3 hrs at 40 °C (regime I). For explanations see legend to Fig. 19. **A**, roots; **B**, leaves; **C**, cotyledon; **D**, ovary; E, anther; **F**, sepals. Bars, 0.5 μm

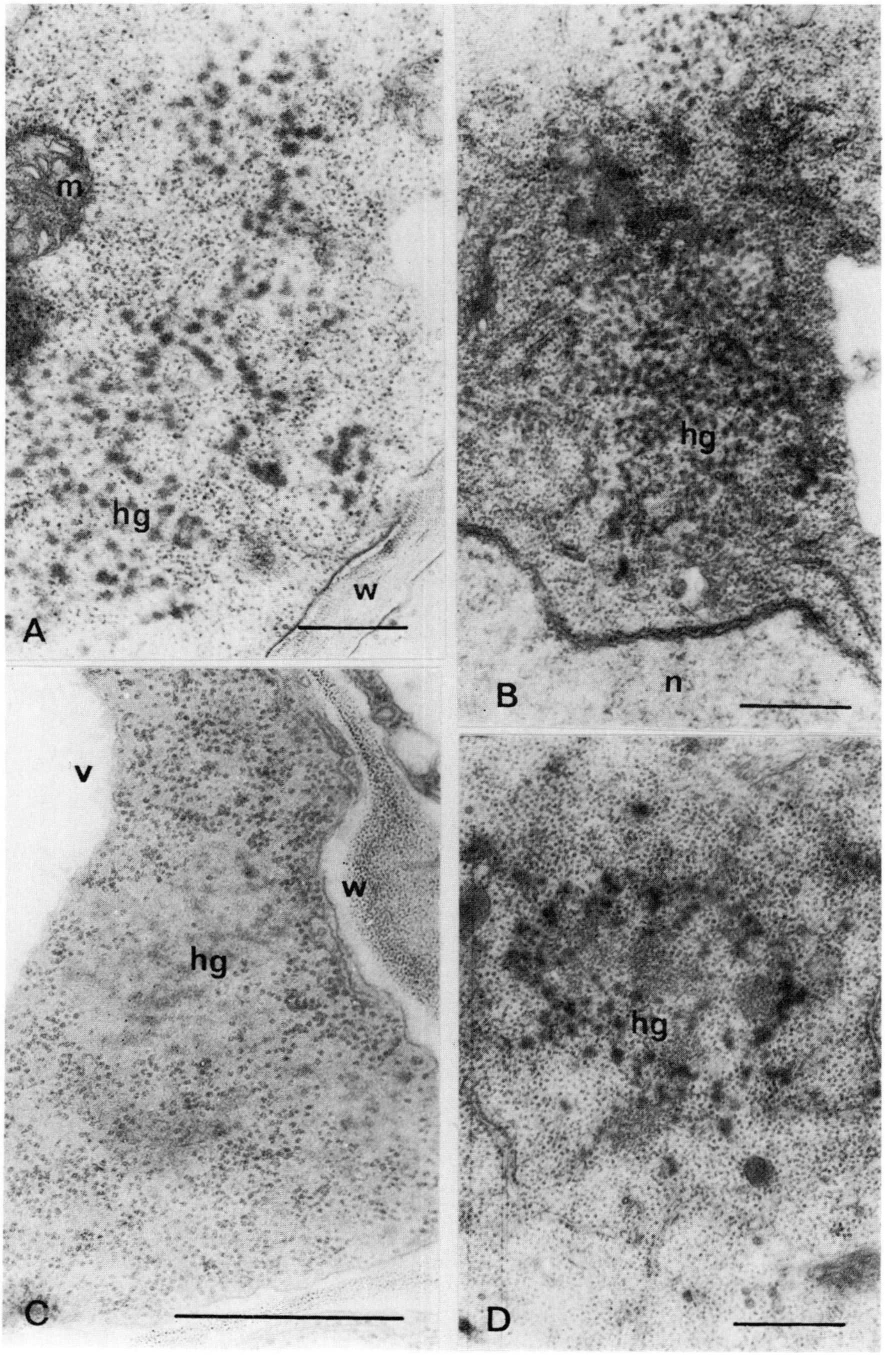

Fig. 25. Heat shock granules in different plants.
Three hours of heat shock, regime I. For explanations see legend to Fig. 19. **A, B, D** roots of *Zea mays* (A, 41 °C); *Lycopersicon peruvianum* (B, 41 °C) and *Glycine max* (D, 43 °C); **C, E–G** and **I–L** leaves of *Arabidopsis thal-*

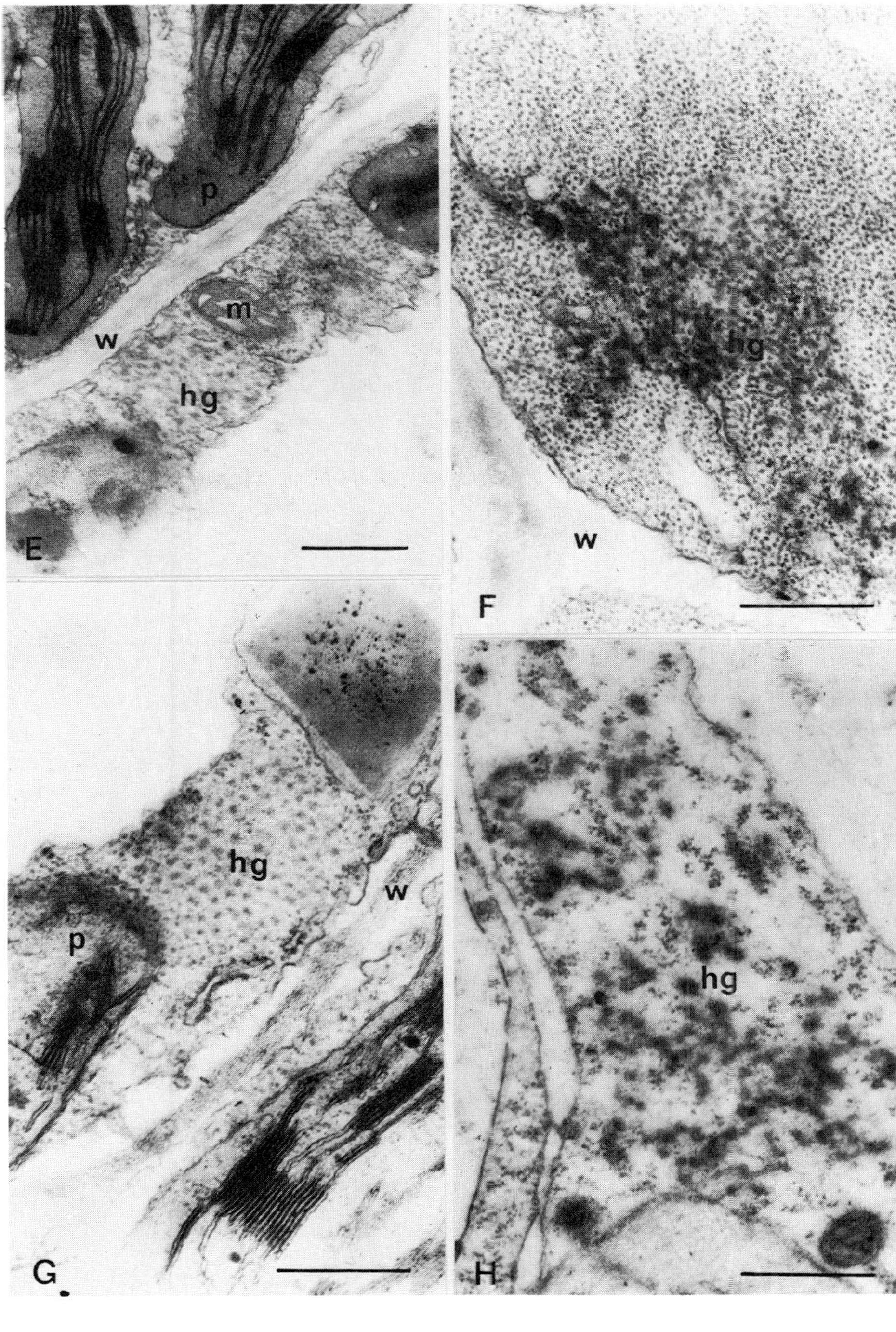

iana (C, 40 °C), *Catharanthus roseus* (E, 40 °C); *Pisum sativum* (F, 40 °C); *L. peruvianum* (G, 41 °C); *Hordeum vulgare* (I, 40 °C); *Cucurbita pepo* (K, 40 °C); *Nicotiana rustica* (L, 40 °C); **H**, *Zea mays* coleoptile (41 °C); **M**, *Glycine max* hypocotyl (43 °C). Bars, 0.5 μm

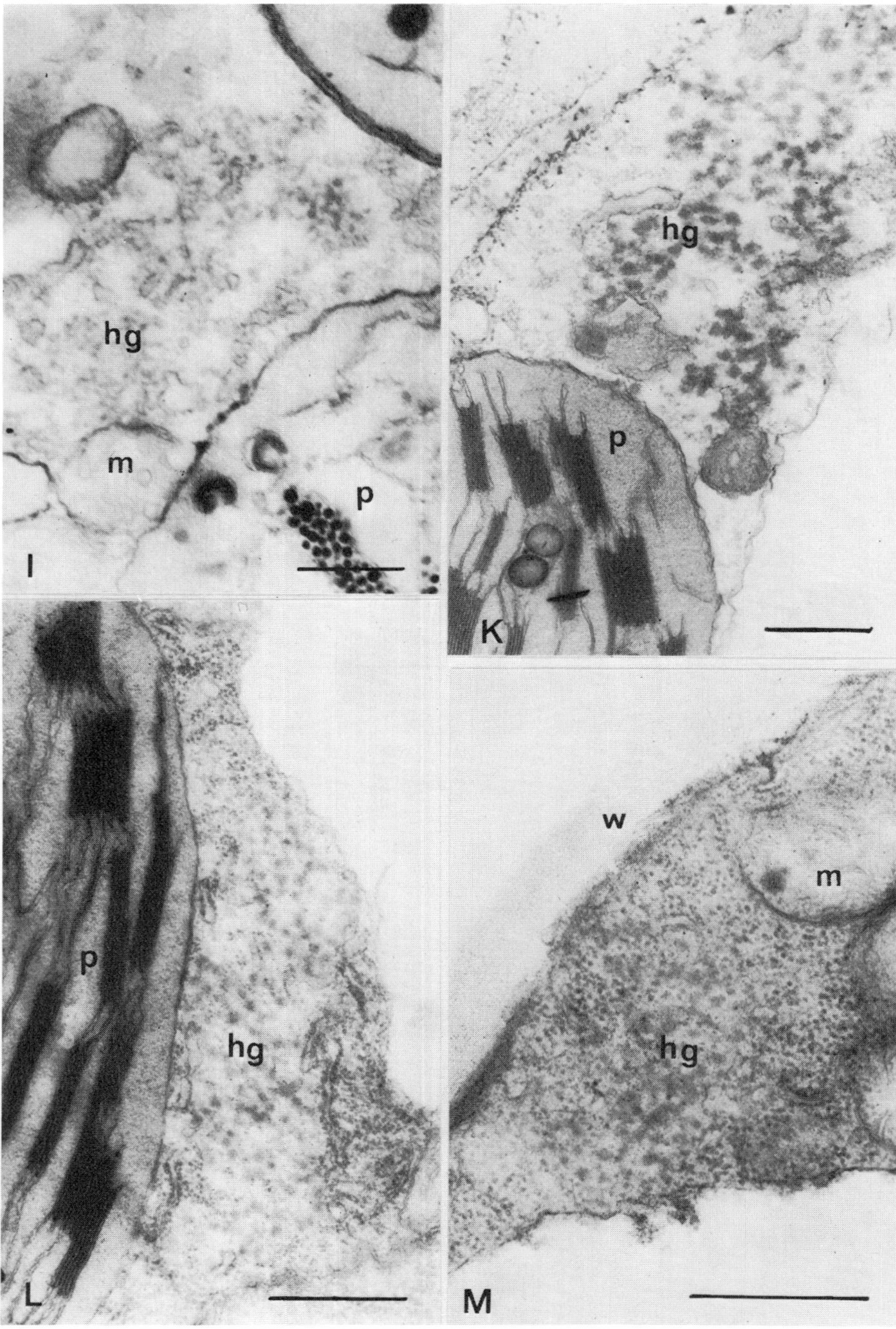
hg
m
p
I
hg
p
K
p
hg
L
w
m
hg
M

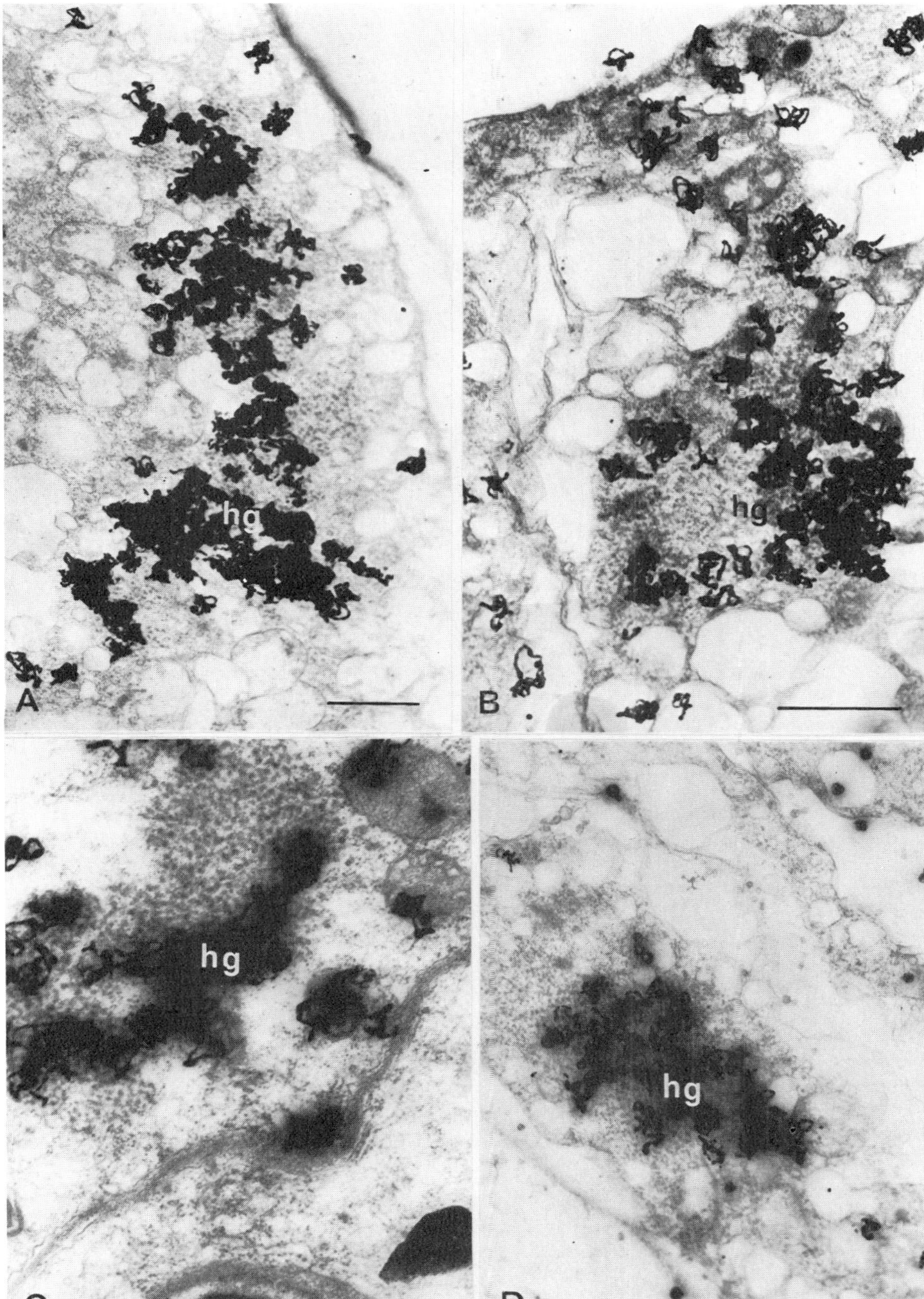

Fig. 26. Preferential accumulation of HSPs in heat shock granules formed in different suspension culture cells. The highly selective labeling of HSPs under conditions of severe hs allows intracellular localization studies by feeding ^{3}H-L-leucine followed by electronmicroscopic microautoradiography (for details see NEUMANN et al., 1984). **A–D**, cell cultures of *L. peruvianum*, **(A)**; *Catharanthus roseus* **(B)**; *Digitalis lanata* **(C)**; *Peganum harmala* **(D)**. Bars, 1 µm

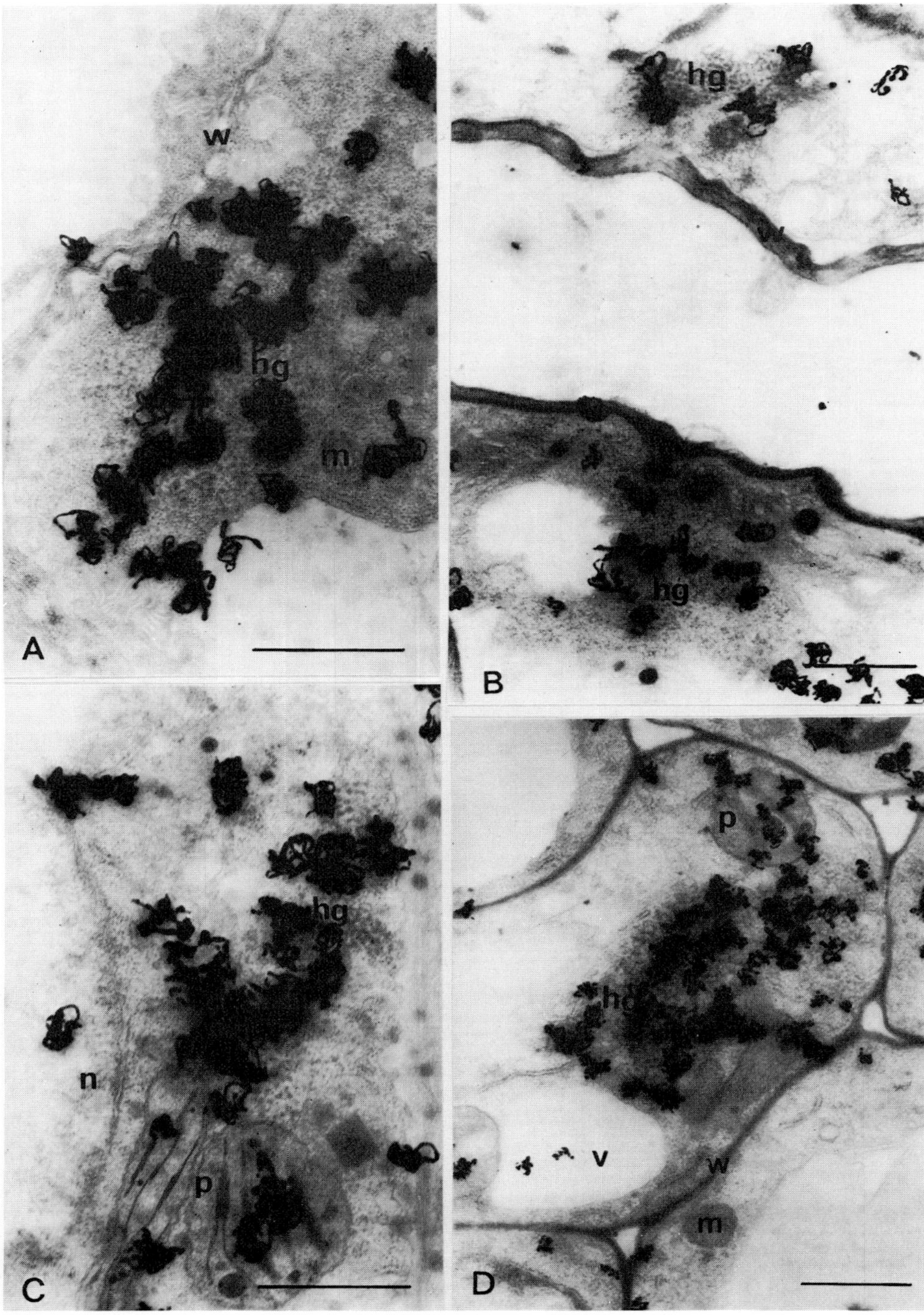

Fig. 27. Preferential accumulation of HSPs in heat shock granules in different organs of tomato *(L. esculentum).*
For experimental details see legend to Fig. 26. **A**, root; **B**, leaf; **C**, ovary; **D**, anther. Bars, 1 μm

lived proteins, whose concentration declined beyond critical levels during the stress period (NOVER et al., 1989).

Studies on the intracellular localization of small HSPs in animal cells led to controversial results. On the one hand, there is evidence for HSG-like aggregates also in *Drosophila* and vertebrates (ARRIGO, 1987; COLLIER et al., 1988; LEICHT et al., 1986). But the close association of the HSG with the perinuclear complex of the intermediate size cytoskeleton (LEICHT et al., 1986) may explain a number of artifacts reported on the nuclear location of small HSPs in *Drosophila* (ARRIGO and AHMAD-ZADEH, 1981; ARRIGO et al., 1988; KLOETZEL and BAUTZ, 1983; LEVINGER and VARSHAVSKI, 1981; SINIBALDI and MORRIS, 1981).

On the other hand, there are new reports in support of a nuclear binding of small HSPs as well. Thus, immunofluorescence data demonstrate a nuclear association of HSP26 in yeast (ROSSI and LINDQUIST, 1989) and of HSP27 in *Drosophila* (BEAULIEU et al., 1989). For *Drosophila,* this result would indicate major differences in the behaviour of the closely related HSP23 (cytoplasmic aggregates, see ARRIGO, 1987; LEICHT et al, 1986) and HSP27 (nuclear locale during hs, see BEAULIEU et al., 1989). Because electron microscopic data are lacking, an aggregation of HSP27 in the perinuclear region cannot be excluded. The results for vertebrate cells are equivocal. COLLIER et al. (1988) reported on the existence of cytoplasmic aggregates of the chicken HSP25, whereas ARRIGO et al. (1988) found pre-HSG-like complexes of the human HSP28 in the perinuclear Golgi complex of HeLa cells under control conditions. After hs large aggregates of the HSP28 are formed in the nuclei. The discrepancy cannot be explained at present because the methods used for analyses of the chicken and the HeLa cells are not comparable.

B.4. Induced thermotolerance and intracellular localization of heat shock proteins

B.4.1. Intrinsic heat sensitivity and different states of thermotolerance

Thermotolerance was introduced as a term by HENLE and DETHLEFSEN (1978) in a review on the application of fractionated hyperthermia in cancer therapy. But many years earlier and long before the detection of heat shock proteins, plant physiologists were familar with phenomena of adaptation to heat stress called acquired tolerance to heat, heat hardening or adaptive heat resistance (ALEXANDROV et al., 1970; ENGELBRECHT and MOTHES, 1964; LANGE, 1961; SAPPER, 1935; YARWOOD, 1961, 1964). It is trivial to state that, unlike animals, plants are usually fixed to their natural habitats and thus, cannot escape stressful conditions. This is the reason for the long lasting interest in stress physiology of plants in general (see Part C) and in the response to heat stress in particular (SACHS, 1864). Many details of these early investigations were summarized by ALEXANDROV (1977) and ALEXANDROV et al. (1970).

With the detection of HSPs and their subsequent extensive characterization, the cellular aspects of the heat stress response and the possible role of the newly formed HSPs for induced thermotolerance became dominant research fields of molecular stress biology of plants and animals (see reviews and books by ATKINSON and WALDEN, 1985; BIENZ and PELHAM, 1987; KEY et al., 1983; KEY and KOSUGE, 1985; KIMPEL and KEY, 1985a; LINDQUIST, 1986; SCHLESINGER, 1986; Schöffl et al., 1984b). For sake of simplicity it is appropriate to discriminate between the intrinsic heat sensitivity of cells (primary thermostability, ALEXANDROV et al., 1970) and the thermotolerance induced by heat shock or chemical stressors. Unfortunately, our knowledge on the molecular basis of both parts is still very fragmentary. There are few well studied examples indicating a stress-related function of HSPs in the protection of subcellular compartments (see below). But it cannot be excluded that either changes of intrinsic heat sensitivity are connected with the hs-independent expression of HSPs (see Sect. B.1.2.1.) or at least affect the velocity or extent of HSP synthesis and thus survival under hs conditions. In addition, there are a number of non-proteinaceous metabolites (proline, polyols, phytohormones), which have a direct or indirect protective effect on the structure and function of cells under stress conditions. They may also act as modulators of the stress response, and their levels may vary independent of the developmental or physiological state and of environmental factors respectively (see Sect. C.2. and HONG-QI et al., 1984; NIETO-SOTELO and HO, 1986). Though

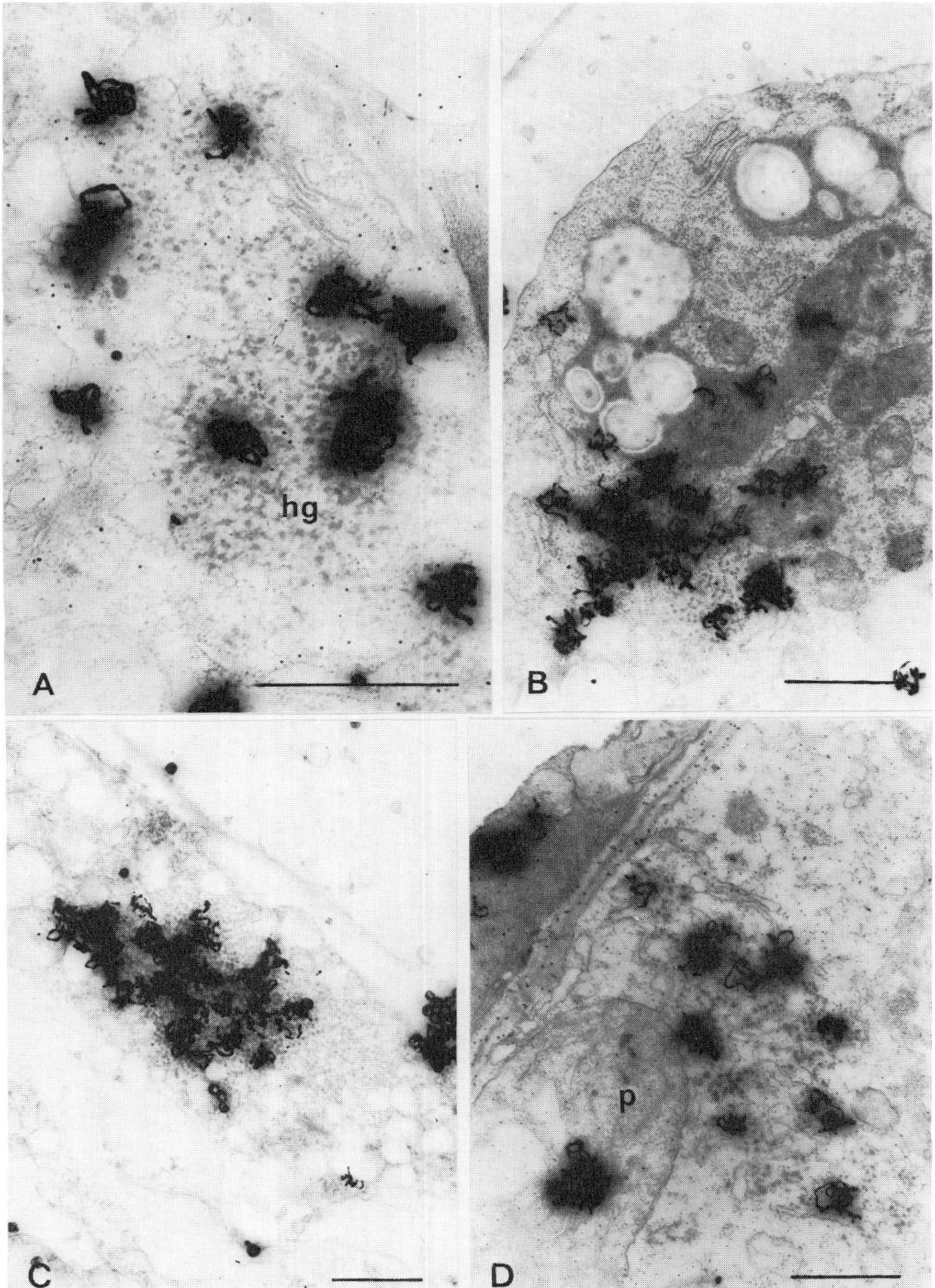

Fig. 28. Preferential accumulation of HSPs in heat shock granules of different plants.
For experimental details see legend to Fig. 26. **A**–**D**, leaves of *Hordeum vulgare* (**A**), *Nicotiana rustica* (**B**); *Cucurbita pepo* (**C**) and *Vicia faba* (**D**). **E**, roots of *L. peruvianum* and **F**, *Zea mays* coleoptile. Bars, 1 μm

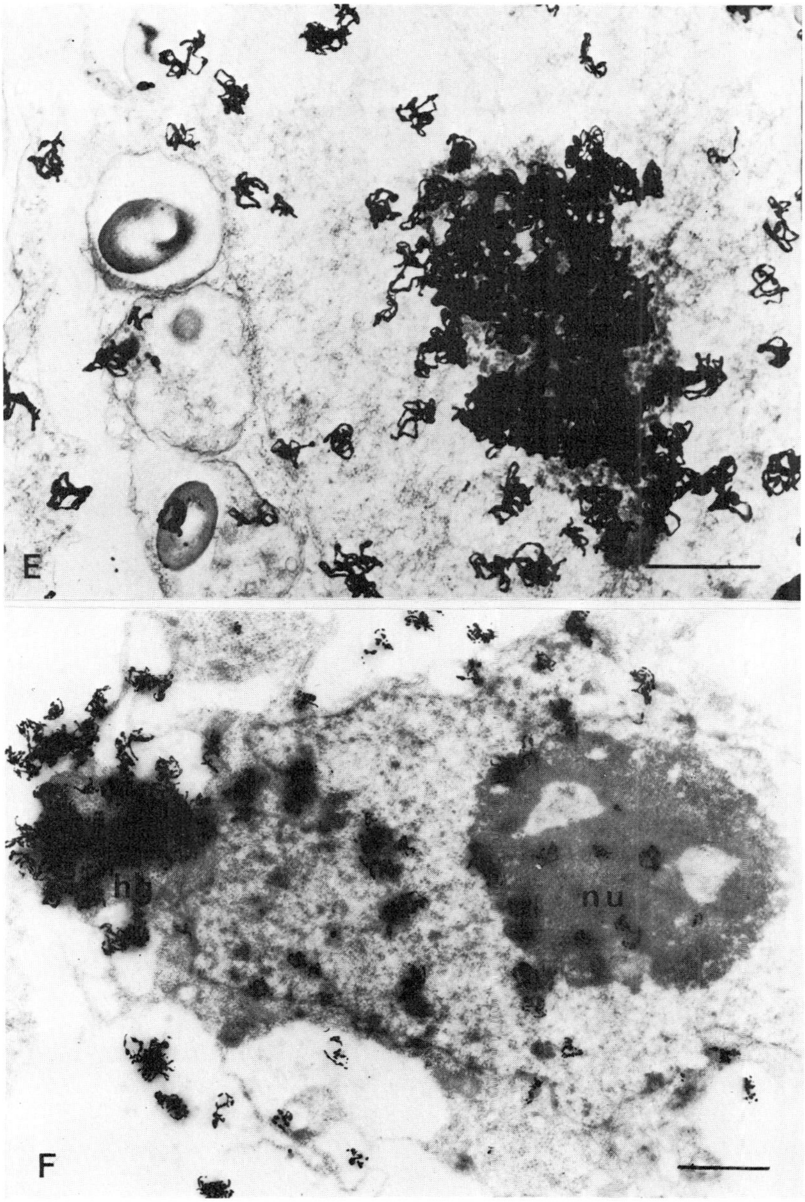

the molecular details are mostly unknown, such changes of the intrinsic heat sensitivity and developmental variation of the heat shock response are well known to plant physiologists (ABERNETHY et al., 1989; ALEXANDROV, 1977; ALEXANDROV et al., 1970; ALTSCHULER and MASCARENHAS, 1985; CHEN et al., 1986; ENGELBRECHT and MOTHES, 1964; KAPPEN and LANGE, 1968; LANGE, 1955, 1961; LANGE and SCHWEMMLE, 1960; SCHWEMMLE and LANGE, 1959; SAPPER, 1935; WU et al., 1984; XIAO and MASCARENHAS, 1985).

The experimental evidence for the existence of more and more stress-related metabolites on the one hand (see Table 4) as well as the synthesis of HSPs or other stress proteins on the other hand

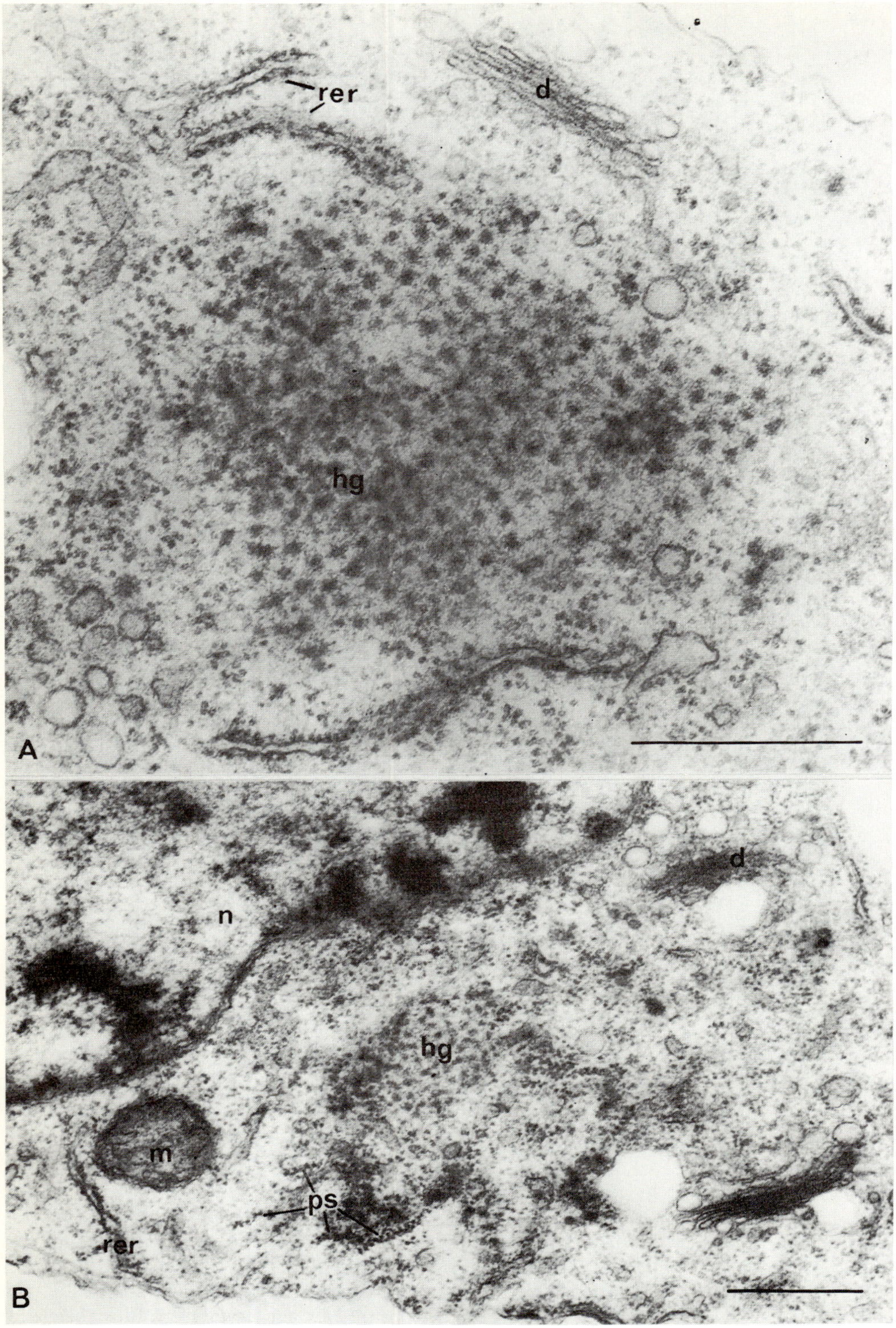
rer
d
hg
A
n
d
hg
m
ps
rer
B

characterize two important parts of induced stress tolerance. It is not suprising that, depending on the method and the heat shock regime applied, different aspects of induced thermotolerance are recorded. Though related to each other, they may not be directly comparable. The differential heat sensitivity of distinct cellular functions/compartments and the particular time kinetics of expression of protective or reparative mechanisms (see summaries by ALEXANDROV, 1977 and NOVER, 1990) explains the evident differences of the thermotolerant states observed either after a short and mild hs (a), after a moderate to severe hs applied for few hours (b) or after a long-term adaptation to growth or at least survival at elevated temperatures (c).

Experimental hs research mostly concentrates on conditions a and b. If both are compared, two different states of thermotolerance can be defined, and there is evidence that HSPs are involved only in the latter (BOON-NIERMEIJER et al., 1986). Compared to this, experimental data on adaptive cellular changes induced by long-term, moderate hyperthermia are rare, though they reflect much more the response under natural surroundings (ALEXANDROV, 1977; ALEXANDROV et al., 1970; DIDDEN-ZOPFY and NOBEL, 1982; DINAR, et al., 1983; KLEIN and FERGUSON, 1987; SÜSS and YORDANOV, 1986; WU and WALLNER, 1984; WU et al., 1984, see review in Chapter 17 of NOVER, 1990). Similar to the relatively slow component of the adaptive response in bacteria (BROCK, 1978, 1985; LJUNGDAHL, 1979), there are changes in the stability, composition and function of the plasma and thylakoid membranes (BENZIONI and ITAI, 1973; DIDDEN-ZOPFY and NOBEL, 1982; KLEIN and FERGUSON, 1987; SÜSS and YORDANOV, 1986; THEBUD and SANTARIUS, 1982; WU and WALLNER, 1984 and reviews by BERRY and BJÖRKMAN, 1980 and LANGRIDGE, 1963). Two remarkable effects of long-term hyperthermia are the destruction of the chloroplast structure connected with bleaching observed in *Euglena gracilis* (ORTIZ and WILSON, 1988) and seedlings of higher plants (see Sect. B.3.2.) as well as the eradication of virus infections from growing meristems of many cultural plants (see Sect. C.10.5.). This hyperthermic cure of diseased material with mass propagation of virus-free plants from meristem cultures ist a convenient method applied world-wide in horticulture (KASSANIS, 1957; WALKEY and COOPER, 1975; WANG, 1985).

B.4.2. Criteria for induced thermotolerance

To induce thermotolerance in plants, tissues or cell cultures, they are either subjected to a continuous, moderate hs, to a preinduction procedure or they are slowly adapted to increasingly severe hs. Two tightly related aspects have to be considered: (i) the protection of cellular fine structure during the stress period and (ii) the effective repair of eventual damages with restoration of the normal metabolic and gene expression activities mainly in the recovery period. The physiology of the repair of non-DNA injury was extensively reviewed by ALEXANDROV (1979). To give a crude estimate of the degree of thermotolerance expressed in a given plant system, it is convenient to measure the relative growth velocity, e.g. of seedlings (LIN et al., 1984; MARMIROLI et al., 1986; ZIVY, 1987), or the temperature of irreversible heat damage to tissues or cells (ENGELBRECHT and MOTHES, 1964; KEE and NOBEL, 1986; MANSFIELD et al., 1988; WU and WALLNER, 1983; YARWOOD, 1961, 1964; XIAO and MASCARENHAS, 1985). But, if analyzed in detail, thermotolerance affects distinct parts or functions of a cell differentially. To approach the molecular basis, it is essential to investigate them separately. If data are available, the references given in this review are restricted to experiments with plants only (for data on animal systems see summary by NOVER, 1990).

Plasma membranes: The particular heat sensitivity and the protection of the plasma membrane are documented by vital staining (KEE and NOBEL, 1986), electrolyte or amino acid leakage (BENZIONI and ITAI, 1973; DANIELL et al., 1969; WU and WALLNER, 1983) and by improved amino acid uptake (NOVER, 1984). Usually, the tonoplast membrane is much more stabile. Leakiness for anthocyanins

---

Fig. 29. Ultrastructure of heat shock granules in the recovery period.
L. peruvianum suspension cultures were treated for 2 hrs at 41 °C following hs regime III and then allowed to recover for 1 hr at 25 °C. Clusters of HSG become more dispersed and are surrounded by organelles indicating active protein synthesis (rER, dictyosomes, polysomes). A and **B** represent different magnifications of hs granules in the recovery period. For explanations see legend to Fig. 19. Bars, 0.5 μm

and suppression of plasmolysis are observed only under extreme hs conditions (SACHS, 1864; ALEXANDROV, 1979). Vacuoles isolated from *Valerianella* leaves were even stable up to 50 °C (WEIGEL, 1983).

Cytoskeleton: Evidently essential for the dynamic response of cells to stress conditions are rapid rearrangements of the cytoskeletal systems. Moreover, the restoration of the original state and function is much improved in thermotolerant cells (HUGHES and AUGUST, 1982; NAPOLITANO et al., 1987; WIEGANT et al., 1987). Investigations in plants are practically lacking. But the reversible block of protoplasmic streaming and its protection by a conditioning treatment were frequently used as experimental criteria (ALEXANDROV, 1979; SACHS, 1864). They reflect hs effects on functions of the actin microfilament system.

Nuclear structure and function: Ultrastructural changes of the nucleus result from the accumulation of unprocessed pre-mRNP as perichromatin granules and the deposition of protein complexes tightly bound to the nuclear matrix. The extent of changes and their rapid reversibility are dependent on HSP, probably on HSP70 synthesis. For details and references see Sect. B.3.2. and Table 2.

Translation: Slow heat adaptation or preinduction can markedly improve translation under severe hs conditions (KEY et al., 1983; SCHARF and NOVER, 1982). A number of modifications of the translation apparatus were reported in microorganisms, plants and animals (see Sect. B.4.3.). For the effective recovery of control protein synthesis the storage of the transiently non-translated mRNAs is essential (ALTSCHULER and MASCARENHAS, 1985; DIDOMENICO et al., 1982b; KEY et al., 1981; NOVER and SCHARF, 1984). A role of the collapsed intermediate filament cytoskeletal system and the protection by HSP aggregates (heat shock granules) is very likely (NOVER et al., 1989; see Sect. B.3.3.).

DNA synthesis and repair: In connection with the application of hyperthermia in cancer therapy, heat shock and thermotolerance effects on DNA synthesis, DNA damage and repair systems were extensively studied in vertebrates (see reviews and books by ANGHILERI and ROBERT, 1986; NOVER, 1990 and OVERGAARD, 1985). Using *Vicia faba* and other plants hs-induced protection from genomic stress by mutagenic agents was studied by RIEGER and coworkers (see Sect. C.11.) and TAYLOR and HOLLIDAY (1984) reported on a decreased recombinational instability of the maize smut fungus *Ustilago maydis*, if exposed to severe hs at 48 °C after a conditioning treatment of 10 min at 42 °C.

Chloroplast structure and function: Similar to plasma membranes the high thermosensitivity of photosynthesis depends mainly on rapid changes of the thylakoid membrane (see Sect. B.5.1.). But chloroplast membranes are the most heat-sensitive (THEBUD and SANTARIUS, 1982; VOLGER and SANTARIUS, 1981). Stabilization was documented by analyzing net CO_2 fixation, electron transport, membrane composition and chlorophyll fluorescence (BAUER, 1972; BERRY and BJÖRKMAN, 1980; PEARCY et al., 1977; SÜSS and YORDANOV, 1986).

B.4.3. The role of heat shock proteins

B.4.3.1. Physiological and genetical evidence

The rapid switch of gene expression from synthesis of normal to heat shock proteins and their marked accumulation during the stress period caused multifold attempts to correlate the acquisition of thermotolerance with the increasing HSP levels. Similar to animal systems (HAHN and LI, 1982, see review by NOVER, 1990), the three phases of hs induction (few minutes), expression (1-3 hrs) and decay (12-48 hrs) of HSPs and thermotolerance are very similar also in plants. The inherent limitations of survival tests were already discussed. Generally, it is not justified to conclude from analyses of growth or survival curves that HSPs are required or not required for thermotolerance (see summary by NOVER, 1990). For example, the observation of XIAO and MASCARENHAS (1985) that growing pollen tubes of *Tradescantia paludosa*, though unable for HSP synthesis, exhibit certain aspects of induced thermotolerance simply demonstrate that other protective agents contribute as well. In the case of growing pollen tubes synthesis of proline may be decisive (HONG-QI et al., 1984). Interestingly, though survival can be improved by a conditioning treatment, there is no comparable effect on the recovery of *Tradescantia* pollen tubes after the heat period (XIAO and MASCARENHAS, 1985).

It is evident that, in view of the complexity of thermotolerance phenomena, survival curves can give only limited informations and cannot substitute for a search of the molecular levels of HSP activities as exemplified in this review. This is also true for attempts to gain insights by manipulating the dosis and/or expression of hs genes. In yeast, site-directed recombination was used to eliminate the four hs-inducible hsp70 and the only hsp26 genes. Deletion of the hsp26 gene had no detectable effect on induction of thermotolerance or ethanol resistance, on sporulation, spore germination or viability after long-term storage at 37 °C (PETKO and LINDQUIST, 1986). Among the four hs-inducible yeast hsp70 genes of the SSA subfamily, two (SSA1 and SSA2) are required and sufficient for growth and sporulation at high-temperatures. There is evidence that they are involved in the translocation of secretory, mitochondrial and microsomal proteins (CHIRICO et al., 1988; CRAIG et al., 1987; DESHAIES et al., 1988a, b; MURAKAMI et al., 1988; WERNER-WASHBURNE et al., 1987). Though the latter is not a priori a stress-related function, the results help to shed light on general properties of proteins of the HSP70 family (see Sect. B.1.2.3.).
Another line was followed by JOHNSTON and KUCEY (1988). Similar to earlier experiments by XIAO and LIS (1986) with *Drosophila* cells, they introduced a *Xenopus* hsp70 promoter fragment inserted downstream of a dehydrofolate reductase gene into Chinese hamster ovary cells. Gene amplification during methotrexate selection led also to an increasing copy number of the Xl-hsp70 promoter fragment, which entrapped the heat shock transcription factor. About 10^4 copies per cell caused 90% inhibition of the hs-induced HSP72 expression, whereas the level of the constitutive HSC73 was not affected. Simultaneously, there was a marked decline of survival at 45 °C and very slow recovery. The results (i) allow an estimate of the HSTF concentration in the range of several thousand molecules per cell and (ii) demonstrate that, despite their close similarity (see Fig. 8), HSC73 cannot substitute for HSP72. But they provide no insight into the molecular details nor on the site of action of HSP72.
An increase of the constitutive HSP70 level was achieved by transforming monkey COS or mouse cells with the *Drosophila* hsp70 gene hooked up to the adenovirus major late promoter (PELHAM, 1984). The heterologous HSP70 correctly binds to the nucleolus during heat shock and evidently contributes to the rapid recovery of nucleolar structure and function after the stress period. Comparable experiments with plants are just beginning. SCHÖFFL et al., (1987a) used the soybean hsp17 gene fused to the cauliflower mosaic virus promoter to achieve high constitutive levels of the hsp17 mRNA in stably transformed tobacco plants (up to 20,000 mol./cell). But the mRNA is relatively short-lived and rapidly decays during a hs period. Synthesis of the corresponding protein and its eventual effects on thermotolerance were not reported so far.
From the point of view of whole plant physiology experiments under field conditions, the use of natural hs by the daily rise of temperature is very important. Exact data on the intracellular temperatures are rare. GATES et al. (1968) investigated daily changes of leaf temperatures in desert plants. For *Artemisia* spec. and grasses they were found to be 2–3 °C higher than the air temperature, whereas in leaves of the succulent *Opuntia,* they exceed the air temperature by 10–15 °C. Data for the root system in the hot soil were not reported. Evidently, the extent of the heat stress is strongly influenced by the water supply, because transpiration can substantially reduce the actual leaf temperature. Thus, expression of hs genes in field grown cotton (BURKE et al., 1985) and soybean (KIMPEL and KEY, 1985b) is markedly enhanced under water deficiency. Both heat and water deficiency stress are the primary factors for the reduction of plant productivity in agriculture (BOYER, 1982; ORT and BOYER, 1985). An interesting application of induced thermotolerance with HSP synthesis was discussed by OUGHAM and STODDART (1986). Cultivation of *Sorghum* in some tropical countries may be limited because of germination arrest at 45 °C. This, however, can be overcome by pretreating 24 hrs seedlings for 30 min at 40 °C.

B.4.3.2. Methods for intracellular localization of heat shock proteins

To approach the molecular biology of induced thermotolerance and the possible role of HSPs, it is essential to study the intracellular localization of these proteins and to attempt a correlation between reversible stress-dependent HSP-binding and the processes protected from hs damage. Different methods are at hand. Because of inherent drawbacks, a reliable analysis requires a combination of several procedures to provide proper controls.

(i) Cell fractionation procedures necessarily disrupt the native state of cellular organization and may be connected with the loss of proteins from sensitive substructures but also with unspecific gain. Though it must be an ultimate goal to isolate intact cell compartments or organelles and to study their protein composition and function *in vitro* these artifactual protein redistributions may seriously affect or even prohibit meaningful experiments in this direction. An illustrative example is the probably contamination of nuclear, mitochondrial, chloroplast or polysomal preparations by cosedimenting heat shock granules suggesting binding of HSP17 and HSP70 to these organelles (COOPER and HO, 1987; LIN et al., 1984; MANSFIELD and KEY, 1988; RESTIVO et al., 1986). At least, if highly purified fractions of chloroplasts and nuclei are carefully inspected in the electron microscope these contaminations can be visualized or detected by immunological methods (NEUMANN and WOLLGIEHN, unpublished). The same is true for polysomes cosedimenting with HSGs. Under appropriate conditions, both types of RNP can be clearly separated (NOVER et al., 1989), and isolated ribosomal subunits do not contain HSPs (MANSFIELD and KEY, 1988; NOVER et al., 1983).

Convenient means to discriminate between adherent proteins and those within an isolated membrane-enclosed cell compartment (chloroplasts, mitochondria, ER vesicles) is the digestion with proteolytic enzymes, which cannot penetrate through the intact membrane. This criterium was also used to document the *in vitro* uptake and processing of chloroplast proteins, including HSPs, translated on cytoplasmic polysomes (see Sect. B.5.2.). But after immobilizing trypsin to an inert matrix, this method was even applied to study the compartmentation of HSP70 with isolated nuclei of chicken embryo fibroblasts (COLLIER and SCHLESINGER, 1986).

(ii) *In situ* localization of HSPs by microautoradiography was pioneered by S. LINDQUIST and coworkers (VELAZQUEZ et al., 1980) using *Drosophila* cell cultures. It is based on the extreme selectivity of HSP synthesis under appropriate heat shock conditions. About 90 % of all newly formed, i.e. radioactively labeled proteins, can be HSPs. Examples using various plant objects are given in Figs. 26–28. Two major sites of HSP accumulation are identified: the nucleoli and the heat shock granules (NEUMANN et al., 1984; NOVER et al., 1983). But more detailed studies with identification of individual HSPs require immunological methods combined with light or electron microscopy (NEUMANN et al., 1987). Crucial for this is a fixation and embedding procedure which retains enough ultrastructural details but also preserves the antigenicity of the HSPs. Formaldehyde fixation and embedding in Lowicryl or LR Gold are well suited for plant materials (NEUMANN et al., 1987).

B.4.3.3. Antibodies to plant heat-shock proteins

The formation of HSP aggregates in plant cells (see Sect. B.3.3.) considerably facilitated the isolation of sufficiently pure HSPs for immunization of rabbits using two-dimensional separation of HSG fractions from heat-shocked tomato cell cultures (Fig. 30). Monospecific antibodies to HSPs 70, 68, 22, 21 and 17 were obtained by this method (NEUMANN et al., 1987 and in preparation). To test the conservation of the corresponding hs antigens in different plant species Western blots of proteins from different control and heat shocked systems were probed with anti-HSP17, anti-HSP70 and anti-HSP68 respectively (Fig. 31). Because there is no detectable cross-reactivity between the latter two antisera reacting with individual members of the HSP70 complex (NEUMANN et al., in preparation), both antisera were probed successively with the same blot.

It is evident that inducible proteins of the HSP17 family are detected in cell cultures or leaves of all plant species tested with the exception of barley. This failure may reflect relatively low amounts of HSP17-type proteins in cereals. But they are detected in 2D-separations of HSPs (Fig. 5D), and heat shock granules are formed as well (Fig. 25I and H). Alternatively, sequence conservation of the primary structure between HSP17 proteins of monocotyledons and dicotyledons may not be sufficient to allow effective cross-reactivity. Sequencing of cereal hsp17 genes can help to clarify this point. The argument in favor of a relatively low sequence conservation even within the group of the dom-

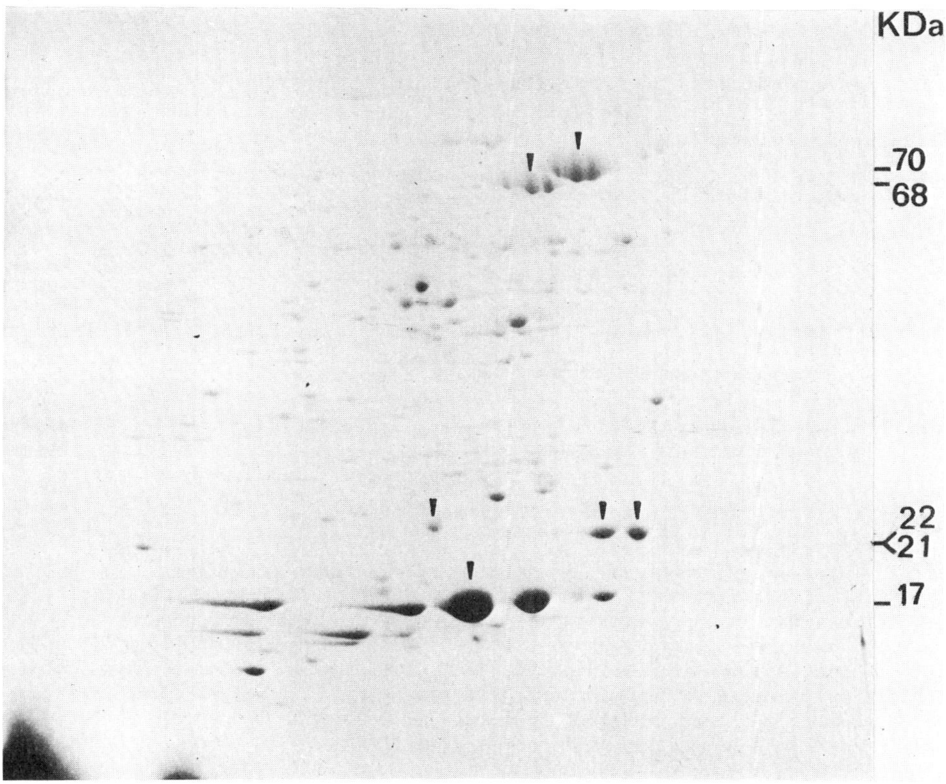

Fig. 30. Protein pattern of heat shock granules of *Lycopersicon peruvianum* cell cultures.
Proteins of purified heat shock granules (Nover et al. 1983, 1989) were separated by two-dimensional electrophoresis and stained with Coomassie blue. For antibody preparation the indicated spots of 10-20 gels were cut out and processed for immunization of rabbits (see Neumann et al., 1987)

inant small plant HSPs of Mr 16-18 kDa is supported by a report of Baszczynski (1986). The maize anti-HSP18 showed practically no cross-reactivity with soybean and pea HSP17. In contrast to this, antibodies against maize HSPs 70 and 89 detected also corresponding proteins in both monocotyledonous and dicotyledonous plants (Baszczynski, 1986). This result reflects the considerably higher conservation in the two families of large HSPs (see Figs. 6 and 8).

As evident from gene analyses (see Sect. B.1.2.5.), in each plant tested so far the dominant small HSPs form a group of closely related proteins coded by a multigene family. Hence, it is not surprising that the antibodies against the tomato HSP17 crossreact with several hs-induced proteins in the 17 kDa region not only of tomato cell cultures (Fig. 32A) but also of cell cultures of *Digitalis lanata* (B), *Peganum harmala* (C) or pea leaves (D). In each case, the major representatives of the HSP17 series, but no other small HSPs, are recognized (compare details in Fig. 32 with the corresponding 2D-patterns of HSPs shown in Figs. 3 and 5E).

Relatively low amounts and structural variability of the HSP21 evidently preclude a detection by Western blotting in protein samples derived from other plants than from tomato. But, at least the combination of electron microscopic and immunological methods demonstrates that cross-reacting HSPs exist in chloroplasts of pea (Fig. 33), tomato and tobacco. The relatively high local concentration of HSP21 in chloroplasts may contribute to this positive result with other plants.

Similar experience with HSP21 antibodies were reported by Vierling et al. (1989). By using fusion constructs between parts of the pea hsp21 gene and the *E. coli* β-galactosidase gene, they generated

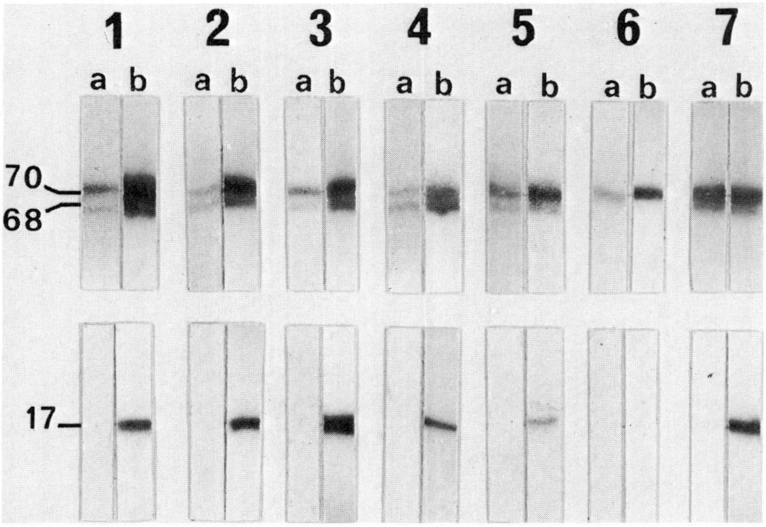

Fig. 31. Cross-reacitivity of antisera against tomato heat shock proteins with corresponding proteins in other plants.
Protein samples from control **(a)** and heat shocked **(b)** plants were separated by 1D-electrophoresis and blotted on nitrocellulose. Positions of hs-induced proteins were detected by binding of specific antibodies followed by incubation with ^{125}I-protein A. Upper part: Because there is no cross-reaction between both antisera, the Western blot was reacted in two consecutive steps with anti-HSP70 and anti-HSP68. In the lower part HSP17 antiserum was used for detection. **1–3**, proteins from cell cultures of *Lycopersicon peruvianum* **(1)**, *Digitalis lanata* **(2)**, *Peganum harmala* **(3)**, **4–7**, leaf proteins from *L. esculentum* **(4)**, *Pisum sativum* **(5)**, *Hordeum vulgare* **(6)** and *Nicotiana rustica* **(7)**. HSP70 and HSP68 are constitutive proteins with increasing amounts after hs, whereas HSP17 is only found in hs cells

antibodies against an N-terminal HSP21 fragment including the leader peptide and a C-terminal fragment, respectively. Only the C-terminal antibodies detected corresponding proteins in other plant species (*Arabidopsis, Petunia* and *Zea mays*), though with relatively low effiency.

B.4.3.4. Survey on the intracellular localization of plant heat shock proteins

Because many aspects of the association of heat shock proteins with heterologous proteins or with distinct cell compartments were already discussed (see Table 2 and Sects. B.1.2. and B.3.), including important data obtained with animal systems, we will only briefly summarize the results with plants which are mainly based on investigations in our laboratories (Table 3). Methods applied are microautoradiography (AR, see Figs. 26–28), cell fractionation combined with analysis of the protein composition (CF, see Fig. 30), immunofluorescence (IF, Neumann et al., 1987) and immune electron microscopy using protein A-gold (I-Au, Neumann et al., 1987; see Fig. 33).
It is apparent that the data are restricted to the macrocompartments or large aggregates only, which can be visualized in the electron microscope and/or purified by cell fractionation. Contrasting to animal systems (Table 2), results on soluble protein/HSP complexes in the cytoplasm or in the organelles are lacking for plants. The same is true for the probable association of HSP70-type proteins with the plant cytoskeleton or ER/Golgi system. When considering the possible role of HSPs for thermotolerance effects, as analysed on the molecular basis, it is essential to investigate also their stress-dependent recompartmentation. At least, the large HSP families (HSP90, HSP70, HSP60) include constitutively expressed members. Thus, discriminating between functions excerted under

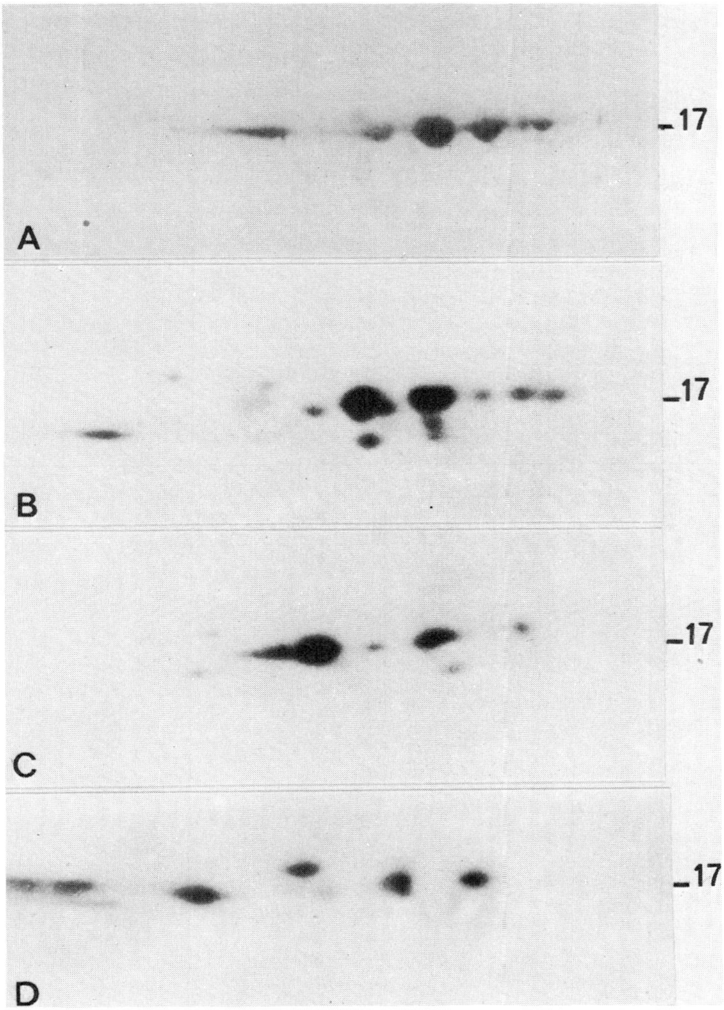

Fig. 32. Two-dimensional Western-blot of proteins of the plant HSP17 family probed with tomato anti-HSP17 serum.
Plant material was subjected to a long-term heat shock (regime II, 41 °C). Total cell proteins were separated by 2D-electrophoresis and blotted on nitrocellulose. **A–C**, protein samples from cell cultures of *Lycopersicon peruvianum* **(A)**; *Digitalis lanata* **(B)** and *Peganum harmala* **(C)**; **D**, from *Pisum sativum* leaves. The figures show only the 17 kDa regions of the 2D-Western blots. In each case the antiserum detects the main HSPs of the HS17 series (see corresponding 2D-pictures in Figs. 3A, 3C, 3F and 5E)

control vs. stress conditions is a prerequisite to find out, why enhanced amounts or new isoforms are required during a hs treatment.
The situation appears relatively clear for HSP17 and HSP70 and their stress-dependent binding to the nucleolus (HSP70) and heat shock granules (HSPs 70 and 17), respectively (Fig. 33, A–C; NEUMANN et al., 1987). In both cases, well defined aspects of induced thermotolerance can be correlated with the presence of the heat shock proteins (see Sects. B.3.2. and B.3.3.). However, the occurrence of HSP68 and HSPs21/22 in the HSG fraction (Fig. 30) cannot be explained at present. It contrasts to the exclusive immunological localization of these proteins in mitochondria (HSP68) and chloro-

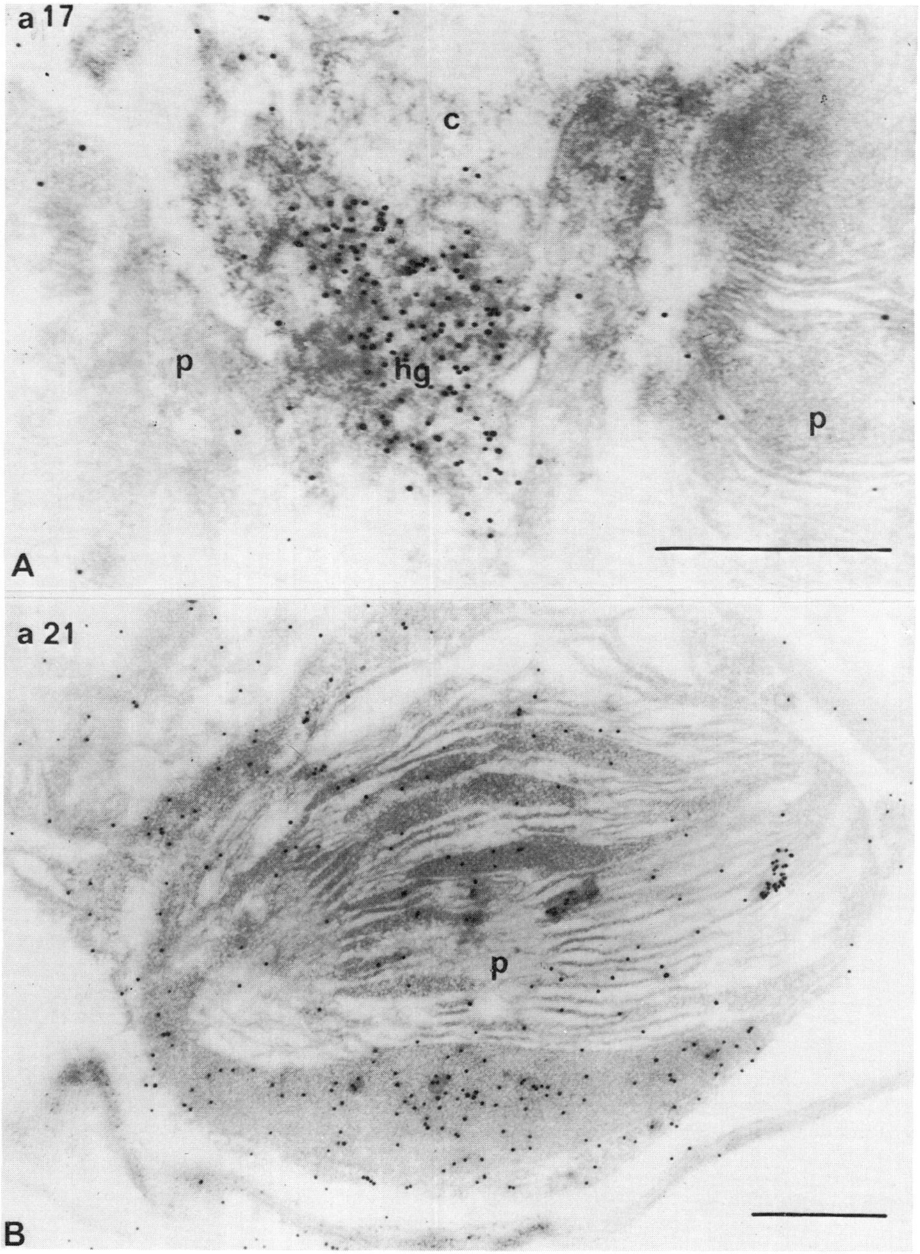

Fig. 33. Immunological localization of plant heat shock proteins.
Heat shocked leaves were fixed in formaldehyde and embedded in Lowicryl. After immune staining with the indicated antisera, antibody binding was visualized by treatment with protein A-gold (16 nm gold particles, for details of antibody preparation and electron microscopic techniques see NEUMANN et al., 1987 and in preparation). **A–C**, *Lycopersicon esculentum* leaf sections stained with anti-HSP17 **(A)**, anti-HSP21 **(B)** and anti-HSP22 **(C)** respectively. **D–G**, *Pisum sativum* leaf sections stained with anti-HSP68 **(D)** and anti-HSP70 **(E–G)** respectively. HSP17 is detected in HSG, HSP21 and 22 are chloroplast HSPs, HSP68 is concentrated in mitochondria and HSP70 is predominant in nucleoli, in the nucleoplasma and in heat shock granules. Bars 0.5 μm

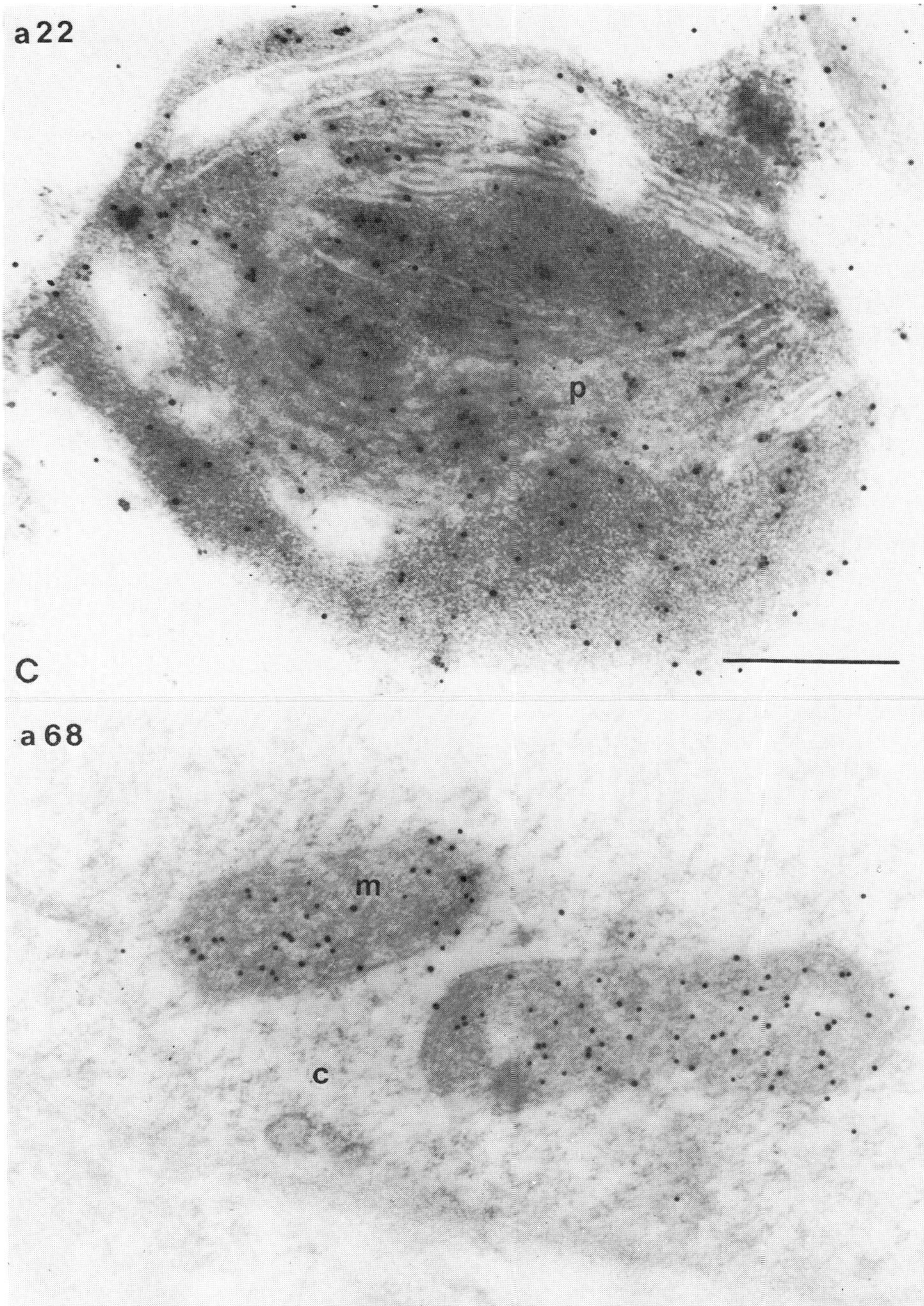
a22
p
C
a68
m
c
D

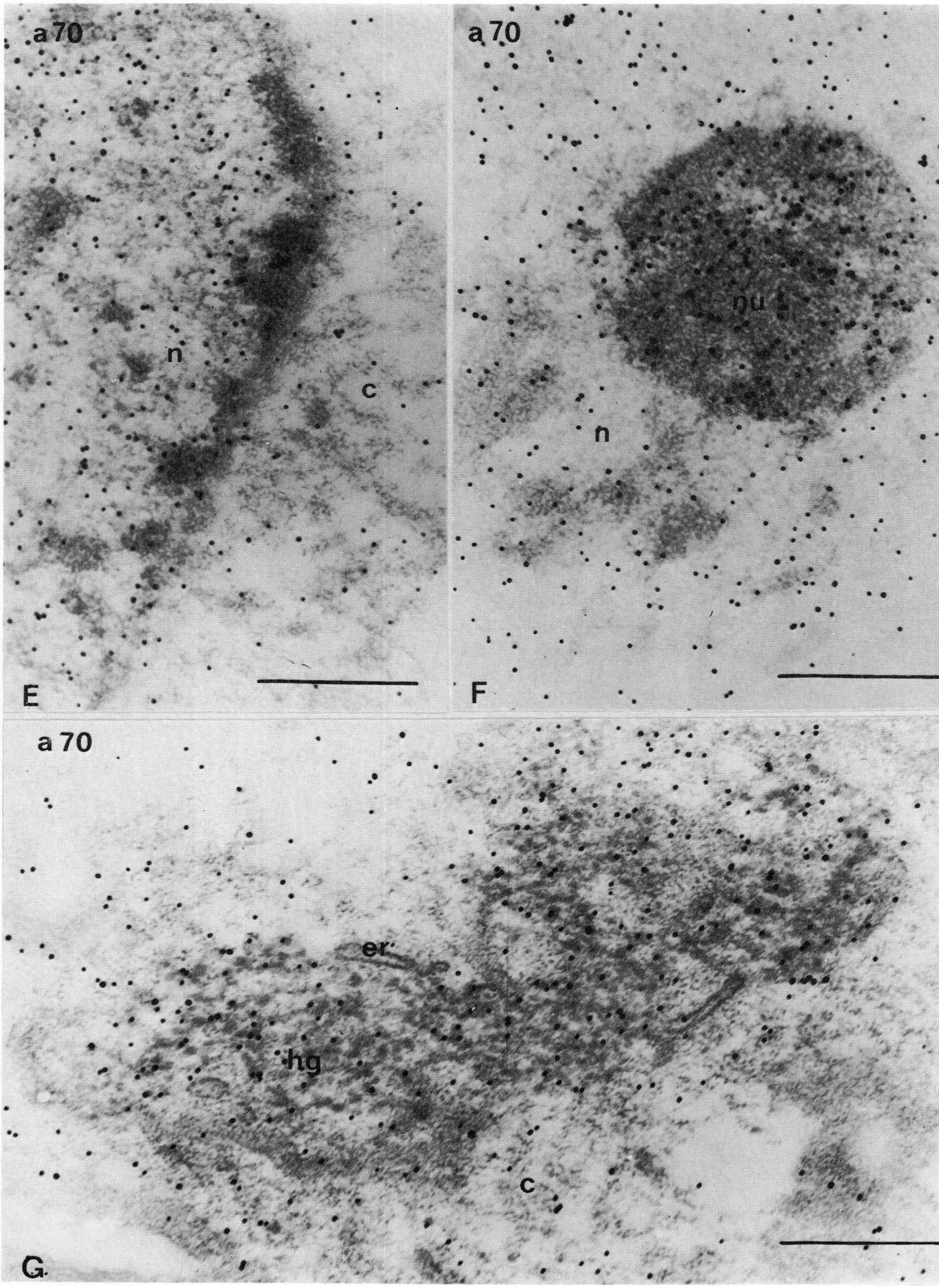
a 70
n
c
E
a 70
nu
n
F
a 70
er
hg
c
G

Table 3. Intracellular localization of plant heat shock proteins

Compartment (Thermotolerance effect)	Stress-dep. recompartm. of HSPs	HSP	Methods[a]	Data in this review	References
Nucleus/nucleolus (protection of synth. and processing of pre-rRNP, recovery of nucleolar functions, see Sect. A.3.2.)	yes	HSP70	AR, IF I-Au	Sect. A.3.2. and Fig. 33	NEUMANN et al., 1984, 1987; NOVER et al., 1986
Heat shock granules (mRNA storage, see Sect. A.3.3.)	yes	HSPs70, 68, 21 and 17[c]	AR, CF IF, I-Au	Sect. A.3.3. and Figs. 28, 30 and 33	MANSFIELD & KEY, 1988; NEUMANN et al., 1984, 1987; NOVER et al., 1983, 1989
Chloroplasts (protection of photosystem II?, see Sect. A.5.2.)	yes/no	HSPs21-25, HSP60	CF, I-Au	Sect. A.5.2., Figs. 33, 34	GLACZINSKI & KLOPPSTECH, 1988; HEMMINGSEN et al., 1988; KLOPPSTECH et al., 1985; NEUMANN et al., in prep.; SCHUSTER et al., 1988; VIERLING et al., 1986, 1988; 1989
Mitochondria (?)	no	HSP68, HSP60 (?)	CF, I-Au	Fig. 33	McMULLIN & HALLBERG, 1988; NEBIOLO & WHITE, 1985; NEUMANN et al., in prep.
ER/Golgi	?	HSP70-like[b]	I-Au	–	COOPER & HO, 1987; NEUMANN et al., 1987
Plasma membrane (protection from solute leakage)	yes	HSPs70 and 15-18	CF	–	COOPER & HO, 1987; LIN et al., 1985

[a] AR, microautoradiography; CF, cell fractionation with analysis of protein composition; IF, immunofluorescence; I-Au, immunogold labeling coupled with electronmicroscopy.
[b] The existence of plant homologs to GRPs94 and 78 (see Sects. A.1.2.2., A.1.2.3.) is not clarified, but we found consistently HSP70-type antigens associated with the Golgi apparatus (NEUMANN et al., 1987).
[c] The presence of HSPs68 and 21/22 in isolated HSG (Fig. 30) could not be verified by *in situ* immune staining (see Fig. 33 and text for discussion).

plasts respectively (HSPs21/22; Fig. 33, D–F; NEUMANN et al., in prep.). Though direct evidence was only provided for HSP21 (see Sect. B.1.2.5.), all three proteins should be produced as precursor with an N-terminal transit peptide. It will be essential to investigate, whether these organellar HSPs become tightly bound parts of the HSG fraction only as a result of the preparation procedure, i.e. after destruction of mitochondria and proplastids respectively. HSP60 is evidently a constitutive protein in mitochondria and chloroplasts, required for effective assembly of oligomeric protein complexes (chaperonin, see Table 2). But it is unclear whether it plays any role in the stress response system of plants or other organisms.

B.5. *Heat shock effects on chloroplasts and mitochondria*

Chloroplasts and mitochondria are semiautonomous organelles of plant cells with their own genetic material and an independent apparatus for gene expression. From the point of view of evolution they probably arose from prokaryotic ancestors incorporated into the ancient prototype of a eukaryotic cell. The high degree of conservation and the ubiquitous occurrence of the hs response from archaebacteria to plant and man prompt the question of how much of the prokaryotic hs program was retained in present day chloroplasts and mitochondria. Unfortunately, investigations in this direction are very fragmentary and mostly restricted to chloroplasts.

B.5.1. *Effects on photosynthesis and gene expression in chloroplasts*

Similar to other stress factors (see Sects. C.3., C.6. and C.7.2.) hs diminishes the photosynthetic activity of chloroplasts (Berry and Björkman, 1980; Rhodes, 1987). Primary site of hs effects are the thylakoid membranes and the ordered reactions connected with the light-induced splitting of water and with photophosphorylation. Generally, the stroma enzymes involved in CO_2 fixation (ribulosebisphosphate carboxylase, Calvin cycle enzymes) are more resistant (Bauer and Senser, 1979; Berry and Björkman, 1980; Volger and Santarius, 1981). But the decrease of ATP generation (Santarius, 1975; Tischner and Lorenzen, 1975) and the dissipation of reducing equivalents by photoreduction of oxygen (Havaux et al., 1987a, b) may contribute to the observed decrease of CO_2 fixation (Weis, 1981). Among the two photosystems system II is the most heat-sensitive (Mohanty et al., 1987; Mukohata et al., 1973; Schreiber and Armond, 1978; Schuster et al., 1988; Süss and Yordanov, 1986; Tischner and Lorenzen, 1975). Both heat stress and high light intensities lead to inactivation of photosystem II connected with aggregation of thylakoid membrane proteins probably as a result of oxygen stress (Sect. C.3.; Schuster et al., 1988). Investigations with pea and *Chlamydomonas* showed that protection of thylakoid membranes and improvement of the recovery of photosystem II activity can be achieved under low light conditions during the stress period and by using hs-preadapted plants and algae respectively.

Our knowledge on details of the synthesis of chloroplast proteins during the stress period is negligible, except the intriguing example given by Vierling and Key (1985). They investigated details of ribulose bishosphate carboxylase (Rubisco) synthesis in green suspension cultures of soybean *(Glycine max)*. It is essential to recall that Rubisco is the most abundant protein in nature making up to 50 % of all protein in green tissues. Hence changes of synthesis and/or amount can be easily analyzed. Labeling of the large and small subunits of Rubisco decrease to 20–30 % of the control values at temperatures of 38–40 °C. For the small subunit, produced as precursor protein in the cytoplasm, this correlates with a decay of its mRNA, whereas for the large subunit, synthesized in the organelle itself, the mRNA levels remain unchanged. Interestingly, the residual pre-SSU is correctly processed and transported to the chloroplast even under hs conditions. Evidently, much further work is needed to better characterize the protein synthesis activity of chloroplasts during hs in general and to investigate the consequences of a non-coordinate synthesis of different parts of multisubunit protein complexes. Particularly interesting in this connection is the role of the Rubisco assembly protein, which belongs to the HSP60 family (see below).

An interesting peculiarity of the hs effects was analyzed in developing plant seedlings (see summary by Feierabend, 1979). Seedlings grown at very mild hyperthermic conditions (32–34 °C) become deficient of chloroplast ribosomes (Feierabend, 1979; Feierabend and Mikus, 1977). Consequently, these leaves are also free of chlorophyll and other components required for chloroplast function. As evident from the example of Rubisco synthesis just mentioned, normal chloroplast development is dependent on the coordinate activities of both gene expression systems of the cell, i.e. of the nucleo-cytoplasmic and of the organellar system. Lack of chloroplast ribosomes blocks the whole process. The temperature-sensitive step of bleaching could not be defined. At least the organellar genome and the corresponding RNA polymerase remain intact (Büngér and Feierabend, 1980; Herrmann and Feierabend, 1980). Thus, assembly and processing of preribosomes are the most likely candidates. In support of this assumption, even a short interruption of the long-term hs at 32 °C, e.g. by one hour at 22 °C, was found sufficient to restore a considerable part of ribosome formation and greening in developing oat seedlings. Interestingly, a number of precursor ribosomal proteins, destined for assembly into chloroplast ribosomes, were found in the cytoplasm of heat-treated rye leaves (FEIERABEND et al., 1988).

B.5.2. Heat shock proteins of chloroplasts and mitochondria

The existence of chloroplast-specific HSPs has been demonstrated by several groups in different objects, including *Pisum sativum* (KLOPPSTECH et al., 1985; VIERLING et al., 1986, 1988, 1989), *Phaseolus vulgaris* (SÜSS and YORDANOV, 1986), *Arabidopsis thaliana* (VIERLING et al., 1989), *Nicotiana plumbaginifolia* (RESTIVO et al., 1986), *Petunia hybrida* (VIERLING et al., 1989) and *Vigna sinensis* (KRISHNASAMY et al., 1988), as well as the green algae *Chlamydomonas* (KLOPPSTECH et al., 1985; SCHUSTER et al., 1988) and *Acetabularia* (KLOPPSTECH et al., 1986). Size and number of HSPs detected in highly purified chloroplasts vary with the plant species investigated. Two examples for pea and tobacco are given in Fig. 34. Besides prominent members of the HSP20 family, frequently large HSPs in the range of 67–90 kDa are found as well (see also KLOPPSTECH et al., 1986; KRISHNASAMY et al., 1988; RESTIVO et al., 1986). The group of large HSPs detected in the pea and tobacco chloroplasts (Fig. 34) is clearly different from the corresponding part of the total HSP pattern (Figs. 5E and 5C). This is an argument against cytoplasmic contamination of the chloroplast fraction. But it should be recalled that intracellular recompartmentation of cytoplasmic non-HSPs is common and may cause the appearance of new proteins in a cell compartment misinterpreted as newly synthesized HSP.

Only in the case of the heat shock proteins in the 21–26 kDa range there is unequivocal proof that they are new chloroplast constituents, whose synthesis is induced by hs. Similar to many other chloroplast proteins (KEEGSTRA, 1989), they are nuclear encoded proteins synthesized in the cytoplasm as precursor forms with a chloroplast-specific N-terminal transit peptide of about 5 kDa, which is removed upon uptake into chloroplasts. This was shown by *in vitro* synthesis and transport using isolated chloroplasts (KLOPPSTECH et al., 1985; VIERLING et al., 1986, 1989). Sequencing of cDNA clones of the pea HSP21 and the soybean HSP22 revealed 79% identity in the amino acid sequence. The pea cDNA clone contains also the 5′ end of the gene with about 140 nucleotides coding for a transit peptide of 46–47 amino acid residues (Fig. 12; VIERLING et al., 1988). Remarkably, no precursor protein was detected for the *Chlamydomonas* chloroplast HSP22 (KLOPPSTECH et al., 1985), and this was recently confirmed by sequencing of a corresponding cDNA clone (GRIMM et al., 1989).

Sequence comparison with other members of the HSP20 family (Figs. 12, 13) shows that homology in the C-terminal part is considerably higher than in the N-terminal part. This fact explains results of an immunological screening of different plants with antibodies against both parts of the pea HSP21 molecule (VIERLING et al., 1989). Antibodies against the N-terminal domain recognized only the pea precursor protein, whereas antibodies against the C-terminus detected related hs-induced chloroplast proteins also in *Arabidopsis*, *Petunia* and *Zea*. The relatively high levels of HSP21–26 in chloroplasts are maintained even many hours after the end of the heat stress period. This may be an essential factor for the protection of membranes (see above).

Besides the members of the HSP20 family, there is a prominent constitutive protein which belongs to the HSP60 family. As detailed in Sect. B.1.2.4. chloroplasts and mitochondrial proteins of this type are homologous to the major *E. coli* HSP, the GroEL protein (GOLOUBINOFF et al., 1989; HEMMINGSEN et al., 1988). The corresponding chloroplast protein was well known before as the Rubisco subunit binding protein (BARRACLOUGH and ELLIS, 1980; MUSGROVE and ELLIS, 1986). It is presumably implicated in the post-translational assembly of the ribulose-1,5-bisphosphate carboxylase from preformed large and small subunits (GOLOUBINOFF et al., 1989; ROY and CANNON, 1988; MUSGROVE and ELLIS, 1986). There is no direct evidence for an increased synthesis of the chloroplast HSP60 in heat-shocked plant cells. But plant HSPs in this size range were detected by KRISHNASAMY et al. (1988) in chloroplast fractions of *Vigna sinensis* and *Sorghum vulgare* and by KISHORE and UPADYAYA (1988), NEBIOLO and WHITE (1985) and SINIBALDI and TURPEN (1985) in mitochondria of pigeon

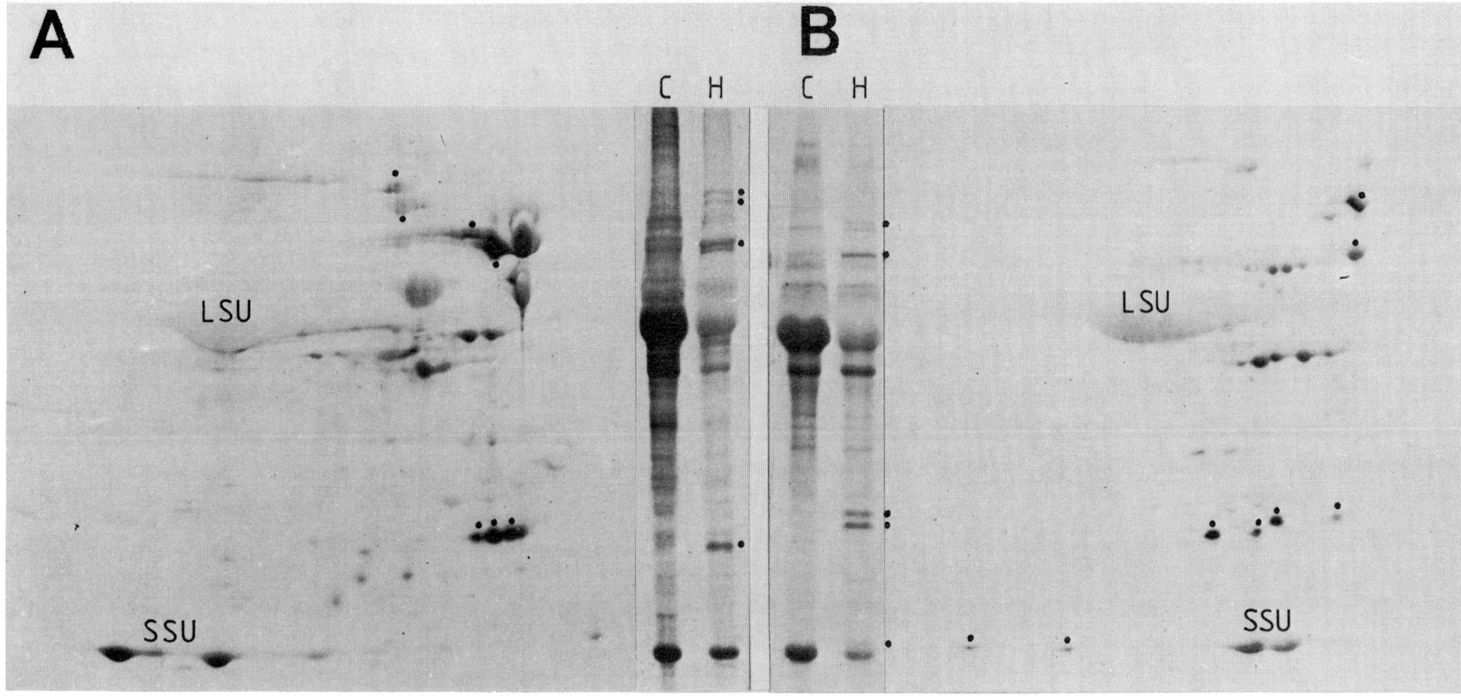

Fig. 34. Heat shock proteins in stroma fractions of purified chloroplasts of pea **(A)** and tobacco **(B)**.
Tips of 7 d old seedlings (pea) or leaf disks of greenhouse plants (tobacco) were preinduced and labeled with ^{35}S-methionine for 4 hrs at 40 °C. Chloroplasts were isolated and purified on discontinuous sucrose gradients. After disruption by osmotic shock stroma proteins were prepared by centrifugation. Autoradiographs from control (C) and heat shock (H) samples are shown as inserts in the 2D-autoradiographs of stroma proteins from the hs samples. Position of the ribulosebisphosphate carboxylase subunits are indicated in the 2D pattern by LSU und SSU, respectively (Mr ~ 50 and ~ 12 kDa). Mr's of the HSPs (dots) are marked from top to bottom: pea 95, 89, 74 and 21 kDa; tobacco 85, 74, 23, 22 and 12 kDa, respectively

pea, maize and cabbage respectively. In all these cases, the authors provide evidence that the proteins are synthesized within the organelles. However, this is not true for the yeast mitochondrial HSP60 and the Rubisco subunit binding protein, which are both synthesized on cytoplasmic ribosomes (CHENG et al., 1989; HEMMINGSEN and ELLIS, 1986). Thus, it remains to be shown, whether both types of plant proteins with Mr 52-61 kDa belong to the same HSP60 family as summarized in Fig. 9.

Remarkable is the exclusive localization of the plant HSP68 in mitochondria under control and hs conditions (Fig. 33). Since antibodies against HSP68 do not cross-react with HSP70, it remains to be shown by cloning and sequencing whether both proteins belong to the same multigene family. Interestingly, a member of the HSP70 family (SSC1) was characterized as mitochondrial HSP in yeast. It is synthesized as precursor protein at cytoplasmic polysomes (CRAIG et al., 1987).

A first step towards an understanding of the function of the chloroplast HSPs are investigations on their localization within the organelle. Depending on the hs conditions, HSPs were found in the stroma fractions or in the thylakoid membranes of the organelles (KLOPPSTECH et al., 1985; KRISHNASAMY et al., 1988; RESTIVO et al., 1986; SÜSS and YORDANOV, 1986; VIERLING et al., 1986). Seemingly contradictory results could be clarified by *in vitro* translation and transport of HSPs 22 and 25 of pea and HSP22 of *Chlamydomonas.* If chloroplasts were taken from control plants, both proteins were found in the stroma fraction. But if chloroplasts were prepared from hs pea plants, they were integrated into the thylakoid membranes (KLOPPSTECH et al., 1985). Moreover, high light intensities applied simultaneously with the heat shock reduced the HSP binding to the membranes, a phenomenon possibly connected with the massive destruction of parts of the thylakoid membranes and photochemical cross-linking of the proteins (GLACZINSKI and KLOPPSTECH, 1988). In *Chlamydomonas* two HSPs of 22 kDa and 29 kDa were also shown to be localized in the grana lamellae (KLOPPSTECH et al., 1985; SCHUSTER et al., 1988). Inhibition of protein synthesis by cycloheximide during the heat shock increases the inactivation of PSII and enhances protein aggregation. These results suggest that in *Chlamydomonas* the chloroplast HSPs may contribute to the protection against damage by photoinhibition during heat shock (SCHUSTER et al., 1988).

It was already mentioned that the representatives of the chloroplast HSP20 family of different plant species are nuclear-encoded and translated as precursor proteins on 80S cytoplasmic ribosomes. The synthesis is inhibited by cycloheximide, but not by chloramphenicol (KLOPPSTECH et al., 1985; VIERLING et al., 1986, 1988). On the other hand, there are results that some HSPs may be encoded in the chloroplast genome itself. (i) In *Acetabularia,* heat shock induces the synthesis of a 70 kDa protein even in the absence of the nucleus, which can be surgically removed in this giant green alga. This HSP70 is localized in an organelle fraction, enriched in chloroplasts. Its synthesis is inhibited by chloramphenicol (KLOPPSTECH et al., 1986). (ii) In isolated chloroplasts of *Vigna sinensis,* the light-dependent protein synthesis *in vitro* responded to heat shock with the induction of four HSPs of 85, 70, 60 and 23 kDa. In chloroplasts of *Sorghum vulgare,* only the three high molecular weight polypeptides are translated. The reaction is inhibited by the transcription inhibitors actinomycin D and rifampicin suggesting that the synthesis of the HSPs is regulated at the transcriptional level (KRISHNASAMY et al., 1988).

Not only the chloroplasts, but also the mitochondria of plants were found to contain HSPs. LIN et al. (1984) isolated a number of high and low mol. weight HSPs (15-18, 22, 24, 69-70 kDa) from gradient-purified mitochondria of heat-shocked soybean seedlings. They also reported evidence for induced thermotolerance connected with the presence of these HSPs: in preconditioned soybean seedlings the tight coupling between O_2 uptake and ATP generation was maintained during hs at 42.5 °C, if soybean seedlings were preconditioned for 2 hrs at 38 °C (CHON et al., 1989). Evidence for the synthesis of HSP60 in mitochondria of pigeon pea *(Cajanus cajan),* corn and cabbage was already mentioned (KISHORE and UPADYAYA, 1988; NEBIOLO and WHITE, 1985; SINIBALDI and TURPEN, 1985). The results were questioned by NIETO-SOTELO and HO (1987), who attributed them to bacterial contamination. At least the GroEL homologous forms of HSP60, identified so far in mitochondria and chloroplasts of eukaryotic microorganisms and plants (see above) seem to be nuclear encoded proteins.

C. Other plant stress response systems

C.1. General stress response and interaction of stressors

In the preceding sections of this review we compiled data on important parts of the hs response in different plants, plant organs or cell cultures. It was emphasized in the discussion that many results, e.g. of gene structure, gene expression regulation, changes of cell ultrastructure or of intracellular localization and function of heat shock proteins are very similar in animals and eukaryotic microorganisms as well. Many aspects of the hs response are highly conserved, making research in this field an important and productive part of molecular cell biology.

The following survey of plant stress response systems in general will concentrate on the cell biological aspects with particular emphasis on the synthesis of stress proteins. The historical and conceptual basis of plant stress physiology in general is remarkably outlined by ALEXANDROV (1977, 1979) and in the two volumes by LEVITT (1980a, b). LEVITT emphasizes the necessity to discriminate between the external stress (stressor) and the internal strain caused by it. The anatomical and physiological organization of whole plants and peculiarities of their life cycles enable them to minimize stress-dependent strain or injury or to escape stressful situations, e.g. by periods of dormancy. However, the description of molecular details of the cellular stress response do not easily fit the highly sophisticated terminology of plant stress physiologists. On the one hand, this results from the apparent difficulties to unequivocally prove a role of a given stress protein or metabolite for the inherent resistance of a cell or its acquired tolerance (see Sect. B.4.). On the other hand, the usually very complex stress response systems represent a continuum of adaptive changes involving preformed and newly formed molecules or structures, respectively, as well as immediate activities to repair damages or to restore the pre-stress functions (see summary by NOVER, 1990).

The increasing number of physical, chemical or biological stressors affecting plants (Table 4) may be related to each other by virtue of a 'natural coupling', by common parts of the responses or signal systems and by the intracellular strain finally generated. The following examples may help to illustrate these points:

1) Heat shock and reduced water supply are a natural pair of stressors with the highest impact on plant productivity (ORT and BOYER, 1985). Evaporation of water from the soil and the plants strongly increases with the environmental temperature and, if the stomata have to be closed to avoid excessive water loss, the leaf temperature rapidly increases above the air temperature. This is documented by increased expression of hs genes in field-grown cotton or soybean, if the water supply of the experimental plants is reduced (BURKE et al., 1985; KIMPEL and KEY, 1985b). In addition to this, water deficiency may result from chilling or freezing and from osmotic or salt stress (see Sects. C.6., C.7.). An increase of abscisic acid (ABA) content was observed in all situations of water deficiency (Table 4), and most parts of the response are mimicked by application of exogenous ABA (Sect. C.2.). In this respect it is an interesting finding that ABA also increases HSP70 mRNA levels in maize mesocotyls (HEIKKILA et al., 1984).
2) In many cases pathogen attack is dependent on or coincides with wounding or leads to local necrosis, e.g. in the case of a hypersensitive reaction. Many parts of both stress response systems are identical or very similar (Table 4). They can be mimicked by application of plant- or pathogen-derived cell wall fragments (elicitors), which frequently represent carbohydrate derivatives (oligosaccharides, see DARVILL and ALBERSHEIM, 1984; RYAN, 1987, 1988) but many also include glycosylated proteins (ASADA and MATSUMOTO, 1987; PARKER et al., 1989; STEKOLL and WEST, 1978). Starting with the local cell death and probable elicitor formation, an unknown wounding signal is spread systemically (see Sect. C.9.) inducing changes or resistance mechanisms in fairly distant parts of the plant.
3) Cellular acidosis is a common result of several stressors. Increased polyamine biosynthesis may help to restore *p*H homeostasis (Table 4, AYRES, 1984). This includes SO_2 stress, unbalanced NH_4 nutrition, K^+-deficiency or acidic soil. But excessive production of H^+ ions is also connected with anaerobic stress (ROBERTS et al., 1984a, 1984b).
4) Starvation or partial nutrient deficiency are frequently encountered, e.g. as a consequence of reduced photosynthetic activity (low light intensities, damage of chloroplasts by photooxidation or

stress-induced destruction of thylakoid membranes, closure of stomata because of drought or salt stress). On the other hand, unbalanced or lacking nutrient uptake by the roots and inefficient distribution within the plant result from nutrient deficits in the soil, salt stress, anaerobiosis by flooding, phytopathogen attack, lacking transpiration, heavy metal intoxication, etc.

5) In many cases the structure and function of SH group-containing proteins is primarily affected, e.g. by heavy metals, many oxidizing compounds (see Sect. C.3.), SH-reactive chemicals (diamide, p-chloromercuribenzoate, N-ethyl-maleimide), heat shock or U.V. irradiation. Increased level of glutathione or of its derivatives may contribute to the restoration of the normal oxidation levels (see Sects. C.3., C.4.). Many stressors in this class were characterized as inducers of HSP synthesis (see Sect. B.2.1.). This close connection of the oxidation stress and heat shock response is also valid for bacteria. MORGAN et al. (1986) and VAN BOGELEN et al. (1987) defined overlapping stress domains, which may be parts of a common regulatory entity (stimulon).

These examples may help to understand the complexity of stress effects and the mutual interaction of stressors typical for plants growing in a natural habitat (LEVITT, 1980a, b). The tight connections between different types of stressors is not only documented by the increased synthesis of common stress metabolites and/or proteins (Table 4) but also by common sensitive sites of the cell primarily affected, e.g. ribosome biosynthesis, (Sect. B.3.2.), polysome decay and reformation (Sects. B.2.6.2., B.3.1.), function of thylakoid membranes (Sect. B.5.1.), SH group containing proteins (Sects. B.2.1., C.3., C.4.) and probably also the ill-defined cytoskeletal systems. Furthermore, acquired stress tolerance may extend to other stressors not used for induction (cross-tolerance, see Sect. B.4.).

The beginning analysis of stress interactions at the cellular level must be complemented by investigations on the role of modulators and phytohormones (Sect. C.2.), on enhancing or inhibitory effects between stressors (BURKE et al., 1985; KIMPEL and KEY, 1985b; MOCQUOT et al., 1987; WALTER, 1989; WALTER and HAHLBROCK, 1986) and on the molecular basis of cross-tolerance (ALEXANDROV et al., 1970, BENZIONI and ITAI, 1975; BONHAM-SMITH et al., 1987; CLOUTIER, 1984; GUY et al., 1986; LIN et al., 1985). Particularly interesting in this respect are many examples of stimulatory or inhibitory effects of hs on the infection or propagation of plant pathogens (see Sect. C.10.5.). The molecular details of this part of the hs response remain to be elucidated. At least one important factor is the high priority of the response to hs when compared to other stressors. Usually it is much faster (see comparison published by BONHAM-SMITH et al., 1988a) and overrides other stress effects. Because of the high selectivity of the gene expression machinery in heat shocked cells, synthesis of other stress proteins is blocked, e.g. starvation proteins in parsley cell cultures (WALTER and HAHLBROCK, 1985), wound-induced proteins in maize (BONHAM-SMITH et al., 1988a), anaerobic proteins in rice (MOCQUOT et al., 1987) or proteins of the pathogenesis-related complex in pea and parsley (DANIELS et al., 1987a; WALTER, 1989).

Table 4. Multivalent stress metabolites and stress proteins (s. S. 80, 81)

The table summarizes experimental evidence for increased synthesis of stress proteins and non-proteinaceous stress metabolites (+). (−), hs inhibits synthesis of PAL and other enzymes of phenylpropanoid metabolism (WALTER, 1989; WALTER & HAHLBROCK, 1985). In many other cases stimulating or inhibitory effects were not analyzed.

[a] Pathogen induction includes fungi, bacteria and/or viruses as well as pathogen-derived elicitors.

[b] The amino acid or aa derivative varies with the plant investigated (JONES, 1984).

[c] In most cases increased ethylene formation is connected with increased synthesis of ACC synthase.

[d] In addition to PAL, several other enzymes of general phenylpropanoid metabolism are induced as well (see Sect. C.10.2.).

[e] SP-26A is a general stress protein of soybean (see Fig. 11). The existence of similar proteins in other plants remains to be established.

[f] Only HSP70 mRNA increased after wounding or addition of ABA.

[g] References are mostly restricted to reviews.

[h] Depending on the cell type investigated, many stress proteins and metabolites mentioned in this table may also be synthesized stress-independently, under developmental control.

Stress metabolites	Wounding	Pathogen infect.[a], elicitor	Osmot. stress	Salt stress	Water deficiency	Heat shock	Low temp. stress	Anaerobiosis
A. Non proteinaceous compounds								
Polyamines, gamma-aminobutyrate		+	+	+				+
Proline, glycine betaine, asparagine[b]		+	+	+	+	+	+	
Sucrose, polyols					+	+	+	
Cell wall material (suberin, cutin, waxes, phenolics, lignin, callose)	+	+			+		+	
Abscisic acid (ABA)		+	+	+	+	+	+	
Ethylene[c]	+	+	+	+	+	+	+	+
Oligosaccharide cell wall fragments	+	+						
Phytoalexins and other secondary metabolites	+	+			+			
B: Proteins								
Hydroxyproline-rich and glycine-rich cell wall proteins	+	+				+		
Phenylalanine-ammonia lyase (PAL)[d]	+	+				−		
Heat shock proteins	+[f]					+		
Stress protein 26-A[e]			+			+		
Proteinase inhibitors	+	+						
Chitinase, glucanase and other hydrolytic enzymes	+	+						
Pathogenesis-related proteins (PRs)	+	+	+					

Heavy metals	Radia-tion	Other inducing conditions[g]	References
		Growth and cell division; cellular acidosis caused by SO_2 stress, NH_4^+ nutrition or K^+ deficiency	AYRES, 1984; BASSO & SMITH, 1974; FLORES et al., 1985; MURPHY et al., 1983; RHODES, 1987; SLOCUM et al., 1984, SMITH, 1984, 1985; YOUNG & GALSTON, 1983 (see Sect. C.2.)
	+	Ozone, SO_2, cell division	AYRES, 1984; DASHEK & ERICKSON, 1981; ERICKSON & DASHEK, 1982; JONES, 1984; RHODES, 1987 (see Sect. C.7.3.)
		–	RHODES, 1987 (see Sect. C.7.3.)
	+	Mineral deficiency, ABA, poly-L-lysine	ASADA & MATSUMOTO, 1987; BAKER, 1974; BLUM, 1975; BONHOFF et al., 1987; DIXON et al., 1986; SCHEEL & HAHLBROCK, 1989; JAFFE & TELEWSKI, 1984; KÖHLE et al., 1984; KOLATTUKUDY & SOLIDAY, 1985
	+	Nitrogen deficiency	DAIE & CAMPBELL, 1981; DEGANI & ITAI, 1978; LALK & DÖRFFLING, 1985; RYAN, 1987 (see Sects. C.2. and C.7.)
+	+	Oxygen stress (O_3), auxin, ripening, senescence	FUHRER, 1982; HOGSETT et al., 1981; MEYER et al., 1987; NAKAGAWA et al., 1988; STERMER & HAMMERSCHMIDT, 1987; YANG & HOFFMAN, 1984 (see Sect. C.2.)
		Agents causing local necrosis	DARVILL & ALBERSHEIM, 1984; GROSS, 1987; RYAN, 1987, 1988
+	+	Agents causing local necrosis ($HgCl_2$, chloroform, Triton, freezing) or reduced growth	DARVILL & ALBERSHEIM, 1984; DICOSMO & TOWERS, 1984; EBEL, 1986; EILERT, 1987; GERSHENZON, 1984; GROSS, 1987; SCHEEL & HAHLBROCK, 1989; HARGREAVES, 1981; KODAMA et al., 1988
		Ethylene, boron deficiency, high nitrate nutrition, glutathione	ESQUERRÉ-TUGAYÉ et al., 1985; ROBY et al., 1985; STERMER & HAMMERSCHMIDT, 1987; WINGATE et al., 1988 (see Sect. C.9.3.)
+	+	Glutathione, vanadate	EILERT, 1987; SCHEEL & HAHLBROCK, 1989; PENDHARKAR & NAIR, 1975; STEFFENS et al., 1989; VAYDA & SCHAEFFER, 1988; WINGATE et al., 1988 (see Sect. C.10.2.)
+		ABA[d]	See Sects. B.1.2. and B.2.1.
+		ABA, 2,4-D	CZARNECKA et al., 1988; HAGEN et al., 1988
		Starvation, ripening of tomato fruits, ethylene, poly-L-lysine	LINCOLN et al., 1987; PEARCE et al., 1988; RYAN, 1987; WALKER-SIMONS & RYAN, 1986 (see Sect. C.9.2.)
		Ethylene	BOLLER, 1985; SHINSHI et al., 1987 (see Sect. C.10.2.)
+		Ageing, flowering, 2,4-D, ethylene or agents causing increased ethylene levels (polyacrylic acid, salicylate, 2-chloroethylphosphonic acid)	GRANELL et al., 1987; SCHEEL & HAHLBROCK, 1989; MÜHLBACH, 1987; OHASHI & MATSUOKA, 1985; RIGDEN & COUTTS, 1988; VAN LOON, 1983, 1985; TRAYLOR et al., 1987; WAGIH & COUTTS, 1981 (see Sect. C.10.2.)

C.2. *Phytohormones and other stress modulators*

C.2.1. *Introduction*

The so-called classic phytohormones (auxins, cytokinins, gibberellins, abscisic acid, and ethylene) and some putative plant growth regulators such as brassinosteroids, jasmonates and polyamines are essentially involved in physiological and developmental processes of plants, organs, and cells. Although several modes of action of these substances are being discussed, increasing evidence points to their primary actions at the level of gene expression (Baulcombe, 1987; Goldberg, 1986; Key, 1987; cf. recent review by Parthier, 1989).
Some of them, especially abscisic acid (ABA), ethylene, cytokinins, jasmonates and a small number of other chemicals may act as stress modulators as defined by Lanks (1986) for animal cells: They act by suppressing or inducing (enhancing) stress responses of the plant tissues or cells in a vastly unknown, complex manner. They might be involved in signal transmission processes turned on in the cells as an early response to stress situations. Under this aspect they can be termed stress metabolites together with a small number of universal metabolites, whose concentration increases as a result of the action of different types of stressors (cf. Table 4). Stress modulators could act directly as internal regulators of certain steps of gene expression. Applied exogenously, these substances can induce physiological changes identical with characteristic parts of the stress responses. In this respect exogenous compounds might replace or amplify the action of the respective internal phytohormones, but other effects cannot be excluded.

C.2.2. *Abscisic acid (ABA)*

As pointed out in several chapters of this review, an increase of internal ABA content serves to protect plants in many if not any situation of water deficiency and other environmental stresses (cf. Sect. B6., B7.). Thus, a central role of this phytohormone in stress response regulation can be assumed. Exogenously applied ABA can mimic the respective processes, e.g. induce freezing resistance in cereals (see Sect. C.6.).
Increase in endogenous ABA levels and higher freezing resistance were found in tobacco plants when they were pretreated under condition of dessication, mineral deprivation and salt stresses (Boussiba et al., 1975). Similar phenomena were observed also for the acquisition of thermotolerance in maize. Seedlings were prestressed either by heavy metal or dessication conditions (Bonham-Smith et al., 1987) or by exogenous application of ABA or triadimefon (Bonham-Smith et al., 1988b). The latter compound 1-(4-chlorophenoxy)-3,3-dimethyl-1,1 (1,2,4-triazol-1-yl)-2- butanone, is a fungicide which causes higher levels of isoprenoids by blocking gibberellin synthesis. Both ABA- and triadimefon-pretreated tissues withstand the effects of a 3 h (40 °C) sublethal or 45 °C lethal heat shock. The acquired thermotolerance was also induced in both roots and shoots, when only the former were pretreated. The protective effect of triadimefon was likewise observed against heat, chilling, drought and ozone injury (Asare-Boamah and Fletcher, 1986; Asare-Boamah et al., 1986; Fletcher and Hofstra, 1985). Similar to ABA, the triazole fungicide was regarded as a plant multiprotectant (Fletcher and Hofstra, 1985). These effects obviously occur without the formation of HSPs (Bonham-Smith et al., 1987). Thus, gene expression may not be involved (see also Sect. B.4.1.).
The physiological role of ABA in stress responses, which is based on changing gene expression of a variety of proteins, will be regarded in more detail in the following example of seed maturation. It is generally assumed that dessication as part of the developmental program of the seed irreversibly terminates the maturation period and, after a period of metabolic rest, redirects the developmental program towards expression of genes necessary for germination and seedling growth. Thus embryogenesis and seed maturation on the one hand and germination on the other seem to be controlled by at least two interdependent factors: low level of water and high concentrations of ABA (Finkelstein et al., 1985, 1987; Fischer et al., 1987). ABA accumulates during early embryogenesis and declines again before dessication. It prevents germination but promotes maturation and the expres-

sion of storage proteins.—ABA synthesized in the ovule is transported into the embryo, where it turns off the translation of mRNAs specific for germination, i.e. usually mRNAs coding for hydrolytic enzymes necessary for the substrate supply of the growing embryo.

In this connection ABA also prevents precocious germination in many seeds (FINKELSTEIN et al., 1985): If immature embryos of *Brassica* are excised from the seeds and cultured *in vitro* in a hormone-free medium, precocious germination is observed, but it is prevented after exogenous ABA application. Further evidence for the regulatory role of ABA is provided by the observation of vivipary in ABA-insensitive mutants, by the addition of ABA inhibitors and by the fact that high concentrations of osmotics in the culture medium substitute for the effect of exogenous ABA, because they produce water deficiency (FINKELSTEIN et al., 1987).

Changes of major seed storage proteins can be used as a marker of ABA-responsive processes in seeds, e.g. cruciferin and napin in *Brassica* (FINKELSTEIN et al., 1985, 1987), vicilin and legumin in *Vicia faba* (BARRATT, 1986), β-conglycinin in soybean (BRAY and BEACHY, 1985), late embryogenesis abundant (lea) proteins in cotton (GALAU et al., 1986; HUGHES and GALAU, 1987; BAKER et al., 1988), or several storage proteins sedimenting as a 13S complex from mustard seeds (FISCHER et al., 1987). All of them seem to be products of gene families. Using storage protein cDNAs as hybridization probes, the steady-state levels of the storage protein RNAs were found to be variably controlled by external ABA. For instance, the hormone transiently enhances the initially low level of cruciferin mRNA but maintains the already high level of napin mRNA (FINKELSTEIN et al., 1985).

In *Brassica*, ABA responses are antagonized by exogenous cytokinin. As mentioned above, high osmotic potentials ("stress" concentration of 12.5 % sorbitol) can substitute for 10 μM ABA and stimulate storage protein synthesis. However, there is a difference: While ABA is effective in maintaining embryonic gene expression only in the predessication stage (embryos younger than 40 days after anthesis), the sorbitol effect is observed at all developmental stages tested (FINKELSTEIN et al., 1987). The authors regard the synergistic action of the two different effectors to be consistent with the hypothesis that ABA acts via restricted water uptake, which affects gene expression in an unknown, development-related manner. For *Sinapis* embryos, FISCHER et al. (1987) assume that "both ABA and osmotic stress elicit maturation development independently of each other", i.e. each of them could maintain the water content of the embryo cells below a critical threshold necessary to prevent germination.

Novel ABA-induced polypeptides wer likewise observed in cereal seeds during embryogenesis (GOMEZ et al., 1988; LIN and HO, 1986; MUNDY and CHUA, 1988; SANCHEZ-MARTINEZ et al., 1986; WILLIAMSON and QUATRANO, 1988). In addition to the inhibitory action of the hormone on the synthesis of several abundant GA_3-induced enzymes such as α-amylase, proteases (NOLAN et al., 1987), or mannase (DULSON et al., 1988), ABA triggers gene expression of more than a dozen of minor polypeptides of largely unknown function in barley aleurone layers (HONG et al., 1988; LIN and HO, 1986; NOLAN and HO, 1988), among them a most abundant 29 kDa protein. These proteins start to show enhanced synthesis 2 to 4 hours after ABA treatment, and their translatable mRNA levels correspond with the levels of *in vivo* synthesized proteins (LIN and HO, 1986). Some of the ABA-induced proteins might function in the conversion of ABA into its catabolite, phaseic acid (UKNES and HO, 1984), others could belong to the wheat-germ agglutinin family (TRIPLETT and QUATRANO, 1982) or other lectins.

Gene expression of an abundant embryo-specific wheat germ protein (Em) was shown to be regulated either at the level of mRNA transcription or through differential mRNA stability (QUATRANO, 1986; WILLIAMSON et al., 1985). MARCOTTE et al. (1988) constructed chimeric genes by translational fusion between 5'-flanking fragments from an Em genomic clone and the bacterial β-glucuronidase gene (GUS) linked to a 3' fragment from the cauliflower mosaic virus 35-S gene. The constructs were introduced into rice protoplasts to serve as a transient expression assay. Gene expression was found to be increased 30-fold within 60 min after addition of 10^{-4} to 10^{-6} M ABA. *In vitro* deletion experiments in the promoter region of 1 800 bp suggest that 650 bp of the Em promoter are sufficient as a *cis*-element for ABA-binding factors, and that a silencer sequence exists between bp -1 800 and -650 (MARCOTTE et al., 1988). Very basic or glycine-rich proteins are the products of rice and maize genes, respectively, which are expressed, when the plant organs or cell suspensions are subjected to ABA or water-stress (GOMEZ et al., 1988; MUNDY and CHUA, 1988). The RAB21 rice

gene-encoded RAB21 protein of mol. wt. 16,529 Da is rich in lysine. It accumulates in roots, leaves and embryos, i.e. RAB gene expression is not tissue-specific. The respective ABA-induced mRNA appears rapidly, in suspension cultures within 15 min after hormone application; it is likewise inducible by 200 mM NaCl treatment, but undetectable after heat shock treatment (MUNDY and CHUA, 1988). The mRNA accumulation depends on the activation of a preformed promoter binding factor(s), since inhibitors of protein synthesis fail to suppress RAB21 transcript formation.
The amino acid sequence of ABA-triggered maize protein (mol. wt. 15,427 Da) was deduced from genomic and cDNA clones. It consists of two domains of different hydrophilicity, of which the C-terminal domain is extremely rich in glycine residues (69.4 %), followed by arginine (12.5 %). Six repetitive motifs gly-gly-tyr-gly-gly and some arg-arg-glu clusters are the characteristic features (GOMEZ et al., 1988). Nothern blot analysis shows that the highest mRNA level for this protein appears 40 days after pollination, when seed dessication starts, but ABA treatment induces a 25-fold mRNA accumulation in premature embryos. A close correlation between water status, ABA concentration and mRNA level could be observed throughout embryo development. Interestingly, the mRNA for this protein was also transiently (2 to 12 hours) induced by wounding but has disappeared after 24 hours. A similar water stress and ABA induced glycine-rich protein of 21–26 kDa was also detected in tobacco and many other plants (SINGH et al., 1987a).

C.2.3. Jasmonates

(-)-Jasmonic acid and its derivatives (e.g. jasmonic acid methylester, JA-Me) are cyclopentanone compounds. They have been regarded as putative plant growth regulators (UEDA and KATO, 1980, 1982; UEDA et al., 1981; reviews: PARTHIER, 1988, 1989; SEMBDNER and KLOSE, 1985). Most remarkably, jasmonates accelerate senescence of isolated leaf tissue (SATLER and THIMANN, 1981; UEDA and KATO, 1982, UEDA et al., 1981; WEIDHASE et al., 1987a, and reviews by PARTHIER, 1988, 1989). Typical senescence symptoms such as loss of chlorophylls, increase of respiration, lowering of photosynthesis, and breakdown of ribulose-1,5-bisphosphate carboxylase/oxygenase were markedly promoted by JA-Me (POPOVA and VAKLINOVA, 1988; POPOVA et al., 1988; WEIDHASE et al., 1987a).
The effects of jasmonates are similar to ABA action, not only with respect to senescence promotion of detached leaves but also in the induced synthesis and accumulation of several abundant protein classes in barley leaf segments (WEIDHASE et al., 1987b). The polypeptides induced by ABA-Me and JA-Me are evidently the same as indicated by immunological cross-reactivity and determination of molecular masses. The most abundant jasmonate-induced proteins (JIPs) of barley have relative molecular masses (M_r) of 10–12, 23, 30, 37, 66 and 110 kDa, but more faint bands or spots could be visualized as well (MUELLER-URI et al., 1988; WEIDHASE et al., 1987b). The three major classes JIP 23, JIP 37, JIP 66, which are immunologically unrelated to each other or to other soluble barley proteins (HERRMANN et al., 1989), consist of several isoforms. They are synthesized on cytoplasmic polyribosomes (WEIDHASE et al., 1987b), but location of at least some of them in organelles and nuclei is highly probable (unpublished results). JA-Me induces the accumulation of species-specific abundant proteins in many plant species (HERRMANN et al., 1989).
The JA-Me-induced massive synthesis of new proteins (JIPs) with repression of normal protein synthesis reminds of the reprogramming of gene expression during heat shock or other stress responses. However, JIPs are neither related to HSPs nor does heat shock of jasmonate treated barley leaf segments affect their synthesis (MUELLER-URI et al., 1988). Wounding or salt stresses (0.5 M NaCl) are unable to induce proteins which cross-react immunologically with JIPs (unpublished results). Thus, in certain aspects jasmonate action resembles that of ABA, but in other aspects, e.g. ABA mimicking salt stress or osmotic shock response (MUNDY and CHUA, 1988), the two substances differ from each other. They can be regarded as natural signals belonging to different but partially overlapping stress domains.

C.2.4. Ethylene

The responses of plant cells to infection by pathogenic fungi, bacteria, or viruses, to elicitors from fungal or plant cell-wall or to wounding stress, is accompanied by an increase in activity of ACC synthase, the key enzyme in ethylene synthesis and subsequent increase in the amount of endogenous ethylene (Table 4). This "stress ethylene" formation is transiently induced, showing a peak of concentration a few hours after infection (Chappell et al., 1984).
Exogenously applied ethylene likewise causes an increase of chitinase and β-1,3-glucanase in uninfected pea pods, the same effects as exerted by fungal elicitors (Boller et al., 1983; Broglie et al., 1986, Felix and Meins, 1987; Mauch et al., 1984), or observed for auxin and cytokinin deficiency in tobacco cell cultures (Felix and Meins, 1985; Mohnen et al., 1985). There are indications that different signals (hormones, elicitors) may act independently and separately (Boller, 1988, 1989; Boller et al., 1983), but the regulation process is coordinated for the two induced enzymes, chitinase and β-1-3-glucanase (Vögeli et al., 1988).
The effect of ethylene on two other plant genes for defense response, which code for phenylalanine ammonium lyase and a hydroxyproline-rich glycoprotein, points into the same direction. Northern blot analysis of RNA from C_2H_4-treated carrot roots revealed a marked increase in the levels of both mRNA species (Ecker and Davis, 1987). Moreover, ethylene-dependent expression of both genes was also shown after transfer of chimeric gene constructs into tobacco (Ecker and Davis, 1987).
Ethylene as a stress metabolite is synthesized under practically all conditions of environmetal stresses (cf. Table 4). It has a triggering and controlling function in fruit ripening (review by Parthier, 1989), and in senescence of flower petals (Mayak and Halevy, 1980) and leaves (Roberts and Tucker, 1985). The latter seems to occur species-specifically or complementary in those plant tissues which respond with senescence on ABA or jasmonate treatments. The metabolic mechanisms in the formation and action of endogenous ethylene is clear as far as the level of ACC synthase is concerned (Hogsett et al., 1981; Imazeki, 1985; Yang and Hoffmann, 1984) and in that autocatalytic steps are involved. But the reactions at the regulatory levels are still obscure. Thimann (1985) described an ethylene-generating system in senescent oat leaves, in which the dehydrogenation of linoleic acids is coupled through Mn^{2+} to the oxidation of ACC, the biosynthetic precursor of ethylene. The rise in ethylene production associated with tissue excision (Kao and Yang, 1983), and the increase in ACC might be wound-induced rather than related to the senescence event per se. It is unclear, whether, in contrast to flower senescence and fruit ripening, a similar increase of ethylene production occurs during foliar senescence on intact plants and whether ethylene formation initiates the process or is rather a consequence of it. The latter would contrast to fruit ripening, in which ethylene contributes causally to the process by inducing specific gene expression for the necessary enzyme equipment (cf. review by Parthier, 1989).

C.2.5. Cytokinins

Together with auxins the cytokinins belong to the essential hormones necessary for a normal cell division and growth as well as organogenesis in plants. This earlier suggestion was verified by a natural gene transfer phenomenon, the T_i plasmid-induced crown gall tumor. It is the result of phytohormone overproduction due to the expression of two genes for auxin biosynthesis and one gene for cytokinin biosynthesis (T-cyt gene) in the T_i-transformed host cells (Morris, 1986; Zambryski et al., 1989). The system opens the possibility to demonstrate the actions of endogenously synthesized cytokinin (and auxin) at a molecular level, especially in gene expression (Klee et al., 1987; Memelink et al., 1988). It will soon replace auxin- and cytokinin-supplemented cell cultures in this respect.
Using the latter system, gene expression (mRNA transcription) for certain proteins was strongly reduced in comparison with cell cultures in auxin- and cytokinin-free medium. Interestingly, the proteins concerned are β-1,3-glucanase and chitinase, which are known as key enzymes induced after phytopathogen attack or wounding stress (Eichholz et al., 1983; Felix and Meins, 1985; Meins et al., 1983; Shinshi et al., 1987). Using corresponding cDNA clones, Mohnen et al. (1985) showed

that exogenously applied auxin and cytokinin in combination block the new formation of β-1,3-glucanase mRNA in tobacco tissue cultured on basal medium (see review by BOLLER, 1988).
Cytokinin overproduction after transfer of the *Agrobacterium* T-cyt gene alone evidently generates stress-like situations. Four of five different cDNA clones corresponding to mRNA species, which were increased in transgenic T-cyt gene tobacco shoots (chitinase, extensin and pathogen-related proteins) are also known to be induced by pathogenic infection or wounding (CHEN and VARNER, 1985b). The mRNAs for these proteins accumulated also in normal shoots in the presence of exogenous cytokinin (MEMELINK et al., 1987a). This positive control by cytokinin could be extended to other more abundant mRNAs in T-DNA transformed shoots. But a negative control (decrease of steady state levels of certain mRNA species) was also noted. Clear differences were found in the relative abundance of 18 among 240 translation products between transformed and normal tobacco RNA preparations translated in a wheat germ system or identified by Northern blot techniques (MEMELINK et al., 1987b, 1988). According to the authors, the T-cyt directed phenotype reflects an obvious stress situation. The increased stress-related mRNAs are translated to putative stress proteins whose detailed composition varies with the tissue analyzed. It will be important to investigate whether the observed effects on gene expression during 'cytokinin stress' are brought about by the hormone itself or by an increased ethylene level connected with it.
Exogenously applied cytokinins (usually as furfurylaminopurine = kinetin, or benzyladenine) have been found to counteract or suppress a number of stress-induced responses including the "normal" and ABA- or jasmonate-triggered leaf senescence process (MOTHES, 1960; PARTHIER, 1988; THIMANN, 1985; cf. part on jasmonates). In an early experiment ENGELBRECHT and MOTHES (1964) observed that tobacco leaves exposed to 49–53 °C for 1 to 2 minutes withstand the lethal consequences of the heat shock when they were sprayed with kinetin before or after the treatment. The lethal heat temperature was increased by several degrees, and the reversibility of the heat stress was explained with the ability of kinetin to strengthen the solute retention capacity of the cell.
This observation points to a membrane-stabilizing effect of cytokinins that has been assumed and discussed frequently without having profound evidence for it. The concept of cytokinin action on membrane integrity (LESHEM, 1984) and thus on transport and retention of substances (MOTHES, 1960) is supported by its counteraction of the membrane-destructive effects of ethylene, ABA and jasmonates. A similar heat shock modulating effect by gibberellic acid has been reported for etiolated mung bean; the exogenously applied hormone helps to survive and to recover the seedlings which, after 40 to 45 °C treatment for 1 hr, would die without hormone-treatment (CHEN et al., 1986).

C.2.6. *Brassinosteroids*

Brassinosteroids are addressed as substances with stress-protecting activity (cf. review by MANDAVA, 1988). However, neither the stress responses nor the protecting actions of the putative hormones seem to be well-defined at present. An interesting observation by KULAEVA et al. (1989) indicates that brassinosteroids (22,23-S-homobrassinolide, 10^{-6} to 10^{-8} M) were able not only to activate protein synthesis in wheat leaves but also to improve the capacity for protein synthesis at heat shock temperatures (43 °C). Several brassinosteroid-specific polypeptides were claimed to be induced. The synthesis of other proteins seems to be enhanced or decreased, but HSP synthesis was not altered by homobrassinolide treatment. The authors suggested an increase in thermoresistance in leaf tissues pretreated with the putative plant growth regulator.

C.2.7. *Polyamines*

Anti-stress effects and antisenescence actions of diamines (putrescine, cadaverine) as well as triamine (spermidine) and tetraamine (spermine) have been a matter of discussion since more than 10 years (for reviews cf. ALTMAN et al., 1982; SLOCUM et al, 1984; SMITH, 1985). Polyamines were found in all plants studied (but likewise in microorganisms). Their concentrations usually rise when growth rates are increased or osmotics such as sorbitol, mannitol, proline, betaine or sucrose are present (see Table 4; FLORES and GALSTON, 1984a, b). Exogenously applied polyamines cause growth-stimulating effects combined with a higher nucleic acid and protein content in the treated tissues. Although stimulation of *in vitro* protein synthesis by polyamines is often noted, the inhibition of protein degradation may well contribute to a high protein level by inactivation of proteases via polyamine binding to the enzymes (KAUR-SAWHNEY et al., 1982). Senescence retardation was suggested to be related to an inhibition of ethylene biosynthesis by polyamines (APELBAUM et al., 1981; SUTTLE, 1981).

One should keep in mind that polyamines have been generally found to stabilize the intramolecular structure of nucleic acids and negatively charged plasma membranes by surface phospholipid binding. This mode of action cannot be excluded for an explanation of their antisenescence function (ALTMAN et al., 1982). The latter effect points to a more general view of polyamine action in plant tissues under stress conditions. FLORES and GALSTON (1982, 1984a, b) observed a rapid polyamine accumulation via activation of arginine decarboxylase in osmotically stressed cereal leaves. Similar observations were made with acid-stressed (*p*H 3.5) peeled oat leaves (YOUNG and GALSTON, 1983). Salinity stress and ion imbalance, protoplast stabilization, wound effects and ethylene formation seem to be related with the metabolic involvement of polyamines (ALTMAN et al., 1982; SMITH, 1985 and Table 4).

C.3. Oxidation stress

A remarkable aspect of evolution was the acquisition of highly sophisticated mechanisms to rapidly remove excess oxygen or oxygen-containing radicals, which are among the most harmful agents to cells. Survival despite the presence of oxygen is an important character of most organisms existing today. Especially in plants, oxidation stress is frequently secondary to other stress conditions, e.g. high irradiation, intoxication with heavy metals or nitrate and stomata closure because of water deficiency (heat, cold or osmotic stress). Studies on plant/pathogen interactions indicate that the generation of active oxygen, followed by lipid peroxidation and induced phytoalexin synthesis are important parts of the defense reaction connected with hypersensitive cell death (BOLLER, 1989; BOSTOCK et al., 1982; LAMB et al., 1989; ROGERS et al., 1988; STELZIG et al., 1983). In addition, damage of DNA coupled to induced repair (BERNSTEIN and JOHNS, 1989; CATHCART et al., 1984; KANEKO et al., 1988) and oxidation of SH-group-containing proteins are major effects of oxygen stress.

Common protective mechanisms in all types of organisms are the synthesis of scavenging enzymes (superoxide dismutase, catalase, glutathione reductase, alkyl hydroperoxide reductase) and probably the increase of glutathione levels (CHRISTMAN et al., 1985; LOVEN et al., 1985; MITCHEL et al., 1983; NIETO SOTELO and HO, 1986; OMAR et al., 1987; SMITH et al., 1985; STORZ et al., 1989). A special aspect of hs-induced oxidation stress was investigated by HARARI et al. (1986, 1989) in mammalian cells. The increased oxidative removal of polyamines leads to toxic levels of hydrogen peroxide and aldehydes and a depletion of glutathione. Consequently, addition of exogenous spermine or spermidine reduces survival of cells under hs conditions (GERNER et al., 1980).

In plants, oxidative stress is particularly prominent in chloroplasts under all conditions when the normal dissipation of the light-induced high energy state is blocked, e.g. because of stomata closure. Thus lipid peroxidation and cross linking of thylakoid membrane proteins are typical consequences of hs plus high light intensities (GLACZINSKI and KLOPPSTECH, 1988; SCHUSTER et al., 1988). Photorespiration may be an important mechanisms to reduce this type of oxidation stress (see summaries by LEVITT, 1980b; RHODES 1987;). A special herbicide (paraquat) is known to generate superoxide radicals in chloroplasts by accepting an electron from photosystem I. Oxidative stress caused by paraquat coincides with induction of all four isozymes of superoxide dismutase in maize, i.e. the two cytoplasmic isozymes and each one isozyme in chloroplasts and mitochondria (MATTERS and SCANDALIOS, 1986a). Interestingly, hs, paraquat treatment and pathogen infection evoke the same modification of the translation apparatus in maize by binding of a 57 kDa preformed protein (WU et al., 1988b), but effects of this modification have not yet been described.

The intimate connections between hs and certain aspects of oxidation stress led to hypotheses about common signal systems. In enterobacteria *(Escherichia coli, Salmonella typhimurium)* partially overlapping sets of stress proteins are induced by hs, SOS response or oxidation stress (JOHNSON et al., 1989; MORGAN et al., 1986; VAN BOGELEN et al., 1987; WALKUP and KOGOMA, 1989). In particular, synthesis of three bacterial HSPs (including the GroEL and DnaK proteins) is also increased after addition of hydrogen peroxide. Other oxidizing agents with similar effects are 1-chloro-2,4,dinitrobenzene, diamide, t-butyl hydroperoxide, N-ethyl-maleimide, iodoacetamide, cadmium chloride and a variety of quinones. Simultaneously, all these stressors cause marked increase of Ap_4A or similar highly phosphorylated dinucleotides. AMES and coworkers suggested that Ap_4A may represent a new class of intracellular stress signal (alarmone), which also triggers the hs response, not only in bacteria (BOCHNER et al., 1984; LEE et al., 1983). In fact, many of the agents mentioned above were also found to induce HSP synthesis in *Neurospora crassa*, insects and vertebrates (see Sect. B.2.1. and summary by NOVER, 1990). But apparently Ap_4A is not required as intracellular signal, neither in eukaryotes (BREVET et al., 1985; MILLER and MCLENNAN, 1986; SEGAL and PECQ, 1986) nor in bacteria (BOCHNER et al., 1986; PLATEAU et al., 1987). Evidence against a causal relation between hs and oxidation stress in plants comes from the observation that induction of HSPs proceeds even under anaerobic conditions (MOCQUOT et al., 1987).
Much more likely as a central link for the induction of HSPs by hs and chemical stressors is the partial denaturation of proteins (see Sect. B.2.1.). Oxidation of SH-group-containing proteins is common under a variety of stressful conditions. These include all oxidative agents mentioned above, but also hs, heavy metals, active oxygen formed by irradiation or smog containing O_3 or peroxyacetylnitrate.

C.4. *Heavy metal stress*

The increasing exposition of plants to heavy metal ions is the consequence of environmental contaminations by industry, mining or sewage disposal. Among the highly toxic metal ions are La^{3+}, Pb^{2+}, Ag^{+}, Al^{3+}, Hg^{2+}, Cu^{2+}, Cd^{2+}, Zn^{2+}, Co^{2+} and Mn^{2+}. Reduced plant growth coincides with damage of the photosynthetic apparatus, impaired uptake or leakage of nutrients and changes of cell ultrastructure (for rewies see FOY et al., 1978; LEVITT 1980b). But some of these ions can also displace Mg^{2+}, Zn^{2+}, Cu^{2+}, Co^{2+} from their binding sites in active centers of enzymes or regulatory proteins. Effects at this level have barely been investigated in detail. Important parameters influencing the availability, uptake and hence toxicity of heavy metal ions are the presence of other mineral nutrients and the complexing capacity and relative acidity of the soil (KLOTZ and HORST, 1988; LEVITT 1980b). Thus, the actual concentration of Al^{3+} may increase by six orders of magnitude, if the *p*H value changes from 7 to 4 (MACDONALD and MARTIN, 1988).
Particularly sensitive targets of transition series metals (Cu^{2+}, Hg^{2+}, Cd^{2+}, Zn^{2+}) are SH group-containing proteins. Their partial denaturation may explain the induction of heat shock protein synthesis in animals (see summary by NOVER, 1990) and plants (Sect. B.2.1.) but also of other general components of the plant stress response (Table 4). This includes synthesis of ethylene, phytoalexins, PAL, pathogenesis-related proteins and others. An interesting peculiarity of Cd^{2+}-stress is the inhibition of intron splicing (see Sect. B.2.5.), which is not observed with other metal ions (Ag^{+}, Co^{2+}, Cu^{2+}, Zn^{2+}; CZARNECKA et al., 1988; EDELMAN et al., 1988; HAGEN et al., 1988; WINTER et al., 1988).
Acquired tolerance to heavy metal stress was frequently observed (BENNETZEN and ADAMS, 1984; JACKSON et al., 1985; LEVITT, 1980b; REESE and ROBERTS, 1984). In animals and *Saccharomyces* increased levels of low molecular weight metal-binding proteins (metallothioneins) are detected under these conditions (see reviews by BRADY, 1982; HAMER et al., 1985; KARIN, 1985). In *Candida* and *Schizosaccharomyces* species, similar functions are exerted by poly(γ-glutamyl cysteinyl) glycine derivatives (phytochelatins, DAMERON et al., 1989; GRILL et al., 1985, 1987, 1988; JACKSON et al., 1987; STEFFENS et al., 1986). These componds are the major chelating agents found in different types of plants and suspension cultured cells after incubation with Cd^{2+}, Pb^{2+}, Zn^{2+}, Ag^{2+}, Ni^{2+}, arsenite, Cu^{2+}, Sn^{2+}, SeO_3, Au^{+}, and others (GRILL et al., 1985, 1987). An interesting observation on the

mutual potentiation of toxic effects of heavy metal ions was reported by JACKSON et al. (1985). When tolerant cells of *Datura innoxia* growing at 250 μM Cd^{2+} were incubated with a non-toxic concentration of Cu^{2+} (100 μM), these cells are killed. Evidently the high affinity binding of Cu^{2+} to phytochelatins displaces part of the much more cytotoxic Cd^{2+} ions. Besides the actual amount of phytochelatins, their binding affinity and, hence, the extent of complex formation and stability may be decisive for the protection against intoxication by heavy metal ions (DAMERON et al., 1989; DELHAIZE et al., 1989).

C.5. Anaerobic stress

Besides heat shock, anaerobiosis belongs to the best studied stress response systems of plants. Major object in this field is maize (see reviews by DENNIS et al., 1987; FREELING and BENNETT, 1985; PRADET et al., 1985 and SACHS and HO, 1986). Anaerobiosis, experimentally established by putting seedlings in a water-saturated nitrogen or argon atmosphere, presumably mimics the stress situation encountered during flooding. Induction of alcohol dehydrogenase (ADH) activity as a result of flooding maize seedlings was first observed by HAGEMAN and FLESHER (1960), SCHWARTZ (1969) demonstrated that flooding tolerance in maize is dependent on a functional adh1 gene, and FREELING (1973) used immunological techniques to analyze the mass accumulation of ADH in roots of maize seedlings between 5 and 72 hours of anaerobiosis.

Later on, careful studies revealed a whole set of about 20 anaerobic stress proteins (ASPs) in maize roots (FERL et al., 1979; OKIMOTO et al., 1980; SACHS and FREELING, 1978; SACHS et al., 1980). In spite of tissue-specific variation in the details, similar sets of ASPs were also found in endosperm, scutellum, anther wall, coleoptile and mesocotyl, but not in leaves of maize (OKIMOTO et al., 1980). Other plant species/organs analyzed for anaerobic induction of ADH or other ASPs are rice roots (BERTANI et al., 1981; MOCQUOT et al., 1981, 1987), barley aleurone layers (HANSON and JACOBSEN, 1984; HANSON et al., 1984), tomato (TANKSLEY and JONES, 1981), *Arabidopsis* (CHANG and MEYEROWITZ, 1986; DOLFERUS et al., 1985), pea (LLEWELLYN et al., 1987), tobacco (ELLIS et al., 1987a), and *Sorghum*, dragopogon and carrot (cited in SACHS et al., 1980) as well as the conifer *Pinus radiata* (HARRY et al., 1988).

An essential aspect of survival of anaerobic stress periods is evidently the control of intracellular *p*H by conversion of lactate to ethanol. Cytoplasmic acidosis due to an ADH deficiency or CO_2 retardation in the medium contributes to anaerobic cell death (ROBERTS et al., 1984a, b). Maize seedlings adapted to low partial pressure of oxygen acquired higher ADH levels and produced more ethanol and ATP under anoxic conditions. Survival of root tips increased from 9 to 22 hrs (SAGLIO et al., 1988); this is indicative of stress tolerance.

The proteins induced by anaerobiosis are not related to HSPs (KELLEY and FREELING, 1982). But similar to hs, the anaerobic response starts with a decay of polysomes as shown for soybean roots by LIN and KEY (1967). They found that the "aerobic" mRNAs are evidently conserved during the stress period, since polysome reformation after return to aerobiosis does not need new synthesis of RNA. This was subsequently confirmed by *in vitro* translation experiments with RNA from anaerobic maize roots (SACHS et al., 1980), i.e. there is a mechanism of translational discrimination between "anaerobic" and „aerobic" mRNAs. This is an interesting parallel to the hs response (see Sect. B.2.6.1.).

Identification of alcohol dehydrogenase (mainly ADH1 and much less of ADH2) as a prominent ASP in maize (ASP40 A and B, ASP42A and B, see FERL et al., 1979, 1980; FREELING, 1973; SACHS et al., 1980) was followed by characterization of other ASPs which represent enzymes of the glycolytic pathway and related parts of the intermediary metabolism. ASP87 is a sucrose synthase (MCELFRESH and CHOUREY, 1988; SPRINGER et al., 1986), ASP64 is pyruvate decarboxylase (LASZLO and LAWRENCE, 1983), ASP55 is glucose-6-phosphate isomerase (KELLEY and FREELING, 1984a) and ASPs 33 and 35.5 are fructose-diphosphate aldolases (KELLEY and FREELING, 1984b).

The delayed formation of ASPs starting only after 5 hrs of anaerobiosis is preceded by new formation of few transition proteins of ~ 33 kDa (SACHS et al., 1980). *In vitro* translation (FERL et al., 1980; SACHS et al, 1980) and later on direct mRNA quantification with specific gene probes (DEN-

NIS et al., 1984, 1985; GERLACH et al., 1982; HAKE et al., 1985; ROWLAND and STROMMER, 1986) demonstrated that increased ASP synthesis results from accumulation of corresponding mRNAs. Differential gene activation was confirmed by nuclear run-off transcription experiments (ROWLAND and STROMMER, 1986). Induced accumulation of ASP mRNAs is not uniform. HAKE et al. (1985) identified three different groups with maximum mRNA levels at 2–3 hrs (aldolase and ASP31 mRNAs), at 6–12 hrs (alcohol dehydrogenase 1 and sucrose synthase mRNAs) and 24 hrs (ASP40), respectively. But there is a coordinate and rapid degradation of "anaerobic mRNAs" after reoxygenation.

The detailed gene structure of the maize Adh gene has been described (DENNIS et al., 1984, 1985; see summaries by DENNIS et al., 1987; FREELING and BENNETT, 1985). 10 exons are separated by 9 introns. The same is true for the pea Adh gene which in its coding parts is 75 % homologous to the maize gene. The 9 introns, though different in length, are in identical positions (LLEWELLYN et al., 1987). The pea Adh bears 80 % homology to the *Arabidopsis* gene. In this case, 3 introns were lost during evolution but the remaining six are again in similar positions as compared to the maize and pea genes (CHANG and MEYEROWITZ, 1986; see DENNIS et al., 1987). 70 % homology was found by comparison of the pine Adh cDNA with the maize Adh sequence (HARRY et al., 1988).

In the 5' flanking region of the maize Adh gene a TATA-box is found at bp -40. Functional analyses of the promoter by deletion and linker scanning mutation was done in fusion constructs with the E. coli *cat* as reporter gene. Using maize protoplasts as a transient assay system, an anaerobic response element (ARE) between bp -81 and -140 was found to be sufficient for effective anaerobic induction (HOWARD et al., 1987; WALKER et al., 1987). Two essential sequence elements were defined in this region with a conserved hexanucleotide motif in each (underlined):

-(-140)CTGCAGCCCGGTTTC-10n-CCGTGGTTTGCTTGCC(-99)-.

Similar hexanucleotide motifs were detected in other Adh genes was well as in aldolase and sucrose synthase genes (see summary by DENNIS et al., 1987). *In vivo* footprinting, analyzing protection to methylation by dimethyl sulfate, provided evidence for the presence of two types of promoter binding factors in the ARE region. Base pairs -109 to -112 and -130 to -137 are always protected, whereas bp -92 to -100 and -178 to -182 are shielded only in the induced state (FERL and NICK, 1987). Interestingly the maize bp -140 to -81 element is also active as ARE in tobacco. But for effective transcription, it needs cooperation with an upstream enhancer, e.g. provided by a 16 bp palindrome derived from the octopine synthase promoter or by parts of the cauliflower mosaic virus 35S promoter (ELLIS et al., 1987a, b).

C.6. Low temperature stress

Since the early investigations on freezing injury of plants (MAXIMOV, 1912; MÜLLER-THURGAU, 1886; WIEGAND, 1906), two important aspects of low temperature stress and related mechanisms of cold acclimation have been distinguished. On the one hand, adaptation means maintenance of sufficient metabolic activities for survival and growth under low temperature conditions (chilling resistance). On the other hand, it concerns survival of and reactivation after frost periods (freezing resistance). Many reviews and books summarize older and newer results from thousands of papers published on this stress domain (ALDEN and HERMANN, 1971; ANDREWS, 1987; BURKE et al., 1976; GRAHAM and PATTERSON, 1982; LEVITT, 1958, 1980a; PATTERSON and GRAHAM, 1987; SINGH and LAROCHE, 1988; STEPONKUS, 1984).

As described by MÜLLER-THURGAU (1886), freezing damage is largely due to the dehydrative and mechanical injury during extracellular ice formation (see reviews by LEVITT, 1980a; SINGH and LAROCHE, 1988 and STEPONKUS, 1984). Massive extrusion of lipid material from the plasma-membrane is associated with the loss of osmotic responsiveness and, finally, expansion-induced lysis (SINGH et al., 1987b; STEPONKUS, 1984). Cold hardening with respect to freezing tolerance evidently includes (i) accumulation of solutes (proline, sucrose, trehalose) protecting cells from excessive water loss and modifying intracellular ice formation (see summary by ALDEN and HERRMANN, 1971), (ii) increase of membrane material, probably in form of cytoplasmic vesicles, which are available for the expansion of the plasmalemma (DOWGERT and STEPONKUS, 1984; JOHNSON-FLANAGAN and SINGH, 1986; ORR et al., 1986) and (iii) adaptation of the plasma membrane composition by increasing the phospholipid/protein ratio, by changes of the protein composition and probably by incorporation of desaturated fatty acids (GRAHAM and PATTERSON, 1982; LYNCH and STEPONKUS, 1987; PATTERSON and GRAHAM, 1987, STEPONKUS et al., 1988; UEMURA and YOSHIDA, 1984; WILLIAMS et al., 1988). This type of low temperature adaptation of membrane fluidity is common to all types of organisms and contrasts to the adaptation by increase of saturated fatty acid content under high temperature conditions. Thus, it may be surprising that in many plants cold hardening, e.g. by low temperature periods in autumn, is connected also with a rise of the heat resistance (results summarized by ALEXANDROV et al., 1970). But this seemingly contradictory result depends on the experimental criterium used to analyze heat resistance. ALEXANDROV and coworkers observed protoplasmic streaming in leaf epidermal cells, which depends on the functioning of the actin cytoskeletal system and not primarily on membrane properties. On the other hand, the accumulation of osmotic solutes (polyhydric alcohols, proline, betaines) is a general part of the stress response, i.e. irrespective of heat, cold, osmotic or water stress (see Table 4). Some of these are well known cryoprotectants, used to improve survival of plant cells after deep freezing (see summaries in EVANS et al., 1983, and REINERT and BAJAJ, 1977).

For the reduction of productivity of crop plants, cold stress effects on the structure and function of chloroplasts are particularly important. On the one hand, chloroplast formation in germinating seedlings can be inhibited or even irreversibly damaged (for review see ANDREWS, 1987). On the other hand, thylakoid membranes are destroyed by lipid peroxidation as a consequence of the lacking dissipation of the photosynthetic potential (oxidation stress, see Sect. C.3., SMILLIE et al., 1988; WISE and NAYLOR, 1988).

There are numerous reports documenting the synthesis of new proteins during or following a cold stress period (see summary by SINGH and LAROCHE, 1988). In all cases where a direct comparison was done, cold stress proteins (CSPs) are different from the heat shock proteins. This is true for spinach (GUY et al., 1985), maize (YACOOB and FILION, 1987) and the grass *Lolium temulentum* (OUGHAM, 1987). But in connection with the acquisition of cold tolerance synthesis of new CSPs was also detected in alfalfa (MOHAPATRA et al., 1987, 1988a, b), rape (JOHNSON-FLANAGAN and SINGH, 1987; MEZA-BASSO et al., 1986), potato (TSENG and LI, 1986), brome grass (ROBERTSON et al., 1987, 1988), wheat (SARHAN and PERRAS, 1987), rye (CLOUTIER, 1984; SINGH and LAROCHE, 1988; UEMURA and YOSHIDA, 1984), barley (CATTIVELLI and BARTELS, 1989) and *Arabidopsis thaliana* (GILMOUR et al., 1988; KURKELA et al., 1988). Inspecting the pattern of CSPs in different plants, no general picture is apparent. But this may primarily result from the insufficient methods of electrophoretic analysis applied. Only in few cases two-dimensional separations were used (CATTIVELLI and BARTELS, 1989; GILMOUR et al., 1988; MEZA-BASSO et al., 1986; ROBERTSON et al., 1987), and direct comparison of CSPs between different plants is totally lacking. A common feature of long-term cold adapted plants may be the accumulation of high molecular weight CSPs (Mr 150–200 kDa) and several CSPs of 45–47 kDa (GILMOUR et al., 1988; KURKELA et al., 1988; ROBERTSON et al., 1988; SARHAN and PERRAS, 1987). Interestingly, a cold shock-induced mRNA of unripe tomato fruits was shown to code for a thiol protease (SCHAFFER and FISCHER, 1988), which is not synthesized during normal fruit ripening.

Following the introductory remark on the two aspects of cold hardiness, analyses are usually based on different methods: (i) extent of protein labeling at low temperatures, (ii) altered protein patterns, (iii) membrane integrity (osmotic responsiveness, solute leakage), and (iv) survival of freeze-thaw cycles. It is remarkable that in several cases abscisic acid was found to evoke cold tolerance connected with the characteristic accumulation of membrane vesicles and changes of protein patterns (CHEN and GUSTA, 1983; JOHNSON-FLANAGAN and SINGH, 1987; ORR et al., 1986; ROBERTSON et al., 1987; TSENG and LI, 1986). The effects reported by CHEN and GUSTA (1983) with cell cultures of winter wheat, winter rye and bromegrass are conspicuous. After 4 days in 7.5×10^{-5} M ABA medium, cells survived treatment at -30 °C, whereas control cells died already at -7 to -8 °C. Similar results were found with alfalfa and carrot but not with cell cultures of *Datura, Catharanthus, Glycine* and *Vicia.*

C.7. *Water deficiency stress*

C.7.1. *Introduction*

Though the term water stress is almost exclusively applied to stress caused by water deficiency, it is not unambiguous and may principally include also flooding stress (LEVITT, 1980b). Water deficiency or desiccation results, directly or indirectly, from a number of stressful conditions, e.g. drought, heat shock, salt and osmotic stress, freezing or phytopathogen attack. The multiplicity of environmental factors involved and the drastic effects on plant development and productivity have been the basis for many years of extensive investigations centered mainly on the organismic aspects (for summaries see HSIAO, 1973; LEVITT, 1980b; MORGAN, 1984; POWLES, 1984; WALTER, 1955).
During evolution plants developed a multiplicity of mechanisms to survive and even grow under conditions of extremely low or frequently changing water supply. More than other environmental factors, access to water contributed to the variability of higher plants with respect to their morphology, growth rhythm, internal water economy and metabolism. The range of water supply conditions covers excess water for plants growing in water or in tropical rain forests, changing water supply in moderate climates and extends to almost constant water deficiency in deserts or permafrost regions.
All environmental factors causing water deficiency were found to markedly increase synthesis of abscisic acid (ABA) and ethylene (see CORNISH and ZEEVART, 1984; HARRIS et al., 1988 and reviews by DAVIES, 1987; RHODES, 1987; YANG and HOFFMAN, 1984), and ABA application can be used to evoke many effects typical for water stress situations (see Sect. C.2.). Experimentally, water stress is frequently produced either by incubating in polyol, sucrose or polyethylene glycol solutions (osmotic stress) or by treating with high concentrations of salt, mainly of NaCl (salt stress). In the latter case, the general effects of water deficiency may be superimposed by specific effects of ion imbalances. However, plants can adapt to relatively high concentrations of sodium chloride (100 mM NaCl) in their cytoplasm without serious effects on their metabolism (BINZEL et al., 1988; LYNCH and LAEUCHLI, 1988; see review by CHEESEMAN, 1988).

C.7.2. *Photosynthetic activity and carbon assimilation*

Closely related to the increasing amount of ABA in guard cells is the closure of stomata under conditions of water deficiency. Though this mechanism is important to reduce water loss, it causes a number of unwanted secondary effects. The block of transpiration severely limits or totally interrupts nutrient supply from the root as well as formation of carbohydrates in leaves. Hence, nutrient starvation is the consequence. On the other hand, the lacking energy dissipation by reductive CO_2 fixation may result in oxidative damage of the photosynthetic apparatus (photoinhibition, photobleaching, see Sect. C.3. and reviews by OGREN, 1984; POWLES, 1984 and RHODES, 1987). These effects are enhanced by the increasing leaf temperature (heat shock). Experimental evidence for a particular sensitivity of the D proteins (P32) in the center of photosystem II was provided by SCHUSTER et al. (1988). Photorespiration, dependent on the oxygenase activity of ribulose-bisphosphate carboxylase, may be a protective mechanism important under these conditions (OGREN, 1984). This complex biochemical pathway is distributed between the chloroplast, peroxisome and mitochondrial compartments and serves as a constant sink for the electron transport chain.
A particular two-step-mechanism of CO_2 fixation observed in C_4 plants, e.g. in *Amaranthus* spec., *Zea mays, Portulacca* spec., *Sorghum* and others, may help to overcome part of the deleterious effects by water deficiency, i.e. starvation of assimilates. This type of plants is capable to fix CO_2 in mesophyll cells by phosphoenolpyruvate carboxylase forming C_4 acids. CO_2, fixed by this reaction during night, can be delivered to bundle sheath cells and, after decarboxylation of the C_4 acids, can be refixed by the usual photosynthetic C_3 pathway using ribulose-bisphosphate carboxylase (see summaries by KLUGE and TING, 1978; OGREN, 1984).
In few species, expression of this "Crassulacean acid metabolism" (CAM) is not constitutive but depends on developmental stages or environmental conditions, e.g. in *Kalanchoë blossfeldiana* (CHAT-

URVEDI and ZABKA, 1977; OTA, 1988) and *Mesembryanthemum crystallinum*. In the latter species CAM can be induced under conditions of water deficiency, e.g. by irrigation with 0.4 M NaCl solution (WINTER and WILLERT, 1972). Increase of PEP carboxylase amount in leaves, but not in C_3 tissues of the same plant, was 40-fold under these conditions, and this correlates with an increased level of its mRNA (HOLTUM and WINTER, 1982; HÖFNER et al., 1987; OSTREM et al., 1987). Recently, these results were extended by BOHNERT and coworkers using cDNA clones for PEP carboxylase and pyruvate-orthophosphate dikinase. Both proteins are in the size range of 100 kDa and, correspondingly, two mRNAs of 2900 and 3300 nucleotides were identified in leaves of NaCl-induced *Mesembryanthemum* (SCHMITT et al., 1988). Their levels rapidly decline after relief from salt stress (VERNON et al., 1988). Similar data on the size of the two proteins and of their mRNAs were derived from experiments with clones for the two light-induced maize genes (HUDSPETH et al., 1986).

C.7.3. *Osmotic adjustment*

A typical aspect of different types of environmental stresses connected with water deficiency is the increase of cellular solute concentrations, e.g. of ions, alcohols, sucrose or amino acids. Though the type of prevalent amino acid varies with the plant species investigated, e.g. aspartic acid, asparagine, glutamic acid or glutamine, the 10- to 300-fold increase of proline concentration up to 40 mg/g dry matter is frequently the dominant feature (HASSAN and WILKINS, 1988; MORGAN, 1984; RHODES, 1987). In a restricted number of plants (Chenopodiaceae, Gramineae, Compositae), proline is replaced by glycine betaine (see summaries by LEVITT, 1980b; MORGAN, 1984; RHODES, 1987). Interestingly, both proline and glycine betaine as well as trehalose are the major factors also in osmoregulation of bacteria (GLAEVER et al., 1988; LERUDULIER and VALENTINE, 1982; LERUDULIER et al., 1984; SMITH et al., 1988).

Proline accumulation results from a markedly increased synthesis from glutamic acid and a corresponding reduction of degradation by proline oxidase, which is presumably localized in mitochondrial membranes (HANDA et al.; 1983; 1986; HANSON and HITZ, 1982; RHODES et al., 1986). In many cases increased synthesis of abscisic acid (ABA) directly precedes the rise of proline synthesis. Both consecutive parts of the water stress response are inhibited by cycloheximide, and application of ABA without water deficiency leads to proline accumulation (STEWART and VOETBERG, 1985; STEWART et al., 1986). Synthesis of glycine betaine involves a two step process starting from choline and involving betaine aldehyde as an intermediate (HANSON and HITZ, 1982). Both enzymes required are localized in chloroplasts, and their activities and/or amounts markedly increase during salt or water stress (HANSON et al., 1985; WEIGEL et al., 1986).

The massive accumulation of amino acids or their derivatives under conditions of water deficiency may be closely related to an increased protein breakdown (DOVER, 1968; KEMBLE and MACPHERSON, 1954; MOTHES, 1928). The early investigations by MOTHES (1928) demonstrated that the loss of nitrogen compounds in older leaves is connected with their translocation to younger leaves. Consequently, after turgor regain there was a rapid senescence of the older leaves. This agrees with the observation that the adaptive flexibilitiy in osmoregulation is much higher in young organs (see review by MORGAN, 1984). Proline, and presumably other amino acids, may serve as storage forms of nitrogen. But in tomato suspension cultures proline was also shown to alleviate growth inhibition after water stress and "osmotic down-shock" (HANDA et al., 1986), Interestigly, the threshold level of water deficit required to stop growth in maize varies in an organ-dependent manner. It increases from stem (-0.5) to leaves (-1.0) to roots (-1.4 MPa) (WESTGATE and BOYER, 1985). Differences in the capability for osmotic adjustment may be the reason. Finally, there is a clear correlation between osmotic adjustment and the maintenance of photosynthetic activity under water stress (BOYER, 1982; CONROY et al., 1988; GUPTA and BERKOWITZ, 1988).

C.7.4. *Desiccation stress proteins (DSPs)*

The close connections between intracellular abscisic acid levels and various aspects of the water deficiency response is also apparent when changes in the patterns of protein synthesis are analysed. In practically all cases investigated, application of exogenous ABA evokes similar effects. Several examples of very specific changes in enzyme synthesis under water stress conditions were already mentioned, e.g. in connection with the increased synthesis of ABA itself as well as of proline but also with the induction of CAM enzymes in *Mesembryanthemum*. These data are complemented by electrophoretic analyses of general desiccation stress-induced proteins (DSPs). Different types of experimental procedures have been applied. Using double labeling techniques DHINDSA and CLELAND (1975) found new proteins synthesized in oat coleoptiles incubated in 0.3 M mannitol or 21 % Carbowax-4000. Following this, FLECK et al. (1982) demonstrated that marked changes of gene expression in protoplasts from *Nicotiana sylvestris* leaves are, at least in part, due to the formation of „osmotic stress“ proteins.

Usually the manifestation of the reprogramming of gene expression under water deficiency takes hours or even days. BARTELS et al. (1988) compared desiccation sensitive, early stages and tolerant, later stages of barley embryo development 12 days and 16 days after pollination. 25–30 new or enhanced proteins were identified in two-dimensional electropherograms. Which of the newly formed proteins are essential for expression of desiccation tolerance is unclear.

The largest group of reports on DSPs is concerned with proteins found in salt-tolerant plants or induced by increasing the NaCl concentration in the extracellular milieu. Experimental objects are tobacco cell cultures (ERICSON and ALFINITO, 1984; SINGH et al., 1985, 1987a), barley roots (HURKMAN and TANAKA, 1987; RAMAGOPAL, 1987), wheat seedlings (GULICK and DVORAK, 1987), embryos, leaves and suspension cultures of rice (MUNDY and CHUA, 1988) and maize callus (RAMAGOPAL, 1986). Effects evoked by other osmotic agents, e.g. by mannitol or polyethylene glycol, are similar but not identical (RAMAGOPAL, 1986; SINGH et al., 1985). Depending on the experimental conditions, as well as the plants and organs investigated, 10–20 new DSPs were detected (GULICK and DVORAK, 1987; HURKMAN and TANAKA, 1987; RAMAGOPAL, 1987).

Most striking are the markedly increased levels of basic, glycine-rich proteins of 21–26 kDa detected in different organs of various plants, irrespective of induction by high salt, PEG or ABA (HURKMAN and TANAKA, 1987; MUNDY and CHUA, 1988; SINGH et al., 1985, 1987a and Sect. C.2.2.). In callus cultures of *Nicotiana tabacum* (Wisconsin), accumulation of this 26 kDa protein is clearly correlated with the acquisition of salt tolerance, which is accelerated by addition of ABA. By immunological methods, similar 26 kDa proteins were identified in different organs of tobacco (stem, leaf, root), their amount varied with the endogenous ABA levels. In other plants, DSP26 was also induced after application of ABA or NaCl, e.g. in potato, soybean, cotton, carrot and millet (SINGH et. al., 1987a).

C.8. *Starvation*

Nutrient starvation and related changes in gene expression were extensively studied in microorganisms. Metabolic reprogramming of *Escherichia coli* and *Salmonella typhimurium* under conditions of glucose, phosphate or nitrogen starvation was found to be associated with synthesis of complex, partially overlapping sets of new starvation proteins (SSPs). Some SSPs were also induced by heat shock (JENKINS et al., 1988; SPECTOR et al., 1986), and in *E. coli* starvation was shown to increase heat resistance (JENKINS et al., 1988).

In eukaryotic microorganisms, starvation can be used to trigger developmental processes, e.g. fruiting body formation in *Dictyostelium* (ALTON and LODISH, 1977; LOOMIS, 1979; ZUKER et al., 1983) or sporulation in the aquatic fungi (BONATO et al., 1987; LEJOHN and BRAITHWAITE, 1984) and yeast (KURTZ et al., 1986). Though in these cases starvation proteins may include sporulation-specific proteins (LEJOHN and BRAITHWAITE, 1984), some evidently belong to the set of HSPs. This is true for HSPs 26 and 89 in yeast sporulating under nitrogen-deficiency (KURTZ et al., 1986), for HSPs 89, 75, 71, 48 and 46 in sulfur-starved yeast cultures entering into the G_0-stage (IIDA and YAHARA, 1984;

Shin et al., 1987), for HSPs 76, 70 and 39 in *Blastocladiella emersonii* sporulating on a Tris-maleate buffer (Bonato et al., 1987), for HSPs 80 and 30 in *Neurospora crassa* on 0.02 % sucrose medium (Roychowdhury and Kapoor, 1988) and for HSPs 80 and 73 mRNAs in *Tetrahymena thermophila* starved on Tris-hydrochloride buffer (Hallberg et al., 1984). Though details of the regulation have not yet been established, it is very likely that the selective synthesis of distinct HSPs in this context does not depend on the hs promoter element (see Sect. B.1.2.1.).
Similar to other stress conditions, nutrient starvation in higher plants leads to polysome degradation (Walter and Hahlbrock, 1985; Webster, 1980; Webster and Van't Hof, 1973), to a drastic reduction of protein synthesis and to a cell cycle block (Amino et al., 1983; Gould et al., 1981). Remarkable is the observation of a new starvation-induced protein of M_r 60–62 kDa in pea roots (Webster, 1980) or in parsely cell cultures cultivated on water or buffer solution (Walter and Hahlbrock, 1985). After 40 hrs starvation, the 62 kDa protein is practically the only protein labelled in parsley cultures. *In vitro* translation experiments show however that many of the non-translated 'control mRNAs' are maintained. If heat shock is applied under these conditions, synthesis of the starvation protein stops and the normal pattern of HSPs appears (Walter and Hahlbrock, 1985).
This contrasts to the situation in microorganisms. Probably as a result of transcription reprogramming, sporulating yeast cells become refractory to hs-induction in nitrogen deficient medium, at least for HSP70 (Kurtz et al., 1986). When cultures of the unicellular ciliate *Tetrahymena thermophila* were starved on Tris-hydrochloride buffer, they became particularly sensitive to a long-term hs at 41 °C. The defect in the normal hs response is not due to the lacking trancription of hs genes but results from a destabilization of ribosomal RNA and a break-down of the translation machinery (Hallberg et al., 1984). Thus, there may be differences of the response of plants and eukaryotic microorganisms to combined starvation and heat stress. But before seriously evaluating these results, interactions between different stress systems need much further investigation.

C.9. Wounding stress

C.9.1. Reprogramming of gene expression

Though frequently neglected in the analysis of stress factors, wounding of plants is a common stross, due to physical factors like wind and hail, to feeding by herbivores or to various kinds of human attacks. In addition, wounding of plants is frequently connected with or even prerequisite for pathogen attack (see Sect. C.10.). Similar to other stress response systems, multifold changes of cell metabolism and gene expression are observed (see summaries by Davies, 1987; Imazeki, 1985; Kahl, 1978). In this connection, it should be recognized that experimental work with plants frequently involves wounding as a necessary step, e.g. for precursor uptake, protoplast isolation or just handling of parts of larger plants. This may certainly influence the outcome of such experiments.
The wounding response includes formation of protective cell wall material (suberization, cutin and callose formation, lignification), the onset of cell division, remarkable changes in the pattern of gene expression, cessation of protoplasmic streaming, the generation of action potentials, formation of stress ethylene and the generation of cell wall fragments acting as elicitors of the wound response.
The generally higher levels of RNA and protein synthesis in regenerating tissues after wounding (ApRees, 1969) is reflected by a corresponding increase of the polysome content and coding capacity (Ishizuka et al., 1981; Oba et al., 1982; Schuster and Davies, 1983; Shirras and Northcote, 1984; Wielgat and Kahl, 1979). Careful analyses of Davies et al. (1986) revealed that the increase of polysome levels is influenced by two stress-dependent factors. Using aged pea epicotyls, they found a „freezing" of preexsting polysomes with strong inhibition of protein synthesis within the first 5 minutes after wounding, whereas after 15–90 minutes protein synthesis is resumed with new formation of polysomes and of wounding-induced proteins (WSPs, Theillet et al., 1982; Pena-Cortes et al., 1988).
Among the WSPs are a number of enzymes whose activity and/or amount became increased. Examples are the key enzyme of ethylene biosynthesis (ACC synthase, see Imazeki, 1985), callose syn-

thase (AIST, 1976), cytochrome P450 (RICH and LAMB, 1977), invertase and ascorbate oxidase (EDELMAN and HALL, 1965; MATSUSHITA and URITANI, 1974), catalase (ESAKA et al., 1983), ribonuclease (PITT and GALPIN, 1971; SACHER et al., 1982; VARIAN and SACHER, 1978), RNA polymerase I and II (KAHL et al., 1982; WIELGAT and KAHL, 1979) and protein kinase. The latter is connected with changes of phosphorylation of chromatin proteins (SCHAFER and KAHL, 1981). Very intriguing are reports on the induction of enzymes involved in plant defense mechanisms, i.e. of phenylalanine-ammonia lyase and chalcone synthase in bean hypocotyls (LAWTON and LAMB, 1987, see Sect. C.10.2.) and on increased levels of HSP70 mRNA in maize (HEIKKILA et al., 1984). The latter result reminds of the trauma-induced HSP70 synthesis in vertebrate cells (CURRIE and WHITE, 1981; HIGHTOWER and WHITE, 1981; HIGO et al., 1987; WHITE and WHITE, 1986).

C.9.2. Induced proteinase inhibitors

A prominent part of the wound response is the mass synthesis of low molecular weight serine proteinase inhibitors stored in the vacuole. In leaves of potato and tomato they may form more than 3 % of the total cellular protein 48 hrs after wounding (GREEN and RYAN, 1972; RYAN and HUISMAN, 1970). Screening of other plants revealed a strong or moderate increase in leaves of alfalfa, tobacco, strawberry, cucumber, squash, clover, broadbean and grape, but no increase in monocots, in lettuce and in members of the Rosaceae family (WALKER-SIMMONS and RYAN, 1977). However, few representatives of the latter group contain high constitutive levels of these inhibitors. The biological significance of the mass accumulation of serine protease inhibitors remains obscure. Their high activity towards trypsin and chymotrypsin was taken as evidence for a protecting role against herbivorous animals.

In tomato there are two types of inhibitors, PI1 and PI2 with Mr 7,800 and 12,300, respectively. Early *in vitro* translation experiments revealed that they are synthesized as preproteins which are processed on their way to the vacuole (NELSON and RYAN, 1980). The 111 amino acid (aa) residue precursor of PI1 contains 42 aa residues, the 148 aa residue pre-PI2 contains 25 aa residues of N-terminal pre-sequences (GRAHAM et al., 1985a, b). Both show highly conserved boxes with sequence homology to similar serine proteinase inhibitors of potato, broad bean and barley, but also of leech.

While the sequence data for the two tomato inhibitors were derived from cDNA clones, genomic clones were used for analysis of the corresponding potato inhibitors (CLEVELAND et al., 1987; THORNBURG et al., 1987). PI1 of potato is coded for by at least six genes. The clone sequenced shows a 107 aa residue open reading frame for the pre-protein with 36 residues of pre-sequence, which is probably removed in two steps. The gene contains two introns of 479 and 417 bp which are inserted in codons 17 and 43, respectively (CLEVELAND et al., 1987). The potato PI2 gene codes for a 153 aa residue precursor with a 47 residue presequence. A single intron of 117 bp is inserted in codon 18 (KEIL et al., 1986; THORNBURG et al., 1987). Hybrid genes constructed from each about 1000 bp of 5' and 3'flanking sequences of the potato PI2 gene and the bacterial chloramphenicol acetyltransferase gene as reporter gene were expressed in transgenic tobacco plants. Both parts of the flanking sequences are required and sufficient for a systemic wounding-induced cat activity in tobacco (AN et al., 1989; THORNBURG et al., 1987).

C.9.3. Hydroxyproline-rich cell wall proteins

Another type of WSPs are hydroxyproline-rich cell wall glycoproteins. Since their discovery by LAMPORT and NORTHCOTE (1960) in sycamore suspension cultures, bean callus and hypoctoyl, several types of these proteins were found to constitute 1–15 % of the cell wall material of dicotyledonous plants (see summary by COOPER et al., 1987). The extensin-type from tomato, soybean seedlings and carrot roots was intensively studied. It includes glycoproteins composed of 35 % protein and 65 % carbohydrate, mainly arabinogalactan. The basic structure of the protein part of carrot extensin was derived from sequencing cDNA and genomic clones (CHEN and VARNER, 1985a, b). An open read-

ing frame of 306 aa residues codes for a 34,225 dalton protein with a 32 aa residue signal sequence. The gene contains an intron in the 3' untranslated region. Extensins from carrot and tomato consist mainly of two repetitive peptides, the major -SPPPP- plus the less abundant -TPVYKYK- oligopeptide motifs (CHEN and VARNER, 1985b; SMITH et al., 1986; see summary by COOPER et al., 1987). At least half of the proline residues are hydroxylated posttranslationally. The structure of the soybean extensin-like protein may be very similar (AVERYHART-FULLARD et al., 1988).
The unmodified extensin precursor, i.e. the non-hydroxylated and non-glycosylated protein lacking the signal peptide, was identified in carrot slices incubated in the presence of α, α'-dipyridyl (SMITH, 1981). The processed extensin monomer of 86 kDa is secreted to the cell wall and insolubilized by formation of isodityrosine cross-links, a process prevented by inhibitors of peroxidase activity (COOPER and VARNER, 1983, 1984).
The particular interest in extensin in this context resulted from the observation that its synthesis and extracellular deposition rapidly increased after wounding (CHRISPEELS, 1969; LAWTON and LAMB, 1987; VAN HOLST and VARNER, 1984). By use of corresponding clones, two mRNAs of 1500 and 1800 nucleotides were identified in carrot tissue, whose abundance increases maximally 100-fold after wounding (CHEN and VARNER, 1985a, b). Interestingly, higher levels of extensin-type proteins were found also after pathogen infection, e.g. *Cucumis melo* infected by *Colletotrichum lagenarium* (ESQUERRÉ-TUGAYÉ et al., 1979) or in *Phaseolus vulgaris* infected by *C. lindemuthianum* or treated with fungal elicitor preparations (LAWTON and LAMB, 1987; SHOWALTER et al., 1985). The physiological significance of extensin formation as a defense mechanism is supported by differences in the kinetics of induced extensin synthesis oberserved between compatible and incompatible host/pathogen interactions. In the latter case a particular rapid response is connected with the hypersensitive 'containment' of the invading pathogen (SHOWALTER et al., 1985). Another type of cell wall protein induced by wounding is glycine-rich. It is characteristic of certain plant organs only, e.g. of pumpkin seed coats, strawberry fruits, bean ovary and vascular tissues and young tissues of *Petunia* (CONDIT and MEAGHER, 1987; KELLER et al., 1988, 1989).

C.9.4. Signals

As evident from Sect. C.10., wounding and the ensuing protective response of plants are important aspects of the interaction with pathogens and herbivores. Generation of plant cell wall fragments as elicitors may be part of the signal system under both stress conditions, and in both cases unknown systemic signals are formed. Thus, increased synthesis of proteinase inhibitor or of extensin is not confined to the directly adjacent tissue, but is observed in other parts of the wounded or infected plant as well (BISHOP et al., 1984; DAVIES and SCHUSTER, 1981; PENA-CORTES et al., 1988; SHOWALTER et al., 1985, see review by BOLLER, 1989). The formation of a systemic signal, rapidly spreading through the plant, was impressively demonstrated by changes of the cell wall composition and/or plasma membrane integrity in distant tissues (WALTER-SIMMONS et al., 1984). The yield of intact protoplasts from wounded plants was found to be markedly reduced, presumably because the protoplasts decay more rapidly than they can be liberated by the enzymic digestion of the cell wall. The nature of the systemic signal is unclear. It moves acropetally and basipetally and is probably not identical to the cell wall oligosaccharide fragments, which are not transported in the phloem (KELLER et al., 1989; PENA-CORTES et al., 1988). Instead, a role of wounding-induced ethylene (IMAZEKI, 1985; LIEBERMAN, 1979), of calcium redistribution or of changing action potentials can be discussed (see summaries by ALLAN and TREWAVAS, 1987 and DAVIES, 1987).
A well established wound signal of plants was analyzed by RYAN and coworkers (see DAVIES, 1987; RYAN, 1987, 1988). A water soluble proteinase inhibitor-inducing factor from autoclaved tomato leaves (RYAN, 1974) was lateron characterized as rhamnogalacturonan released from cell walls of sycamore, tomato or potato by endogenous galacturonase activity or by a corresponding enzyme of a fungal pathogen (BISHOP et al., 1984; RYAN et al., 1981). Optimum inducing activity is found with a hexagalacturonide (BISHOP et al., 1984). But formation of oligosaccharide fragments may also be responsible for the accumulation of proteinase inhibitor in nutrient-deprived tomato cell suspension cultures (WALKER-SIMMONS and RYAN, 1986) or in ripening tomato fruits (LINCOLN et al.,

1987). In addition to the oligogalacturonide, RYAN and coworkers presented preliminary evidence for the generation of a heat-stable, non-carbohydrate inducer, which is at least 2 orders of magnitude more potent in the induction of proteinase inhibitor and which may reprensent the systemic signal compound (FARMER et al., 1989). Both the oligogalacturonide and the new superinducer stimulate phosphorylation of plasma membrane proteins with 29 and 34 kDa when added to *in vitro* protein kinase assays with isolated membrane fractions. The significance of protein phosphorylation in the signal transduction process remains to be established.

A special type of wound signal was described for the interaction of plants with *Agrobacterium tumefaciens.* Release of phenolic compounds, e.g. of acetosyringone and its α-hydroxyderivative (STACHEL et al., 1985), activate genes of the *vir* region of the bacterium, whose gene products initiate the infection cycle and finally leads to the transfer of the Ti-plasmid into the plant cells. Wound healing connected with cell division in adjacent tissues is prerequisite for the successful genetic colonization of plants by *Agrobacterium* (LIPETZ, 1965; for summary see ZAMBRYSKI et al., 1989).

C.10. Biogenic stress by pathogen attack

C.10.1. Complexity of the defense response systems

Depending on the virulence of the pathogen, on environmental factors and on the genetic and physiological state of the plant, the interaction with fungi, bacteria or viruses can have very different effects. At one end of a continuous scale are infections with cryptic viruses existing almost undetected in many plants. Clearly visible as disease symptoms but with minor impact on the whole plant are localized foci of hypersensitive cell death. The development of the pathogen, in particular its systemic propagation, is stopped by a programmed death of the infected and immediately adjacent cells. At the other end of the disease scale are highly aggressive pathogen attacks with severe damage or even death of the host plant. Analyses of the latter type of interactions but also of the hypersensitive reaction easily lead to the concept of the pathogen as a biological stressor to be considered together with physical and chemical stressors. A number of reviews and books summarize the mass of relevant data on the pathophysiology and genetics of plant-microbe interactions (AYRES, 1984; COLLINGE and SLUSARENKO, 1987; DAY and JELLIS, 1987; HORSFALL and COWLING, 1977-1980; KOSUGE and NESTER, 1987; KUC, 1983; MÜHLBACH, 1987; VAN LOON, 1983). In the context of this specialized review we will mainly concentrate on the molecular cell biologiy, i.e. on the reprogramming of gene expression including the synthesis of phytoalexins and other secondary products, on possible mechanisms of the pathogen defense and on the interaction of pathogens with other stressors.

In contrast to the general response of plants to abiogenic stressors, many aspects of the interaction with pathogens are much more specific, since two independent genetic systems are involved. For molecular biologists, the complicate experimental analysis was greatly facilitated, when essential parts of the response could be mimicked by wounding or application of biological elicitors to whole plants, to parts of them, to cell cultures or even to protoplasts (for reviews see BELL, 1981; DARVILL and ALBERSHEIM, 1984; DIXON, 1986; EBEL and GRISEBACH, 1988; ESQUERRÉ-TUGAYÉ et al., 1985; SCHEEL and HAHLBROCK, 1989; LAMB et al., 1989; RYAN, 1987, 1988). The close similarity between important parts of the pathogen and wounding responses results from two factors: (i) pathogen penetration can depend on wounding, and (ii) in the case of hypersensitive reactions the pathogen is entrapped by local necrosis of the infected tissue. It is very likely that in both cases formation of cell wall-derived elicitors trigger characteristic parts of the response.

Similar to other stress situations pathogen infection or elicitor treatment results in a complex reprogramming of gene expression. When analyzed at the transcription level, the earliest parts of the response can be detected 2-3 minutes after addition of the elicitor (HEDRICK et al., 1988; KOMBRINK et al., 1988; LAWTON and LAMB, 1987; SOMSSICH et al., 1986 and reviews by DIXON et al., 1986; SCHEEL and HAHLBROCK, 1989; LAMB et al., 1989). In fact, though never described in its whole complexity, there is evidently a characteristic temporal program of changes in gene expression. An illustrative example was given by CHAPPELL et al. (1984) analyzing the increase of enzyme activities in

parsley cell cultures induced by a fungal elicitor. The immediate early increase of ACC synthase activity (ethylene synthesis) was followed by an increase of PAL (phenylpropanoid metabolism) and much later by chitinase increase (degradation of fungal cell wall). Other proteins with delayed expression are methyltransferases involved in late steps of phenylpropanoid metabolism or the hydroxyproline-rich glycoproteins incorporated into the cell wall (for details see reviews by SCHEEL and HAHLBROCK, 1989 and LAMB et al., 1989).
There is evidence that the time course of enzyme synthesis may depend on the type of interaction between plant and pathogen, i.e. incompatible vs. compatible interaction and that these differences may disappear if elicitor is used for induction (HABEREDER et al., 1989). Formation of elicitor is one element in the temporal organization of the defense program, but others are mostly unknown. Important new insights into the spreading of the defense reaction from the infection site into the adjacent tissues of a plant organ came from *in situ* hybridization data with antisense mRNA and from immune histochemical studies (CUYPERS et al., 1988; JAHNEN and HAHLBROCK 1988a, b; SCHMELZER et al., 1988; SOMSSICH et al., 1988; see review by SCHEEL and HAHLBROCK, 1989).
The considerable progress in the description of essential parts of the plant defense system against pathogens was achieved with few experimental systems only (see Table 5) and reviews by SCHEEL and HAHLBROCK, 1989; LAMB et al., 1989). The compilation of data in Table 5 indicates that even among the preferred objects the state of our knowledge is unequal. Only in few cases, complex studies on newly formed proteins, on their mRNAs and coding genes as well as on regulatory elements involved and on the tissue-specific organization (regionalization) of the response led to remarkable new insights into details of plant interactions with pathogens (see nos. 3, 7, 10 in Table 5) and reviews by SCHEEL and HAHLBROCK, 1989; LAMB et al., 1989).
By use of *in vitro* translation and cDNA cloning, between 7 and more than 20 new mRNAs were detected after pathogen infection or elicitor induction. With respect to the onset of transcription and the kinetics of mRNA accumulation they fall into different classes. Only few of them could be assigned to known proteins of the pathogen defense system (COLLINGE et al., 1987; SOMSSICH et al., 1989). But the overall complexity is not surprising, since even the synthesis of different phenylpropanoid compounds (phenolics of the cell wall, phytoalexins) may require more than 20 enzymes, i.e. the number of isolated cDNA clones certainly represents only part of the whole set of newly formed gene products.

C.10.2. Basic elements of the plant defense system

The pathogen- or elicitor-induced proteins can be roughly ordered into four main parts of the response. Though definite proof is lacking in practically all cases, there is good evidence that they are somehow involved in the whole set of defense reactions (for reviews see BOLLER, 1985, 1987, 1988; COLLINGE and SLUSARENKO, 1987; DIXON et al., 1986; EBEL and GRISEBACH, 1988; SCHEEL and HAHLBROCK, 1989; KOSUGE and NESTER, 1987; RIGDEN and COUTTS, 1988; VANCE et al., 1980; VAN LOON, 1985).
Cell wall modification: Several components contribute to the reinforcement of the cell wall. In many plants, increased synthesis of lignin, suberin or incorporation of other phenolic compounds coincides with induction of phenylalanine-ammonia lyase and several other enzymes of phenylpropanoid metabolism (ASADA and MATSUMOTO, 1987; BELL, 1981; DIXON et al., 1986; SCHEEL and HAHLBROCK, 1989; KÖHLE et al., 1984; VANCE et al., 1980). On the other hand, the deposition of callose between cell membrane and wall in close vicinity to viral infection sites or growing fungal hyphae evidently involves activation of preformed membrane-bound callose synthase (AIST, 1976; BONHOFF et al., 1987; KÖHLE et al., 1984; 1985). Finally, depending on the plant species, the content of hydroxyproline-rich (extensin) or glycine-rich cell wall glycoproteins is markedly increased (CORBIN et al., 1987; ESQUERRÉ-TUGAYÉ et al., 1979; SHOWALTER et al., 1985, see also Sect. C.9.3. on wounding response). Analyses in French bean demonstrated that, together with the induction of corresponding extensin mRNA, the pathogen response is also connected with increased levels of enzymes required for processing of the extensin precursor, e.g. prolyl hydroxylase and protein arabinosyl transferase (see review by DIXON et al., 1986).

Table 5. Pathogen or elicitor-induced gene expression
The table presents a survey of experimental systems mainly used for studies on gene expression induced by interaction with pathogens or biological elicitors. Protein groups investigated are: 1, pathogenesis-related proteins; 2, enzymes of phenylpropanoid metabolism; 3, hydrolytic enzymes (chitinase, glucanase); 4, hydroxyproline-rich cell wall glycoproteins; 5, ACC synthase (key enzyme of ethylene biosynthesis). Complex analysis in systems 3, 4, 7, 8, 10 includes measuring of newly formed proteins and mRNAs, cDNA and gene cloning, sequencing and, in some cases, functional dissection of promoters (for details see text).

Host plant	Pathogen/elicitor	Method, identification of proteins	References
1. *Hordeum vulgare* (barley)	*Erysiphe graminis*	cDNA cloning	Davidson et al., 1987
2. *Pisum sativum* (pea)	*Fusarium solani*, elicitor	cDNA cloning, protein groups 2, 3	Boller, 1985; Daniels et al., 1987a, b; Riggleman et al., 1985
3. *Phaseolus vulgaris* (French bean)	*Colletotrichum lindemuthianum*, elicitor, wounding	Complex analysis, protein groups 2–4	Corbin et al., 1987; Dixon et al., 1986; Hedrick et al., 1988; Lamb et al., 1989
4. *Glycine max* (soybean)	*Phytophthora megasperma*, elicitor	Complex analysis, protein group 2	Ebel & Grisebach, 1988; Parker et al., 1989
5. *Brassica campestris* (rape?)	*Xanthomonas campestris*	in vitro translation	Collinge et al., 1987
6. *Cucumis melo* (melon)	*Colletotrichum lagenarium*	in vitro translation, protein analysis, protein groups 3, 4	Esquerré-Tugayé et al., 1979; Roby & Esquerré-Tugayé, 1987
7. *Solanum tuberosum* (potato)	*Phytophthora infestans*, elicitor, wounding viroid	Complex analysis, protein groups 1, 2, 3	Conejero et al., 1979; Scheel & Hahlbrock, 1989; Kombrink et al., 1988; Marineau et al., 1987
8. *Nicotiana tabacum* (tobacco)	Tobacco mosaic virus	cDNA cloning, complex analysis, protein groups 1, 3	Carr et al., 1987; Jamet & Frittig, 1986; Legrand et al., 1987; Shinshi et al., 1987; Vögeli-Lange et al., 1988
9. *Lycopersicon esculentum* (tomato)	Viroid, *Verticillium alboatrum, Cladosporium fulvum*	Protein analysis, protein groups 1, 3	Boller, 1985; DeWit & Bakker, 1980; Lucas et al., 1985; Pegg & Young, 1982; Vera et al., 1988
10. *Petroselinum hortense* (parsley)	Pmg elicitor	Complex analysis, protein groups 1–3, 5	Chappell et al., 1984; Scheel & Hahlbrock, 1989; Somssich et al., 1986, 1988, 1989

A new type of low molecular weight (5 kDa) sulfur-containing polypeptides (thionins) in cell walls of barley leaves was described by Bohlmann et al. (1988). They are highly toxic to pathogenic fungi. Up to 100 different genes coding for the 15 kDa thionin precursor were detected on chromosome 6. Expression is usually under circadian control. It is very high in the dark and negligible in the light period. But thionin synthesis is also induced by infection with barley mildew. Thionins are structurally homologous to viscotoxin isolated from *Viscum album*.

Synthesis of phytoalexins and other soluble phenylpropanoid compounds: Plant 'antibiotics' (phytoalexins) are wide-spread, plant-specific secondary products, whose synthesis is triggered by pathogen attack (for reviews see Dixon et al., 1986; Ebel, 1986; Ebel and Grisebach, 1988 and

SCHEEL and HAHLBROCK, 1989). Depending on the type of phytoalexin, the biosynthetic chain can involve few or many newly formed enzymes. Central to all pathways, including the synthesis of cell wall phenolic compounds, is the general phenylpropanoid metabolism catalyzed by phenylalanine-ammonia lyase (PAL), cinnamate-4-hydroxylase and 4-coumarate: CoA ligase (4-Cl). Subsequent to this 'general part', chalcone synthase (CHS) is responsible for the condensation of 4-coumarate: CoA with 3 molecules of malonyl-CoA to form naringenine chalcone, the first product of the special branch leading to many phytoalexins and other flavone derivatives (for details see COLLINGE and SLUSARENKO, 1987; SCHEEL and HAHLBROCK, 1989).

A great number of papers deal with the induced expression of PAL, 4-Cl, CHS and additional enzymes of phytoalexin synthesis. Besides pathogen infection, fungal elicitors, wounding or chemical stressors were used for induction (see ref. in Table 4). In few experimental systems (preferentially bean, parsley, soybean, potato, see Table 5) the initial analyses of enzyme accumulation was complemented by investigations on mRNA levels, altered transcription patterns and, in recent years, by cloning of the corresponding genes and attempts towards a functional dissection of promoter structures (CHAPPELL and HAHLBROCK, 1984; CRAMER et al., 1985; DANGL et al., 1987; DOUGLAS et al., 1987; DRON et al., 1988; HAGMANN et al., 1984; HERRMANN et al. 1988; KUHN et al., 1984; LAWTON and LAMB, 1987; LIPPHARDT et al., 1988; MEHDY and LAMB, 1987; RYDER et al., 1984; SCHRÖDER et al., 1988; see reviews of DIXON et al., 1986; EBEL and GRISEBACH, 1988; SCHEEL and HAHLBROCK, 1989; LAMB et al., 1989).

A role of phenylpropanoid compounds in the plant defense system is suggested by three types of evidence: (i) Inhibition of PAL activity by L-2-amino-oxy-3-phenylpropionic acid in potato (SCHEEL and HAHLBROCK, 1989) and soybean (MOESTA and GRISEBACH, 1982) prior to infection with incompatible races of *Phytophthora* spec. prevented the expression of early resistance mechanisms and led to a quasi-compatible interaction. (ii) Comparing incompatible vs. compatible interactions between soybean roots and *Phytophthora megasperma* HABEREDER et al. (1989) found characteristic differences. The early block of fungal development in the case of infection with the incompatible race 1 corresponds to a rapid increase of PAL and chalcone synthase mRNAs. But the response is delayed by almost 10 hrs during the compatible interaction with *P. megasperma* race 3. (iii) Direct proof comes from experiments of KISTLER and VAN ETTEN (1984). Resistance of pea leaves against infection with the fungal pathogen *Nectria haematococca* depends on pisatin levels, and fungal strains with an inducible pisatin demethylase system can override the inhibitory effect exerted by the phytoalexin.

Excretion of hydrolytic enzymes: Chitinases and glucanases can be prepared from a wide range of plants (see review by BOLLER, 1985, 1987, 1988). The early purification from bean leaves connected with the finding that they are ethylene-inducible (ABELES et al., 1970) suggested a role in pathogen response. This has been verified by characterization of elicitor or pathogen-induced chitinases and glucanases from bean (HEDRICK et al., 1988), parsley (KOMBRINK et al., 1988), pea (MAUCH et al., 1984, 1988 a), melon and cucumber (METRAUX et al., 1988; ROBY and ESQUERRÉ-TUGAYÉ, 1987; ROBY et al., 1986), tobacco (LEGRAND et al., 1987; SHINSHI et al., 1987; VÖGELI-LANGE et al., 1988) and tomato (PEGG and YOUNG, 1982). Many of these hydrolytic enzymes are basic proteins with molecular weights of 25-35 kDa (see summaries by BOLLER, 1985, 1987, 1988). They are contained in the vacuoles or in the cell wall (BOLLER, 1985; BOLLER and METRAUX, 1988; KOMBRINK et al., 1988).

Details of the mechanism involved in the intracellular or intercellular localization of these enzymes remain to be established. At least in the case of the tobacco basic β-glucanase a complex postsynthetic processing of a preproenzyme was elucidated. In a first step an N-terminal extension of 21 aa residues is removed, and the C-terminus of the proenzyme is modified by N-glycosylation. The second step involves the loss of the C-terminal extension of 22 residues with the carbohydrate side chain (SHINSHI et al., 1988).

The induction ethylene may be independent of the pathogen effects (MAUCH et al., 1984; see reviews by BOLLER, 1988; COLLINGE and SLUSARENKO, 1987). Interestingly, ethylene-stimulated synthesis of chitinase and glucanase in leaves is inhibited at the mRNA level by simultaneous addition of cytokinin and auxin. This result argues for an interaction of different phytohormones in the control of at least parts of the defense response, which may be significant for the outcome of the warfare between plants and pathogens (FELIX and MEINS, 1987; SHINSHI et al., 1987; see also Sect. C.2.). There is some evidence that different members of small gene families may respond differentially to pathogen stress, hormones or other developmental signals (MAUCH et al., 1988a; SHINSHI et al., 1987).

Chitinases alone or in combination with glucanases are potent inhibitors of fungal growth. It is tempting to speculate that they play an important part in the plant defense system (BOLLER, 1985, 1988; COLLINGE and SLUSARENKO, 1987; MAUCH et al., 1988b; ROBY and ESQUERRÉ-TUGAYÉ, 1987; SCHLUMBAUM et al., 1986). However, using tobacco plants infected with viral, fungal or bacterial pathogens, MEINS and coworkers demonstrated that a marked increase of chitinase and glucanase synthesis was restricted to interactions with hypersensitive reactions; it was not observed during systemic infections. This result may support a direct role of these enzymes as local 'antibiotic proteins'. Alternatively, it may simply reflect the generation of inducing signals by the necrotic reaction (MEINS and AHL, 1989).

Pathogenesis-related proteins (PRs): this is an operational term for pathogen- or elicitor-induced proteins now mostly applicated to a characteristic group of basic or acidic low molecular weight proteins of Mr 10–20 kDa (see reviews by RIGDEN and COUTTS, 1988 and VAN LOON, 1983, 1985). Sometimes other groups of elicitor-induced proteins are included as well, e.g. chitinases and glucanases (KOMBRINK et al., 1988; LEGRAND et al., 1987; METRAUX et al., 1988; RIGDEN and COUTTS, 1988) or an alkaline endoproteinase of Mr 69 kDa purified from viroid-infected tomato leaves (DEWIT and BUURLAGE, 1986; GRANELL et al., 1987; VERA and CONEJERO, 1988). PRs and the hydrolytic enzymes share the common properties that they are frequently secreted to vacuoles or to the extracellular space and that, in many instances, their induction is connected with the generation of necrotic lesions by pathogens or abiotic elicitors (JAMET and FRITIG, 1986; TRAYLOR et al., 1987; VAN LOON, 1983; VERA et al., 1988; VÖGELI-LANGE et al., 1988).

The early evidence for the extracellular localization of PRs in tobacco, cowpea, cucumber and *Chenopodium quinoa* (PARENT and ASSELIN, 1984; WAGIH and COUTTS, 1981) was confirmed and detailed for the PR1 of *Nicotiana tabacum*: (i) PR1 is synthesized as a preprotein on membrane-bound polysomes (CARR et al., 1985). (ii) Protoplasts do not contain PR1 but are capable of synthesis and secretion of PR1 (CARR et al., 1987). (iii) By use of antibodies the protein was localized in the intercellular spaces of salicylate-induced tobacco leaves (HOSOKAWA and OHASHI, 1988). (iv) The PR1 gene codes for a protein with a 30 amino acid residue leader sequence preceding the 138 aa residue coding region for the mature PR1 (PFITZNER et al., 1988). No evidence for a pre-protein and a leader sequence was found for the parsley PR1, whose gene sequence was reported recently (SOMSSICH et al., 1988). Though immunological investigations detected related PRs in maize, barley, *Chenopodium*, tobacco, tomato, and potato sequencing of three PR genes from tobacco, parsley and tomato, revealed a considerable heterogeneity (CORNELISSEN et al., 1986, 1987; CUTT et al., 1988; LUCAS et al., 1985; PFITZNER et al., 1988; SOMSSICH et al., 1988).

The list of inducers of PR synthesis is particularly long including all types of pathogens (fungi, bacteria, viruses, viroids), plant hormones or growth regulators (ethephon, benzyladenine, indolyl acetic acid, 2,4-D), heavy metals (Ba^{2+}, Mn^{2+}, Co^{2+}, Ag^{+}, Hg^{2+}), polyacrylic acid and salicylate as well as osmotic stress by mannitol or wounding. In addition, levels of some PRs increase during flowering or senescence (see Table 4). In many instances, induction is connected with irreversible cell damage. Hence, PRs were referred to as necrogenic proteins (JAMET and FRITIG, 1986; OHASHI and MATSUOKA, 1985; TRAYLOR et al., 1987; VAN LOON, 1983, 1985; VERA et al., 1988; VÖGELI-LANGE et al., 1988). The generation of necrotic lesions with cell wall incrustation by phenolic compounds and lipid peroxidation is a frequently encountered, effective mechanism for pathogen containment (COLLINGE and SLUSARENKO, 1987; ROGERS et al., 1988). Though accumulation of PRs is predominant in cell layers surrounding these local lesions (ANTONIW and WHITE, 1986; DEWIT and VAN DER MEER, 1986; SOMSSICH et al., 1988) this must not be connected with any function in the pathogen defense but rather indicate high levels of inducing elicitor in these regions.

C.10.3. *Genes and promoters*

It was already discussed in the introduction (Sect. A.1.) that many stress-related genes are not only part of different stress domains (see Table 4) but they are also under developmental control. This is also true for genes identified as parts of the defense-related complex (BOHLMANN et al., 1988, BOLLER, 1985; COLLINGE and SLUSARENKO, 1987; JAHNEN and HAHLBROCK, 1988b; KNOGGE et al., 1987; LAMB et al., 1989; SHINSHI et al., 1987; VERA et al., 1988; VAN LOON, 1985). Though detailed analyses are just beginning the integration into different gene expression programs is usually brought about by multifunctional promoters and by multigene families with differential regulation of individual members. Examples are genes coding for phenylalanine-ammonia lyase (SCHEEL and HAHLBROCK, 1989; LAMB et al., 1989), chalcone synthase (DIXON et al., 1986; RYDER et al., 1987), hydroxyproline-rich cell wall proteins (CORBIN et al., 1987), PR proteins (CORNELISSEN et al., 1986, 1987) and leaf thionins (BOHLMANN et al., 1988). The currently available data are summarized by SCHEEL and HAHLBROCK (1989) and LAMB et al. (1989). These reviews also document plant-specific differences in the multiplicity of these gene families.

With proceeding analysis of more and more defense-related genes functional dissection of the corresponding promoters becomes feasable. Three different methods were applied. (i) Fusion constructs of the French bean and *Antirrhinum majus* chalcone synthase promoter with reporter genes were analyzed in soybean and parsley protoplasts, respectively (DRON et al., 1988; LIPPHARDT et al., 1988). A UV-responsive element and enhancer and silencer regions were defined by deletion mapping. (ii) Direct access to corresponding protein binding sites of the parsley CHS promoter was achieved by *in vivo* footprinting (SCHULZE-LEFERT et al., 1989). One of the two sequence elements identified is conserved between different UV-light inducible genes also in other plant species (see review by SCHEEL and HAHLBROCK, 1989). Its functional significance was supported by site-directed mutagenesis and tests in fusion constructs with an appropriate reporter gene (SCHULZE-LEFERT et al., 1989). (iii) Sequence comparison between different promoters of PR genes of tobacco and parsley indicated motifs similar to the hs element (see Sect. B.2.4.). In the parsley gene, two isolated, bad monomers of imperfect homology located far upstream of the TATA box are presumably nonfunctional (SOMSSICH et al., 1988), whereas in the tobacco genes a repetitive cluster of distorted GAA2nTTC motifs is found directly upstream of the TATA box (CORNELISSEN et al., 1987; PFITZNER et al., 1988):

(-72)TAC2nGAA2nATC6nAAA2nTTC2nGAT7nTATAAA(-29).

Though functional analyses of PR promoters are lacking, it is conceivable that, similar to hsc genes, this promoter region helps to maintain transcription during hs, if the PR genes are in an induced state (see Sect. B.2.4.).

C.10.4. *Signals*

The multiplicity of induced defense mechanisms including many different proteins and their stress metabolites, respectively, leaves many important questions unanswered. Similar to the situation with the heat shock response (see Sect. B.2.1.), much experimental efforts now concentrate on the signal systems involved. On the one hand, this means sequencing and functional analyses of promoters and of their binding proteins (see Sect. C.10.3.). On the other hand, there must be a whole network of primary and secondary signals which trigger the rapid and delayed steps of gene expression. In addition, systemic signals may extend the response also to other organs of the plants located at considerable distance from the infection site (CHAPPELL et al., 1984; DIXON et al., 1983; LAMB et al., 1989; KUC, 1983; LAWTON and LAMB, 1987, SHOWALTER et al., 1985, SOMSSICH et al., 1989). As a direct consequence of the pathogen attack, biological elicitors are produced which are fragments of the pathogen or plant cell walls. Their formation may be an intimate part of the wounding response (Sect. C.9.4.). Oligosaccharides of the β-glucan, chitosan and oligogalacturonide types are well characterized (DARVILL and ALBERSHEIM, 1984; RYAN, 1987, 1988). In addition, fatty acid derivatives,

e.g. arachidonic acid (Bostock et al., 1982; Stelzig et al., 1983) and protein constituents (Parker et al., 1989) may be active as well.
The search for secondary signals, possibly involved in the systemic induction of defense genes was not very successful. Ethylene, formed rapidly after pathogen attack and wounding, might be a good candidate (Abeles et al., 1970; Chappell et al., 1984). At least, ethylene or ethylene-generating compounds (ACC, ethephon) are inducers of PR and chitinase synthesis. But inhibitors of ethylene synthesis (Co^{2+} ions or aminoethoxy vinylglycine) applied together with pathogen or elicitor did not prevent PAL induction and glyceollin synthesis in soybean (Paradies et al., 1980) or chitinase and β-1,3-glucanase induction in pea pods (Mauch et al., 1984). Thus, ethylene is more an indicator of the stress situation than an integral part of the signal system. But a final decision about the role of ethylene requires investigations of all parts of the pathogen response of whole plants including the gene expression found in distal organs. Important new impulses as to the characterization of receptors and the generation of secondary signals come from several observations (see also reviews by Boller, 1989; Scheel and Hahlbrock, 1989): (i) Parsley protoplasts retain their capability to respond to fungal elicitor (Dangl et al., 1987), and soybean protoplasts exhibit a binding site for *Phytophthora megasperma*-derived β-glucan elicitor (Ebel and Grisebach, 1988). (ii) 0.01–1 mM concentrations of reduced glutathione stimulate synthesis of hydroxyproline-rich glycoproteins, PAL and chalcone synthase in bean (Wingate et al., 1988). (iii) Redistribution of Ca^{2+} ions (probably Ca^{2+} influx) may be crucial for signal transformation in soybean (Ebel and Grisebach, 1988; Staeb and Ebel, 1987). At least Ca^{2+} depletion by addition of EGTA or low levels of La^{3+} ions abolished the inducing activity of Pmg elicitor. (iv) The role of membrane potentials or ion transport in signal systems is suggested by the observation that 50 μM vanadate induces synthesis of PAL, cinnamate-4-hydroxylase and stilbene synthase in peanut suspension cultures to the same extent as observed after application of cell wall elicitor. The concentration of vanadate corresponds to that required for optimum inhibition of membrane ATPase (Steffens et al., 1989). (v) Together with the isolation of a potential systemic signal compound from wounded tomato, Farmer et al. (1989) presented evidence for changing phosphorylation of membrane proteins detected in an *in vitro* kinase assay (see Sect. C.9.4.). The significance and possible connections between these seemingly unrelated results in the frame of signal systems involved in the plant-pathogen interactions remain to be established. The complexity of the response certainly requires several independent signal transduction chains.

C.10.5. Interactions between heat shock and pathogen stress

Disregarding the interesting question as to the possible role of heat shock elements in promoters of PR genes, the decisive influence of high or low temperatures on the infection and development of a pathogen and on the induced defense mechanisms is well known to plant pathologists. There is a broad field of stress interactions barely studied so far at the molecular level. To illustrate the diversity of responses we will only briefly review examples of heat shock effects on plant pathogens or disease resistance. Similar to the situation with animal viruses or non-viral pathogens (see summary in Nover, 1990), there are various types of mutual interactions. Their outcome is influenced by the type of the host/pathogen pair (host vs. non-host plants, compatible vs. non-compatible interactions).

1) High temperature treatment for virus eradication: Since many years, long-term, moderate heat shock treatments have been used to stop virus multiplication in meristems of cultivated plants. *In vitro* regeneration of virus-free plants from meristem cultures is a broadly applicated method to maintain or improve the productivity of these plants. With improving analytical facilities and procedures for mass propagation of plantlets, more and more cryptic viruses are detected in cultivated plants, and the range of application of this type of hyperthermic cure extends (Kassanis, 1957, Walkey and Cooper, 1975; Wang, 1985).
   The molecular mechanisms of virus eradication are barely investigated. The only example studied so far cannot give a plausible explanation. This is due to the differences of hs treatments, i.e.

long-term, moderate hyperthermia for days or even weeks for virus eradication vs. 1 hr severe hs for experimental purposes. Thus, DAWSON (1976, 1983) analyzed hs effects (1 hr 40 °C) on tobacco mosaic virus RNA and protein synthesis. There was an immediate block of TMV single strand RNA synthesis, whereas formation of the double-stranded replicative intermediate continued and was even stimulated at 37 °C. Extent of viral protein synthesis correlates with the ongoing formation of dsRNA. Evidently the hs-induced block of translation of control proteins (see Sect. B.2.6.) is not effective for viral proteins (DAWSON and BOYD, 1987). Recent data by SCHÖFFL et al. (1989) indicate that, similar to the hs mRNAs, viral RNAs contain leader sequences which allow an internal, cap-independent start of translation (see Sect. B.2.6.2.).

2) Heat shock-induced pathogen resistence: STERMER and HAMMERSCHMIDT (1984, 1985, 1987) investigated long term changes induced by pulse heat shock treatments (40 sec. 50 °C) of cucumber seedlings. Infection with fungal parasites, e.g. *Cladosporium cucumerinum* and *Colletotrichum lagenarium,* was tested 12–96 hrs after the hs. Increased resistance was observed with an optimum after 48 hrs. The effect correlates with induced synthesis of ethylene, cell wall-bound peroxidase and extensin.
3) Increased susceptibility to pathogen attack: There is no doubt that the effective defense of a plant against an invading pathogen requires high metabolic and gene expression activities. Consequently, hs can inhibit especially the early phases of a pathogen response, including phytoalexin synthesis and hypersensitive cell death (HAZEN and BUSHNELL, 1983; HEATH, 1979; JEROME and MÜLLER, 1958; TOMIYAMA, 1967; YARWOOD, 1956). Using *in vitro* translation assays and mRNA analyses by cDNA clones, DANIELS et al. (1987a) and HADWIGER and WAGONER (1983) discriminated between hs and pathogen-induced gene expression in pea. Heat shock (2 hrs 40 °C) inhibited the disease response induced either by infection with *Fusarium solani f. sp. phaseoli* or by treatment with a chitosan elicitor. In support of this, WALTER (1989) reported on an immediate block of PAL and CHS mRNA accumulation after hs treatments of UV- or elicitor-induced parsley cell cultures.
4) Heat shock promotes formation of infection structures: Upon contact with their host plants, stomata-penetrating rust fungi differentiate an appressorium and an infection hypha, required for invasion of the leaf or stem. Artifical collodion membranes or brief heat shock treatments (1.5 hr at 30 °C) are also effective to trigger this differentiation process in uredospore germlings of *Puccinia graminis tritici* and *P. coronata.* The artificially induced infection structures are normal, and it can be speculated that supranormal temperatures under natural conditions accelerate infection by rust fungi (DUNKLE et al., 1969; MENDGEN and DRESSLER, 1983; STAPLES et al., 1983; WANNER et al., 1985).

C.11. Stress-induced protective effects against DNA damage

The confrontation of cells and organisms with a multitude of physical and chemical factors, including heat shock, may exert "genomic stress" (MCCLINTOCK, 1978; 1984) eventually resulting in mutations, chromosome structural changes, reshuffling of transposable genetic elements or cell death. Various constitutive and inducible functions (for review see WALKER, 1985) may counteract genomic stress and render prokaryotic and eukaryotic cells more resistant or tolerant to the adverse effects of environmental genotoxic factors, i.e., stressors may induce genetic damage on the one hand, and they may trigger protective functions on the other, the resultant being dependent on the quality and quantity of stress impact.

Cells exposed to a nonlethal heat shock (hs) acquire a transient resistance to subsequent heat challenge. The phenomenon is called induced thermotolerance (see Sect. B.4.). Thermotolerance may also be induced in response to other environmental insults such as oxidative stress, metabolic inhibition, anoxia, application of amino acid analogues and heavy metals. On the other hand, mutagen pretreatment and also hs pretreatments can protect plant cells from damage by genomic stressors.

Early indications (e.g., CALDECOTT and SMITH, 1952; SWANSON et al., 1953; RIEGER and MICHAELIS, 1965; BROERTJES, 1972; SYBENGA and KLEIJER, 1976) of the existence of such protective functions in plants which are inducible by various stressors and help to alleviate genetic damage induced by various mutagens (clastogens) remained more or less un-noticed. This changed, wheen the studies of hs effects in bacteria, animal cells, and plants aroused new interest in inducible stress tolerance in general. At least in animal systems, this interest is tightly connected with the application of hyperthermia in cancer therapy. Heat shock-induced DNA damage and deficiency of repair mechanisms are major objects of research in this field (for reviews see ANGHILERI and ROBERT, 1986; NOVER, 1990).

For higher plants, the most extensive data on stress-induced protective functions against the generation of chromosome structural changes (chromatid aberrations) are available from experiments of RIEGER and coworkers (for review see RIEGER et al., 1986a; HEINDORFF et al., 1987c). These authors pretreated root tip meristems of *Vicia faba* with low mutagen (clastogen) concentrations, with ethidium bromide (EB) or nalidixic acid (NA) and, after some time, with challenge concentrations of various alkylating or nonalkylating agents. They found that the yield of chromatid aberrations (chromatid and isolocus breaks, duplication-deletions, intercalary deletions, and chromatid translocations) observed in first metaphases after consecutive conditioning and challenge treatments was usually significantly lower than that obtained after challenge treatment alone. Similar observations were made when, instead of low mutagen concentrations or the inhibitors EB and NA, heat shocks or various inorganic salts were used for the conditioning treatment. The data now at hand (Tables 6-8) allow to infer that a variety of stress factors administered prior to mutagen challenging may, under appropriate conditions, trigger protective functions which counteract the induction of DNA damage caused by the challenge treatment and thus reduce the yield of chromosome structural changes. The main conclusions from these observations may be summarized as follows:

1) Conditioning pretreatments with low concentrations of certain mutagens, metabolic inhibitors, heat shock and inorganic salts may result in adaptive responses, which reduce the genotoxic effectivity of mutagens inducing chromatid aberrations (Table 6 and 8).
2) In the case of conditioning pretreatment with mutagens, "cross-protection", i.e., protection after pretreatment with one mutagen and a subsequent challenge treatment with another, depends on the mutagens being used. No "cross-protection" was observed after pretreatment with nonalkylating agents and challenge treatment with alkylating agents, and *vice versa.* All alkylating agents tested proved to be able to substitute for each other (Table 7), their use for conditioning and challenge treatment resulted in what may be called 'clastogenic adaptation'. Presence and absence, respectively, of cross-protection (Fig. 35) in dependence on the mutagens (clastogens) used for conditioning and challenge treatments allow to infer that protective functions triggered

Table 6. Clastogens tested in *V. faba* with respect to their capability to trigger protective functions against induction of chromatid aberrations

Types of clastogens tested	Conditioning treatment	Challenge treatment	Protection[1]	Reference
a) S-phase dependent agents alkylating	MNU[2]	MNU	+	RIEGER et al. (1982)
	Trenimon	Trenimon	+	unpublished
	TEM	TEM	+	RIEGER et al. (1982)
non-alkylating	MH	MH	+	RIEGER et al. (1982)
	Ethanol	Ethanol	−	RIEGER et al. (1982)
b) S-phase independent agents	Bleomycin	Bleomycin	+	HEINDORFF et al. (1987b)
	X-rays	X-rays	+	HEINDORFF et al. (1987b)

[1] + = significant reduction (by more than 50%) of aberration yield due to the conditioning pretreatment
− = no protection

[2] Abbreviations: EMS, ethyl methanesulfonate; MH, maleic hydrazide (1,2 dihydropyridazine-3,6-dione); MNU, N-methyl-N-nitrosourea; TEM, triethyleneiminotriazine

Table 7. Combinations of agents tested in *V. faba* with respect to the occurrence of cross-protection

Combination of agents	Conditioning treatment	Challenge treatment	Cross-protection[1]	Reference
a) Alkylating vs. alkylating agent	MNU[2]	TEM	+	Rieger et al. (1982)
	TEM	MNU	+	Rieger et al. (1982)
	EMS	TEM	+	unpublished
b) Alkylating vs. non-alkylating agent	EMS	MH	−	unpublished
	MNU	MH	−	Rieger et al. (1982)
	MH	MNU	−	Rieger et al. (1982)
	Bleomycin	Trenimon	+	Heindorff et al. (1987c)
	Bleomycin	TEM	+	Heindorff et al. (1987c)
	TEM	Bleomycin	−	Heindorff et al. (1987c)
c) Non-alkylating vs. non-alkylating agent	Bleomycin	MH	+	Heindorff et al. (1987c)
	MH	Bleomycin	+	Heindorff et al. (1987c)
	Bleomycin	X-rays	+	Heindorff et al. (1987b)
	X-rays	MH	+	Heindorff et al. (1987c)
d) Peroxides or hydrazines vs. non-alkylating agent	Hydrogen peroxide	MH	+	Heindorff et al. (1987a)
	tert-Butyl hydroperoxide	MH	+	Heindorff et al. (1987a)
	N,N′-diformyl hydrazine	MH	+	Heindorff et al. (1987b)
e) Peroxides or hydrazines vs. alkylating agent	Hydrogen peroxide	TEM	−	Heindorff et al. (1987a)
	tert-Butyl hydroperoxide	TEM	−	Heindorff et al. (1987a)
	N,N′-diformyl hydrazine	TEM	−	Heindorff et al. (1987b)
f) Metabolic inhibitors vs. non-alkylating agent	Nalidixic acid (NA)	MH	+	Michaelis et al. (1988)
	Ethidium bromide (EB)	MH	+	Michaelis et al. (1988)
g) Metabolic inhibitors vs. alkylating agent	Nalidixic acid (NA)	TEM	−	Michaelis et al. (1988)
	Ethidium bromide (EB)	TEM	−	Michaelis et al. (1988)

[1] + = cross-protection observed
− = no cross-protection

[2] Abbreviations: EMS, ethyl methanesulfonate; MH, maleic hydrazide (1,2 dihydropyridazine-3,6-dione); MNU, N-methyl-N-nitrosourea; TEM, triethyleneiminotriazine

by conditioning pretreatment are dependent on the agent being used for conditioning. A similar conclusion may be drawn from the differential effects of pretreatment with ethidium bromide and nalidixic acid (Michaelis et al., 1988). In this case protection occurred against MH challenging but not against TEM challenge treatment.

3) Protective effects (clastogenic adaptation) were found to be dependent on the mutagen concentration used for conditioning and on the time span between conditioning and challenge treatment (Rieger et al., 1985). This indicates that a certain minimal conditioning dose is necessary to trigger protection and that the protective effect decays with time.
4) Protection by conditioning with mutagens or heat shock is prevented by inhibition of protein synthesis by cycloheximide prior to the conditioning pretreatment (Rieger et al., 1985; Nico-

Table 8. Protection by conditioning treatment with heat shock or inorganic salts against aberration induction in *V. faba* root tip meristems and the effect of benzamide (BA) pretreatment on protective effects against MH and TEM

Conditioning treatment	Challenge treatment	Protection[1] no BA	Protection[1] BA[2]	Reference
hs	Bleomycin	+	nt[3]	HEINDORFF et al. (1987a)
hs	X-rays	+	nt	HEINDORFF et al. (1987a)
hs	Trenimon	+	nt	unpublished
hs	MH	+	+	RIEGER and MICHAELIS (1988a)
hs	TEM	+	−	RIEGER and MICHAELIS (1988a)
NH_4Cl	MH	+	+	RIEGER and MICHAELIS (1988a)
NH_4Cl	TEM	+	−	RIEGER and MICHAELIS (1988a)
$Cd(NO_3)_2$	MH	+	+	RIEGER and MICHAELIS (1988a)
$Cd(NO_3)_2$	TEM	+	−	RIEGER and MICHAELIS (1988a)
$ZnSO_4$	MH	+	+	RIEGER and MICHAELIS (1988a)
$ZnSO_4$	TEM	+	−	RIEGER and MICHAELIS (1988a)

[1] + = protection observed
− = no protection

[2] BA, benzamide (5×10^{-3} M for 1 h) prior to conditioning

[3] nt, not tested

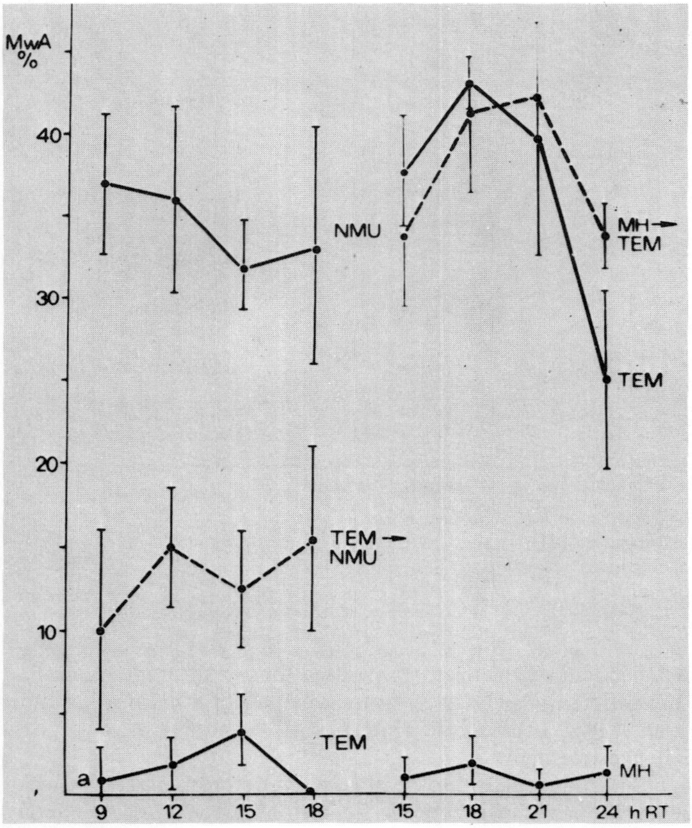

Fig. 35. Cross-protection after conditioning with TEM (0.5 h, 10^{-5} M) and, two hours later, challenge treatment with MNU (1 h, 10^{-3} M), and *vice versa* (left side curves), absence of cross-protection after conditioning with MH (0.5 h, 5×10^{-5}) and challenge treatment with TEM (0.5 h, 10^{-4} M), and *vice versa* (right side curves). MwA %: percentage of metaphases with chromatid aberrations observed after recovery times (RT) from 9 to 24 h

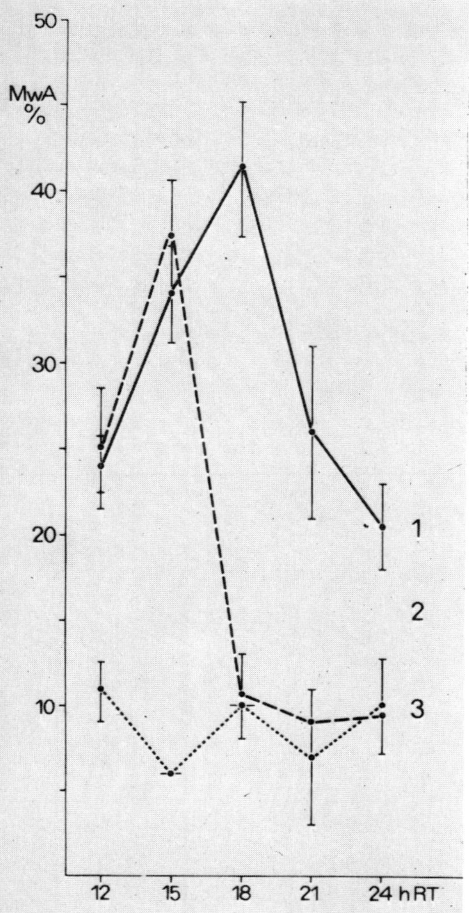

Fig. 36. The yield of metaphases with chromatid aberrations (MwA %) as affected by heat shock (hs) pretreatment for 1 min at different temperatures followed, after 1 h intertreatment time (IT), by challenge treatment with MH.
Curve 1: 0.5 h 5×10^{-4} M MH (MH challenging);
curve 2: 1 min 38 °C hs – 1 h IT – 0.5 h 5×10^{-4} M MH;
curve 3: 1 min 40 °C hs – 1 h IT – 0.5 h 5×10^{-4} M MH

LOFF et al., 1985), i.e., unimpaired protein synthesis during conditioning is a prerequiste for protection to occur.

5) The protective effect of heat pretreatment (40 °C for 1 to 30 min) is dependent on the temperature and duration of hs as well as on the time span between hs and challenge treatment (RIEGER and MICHAELIS, 1988a, b; RIEGER et al., 1986a, b). Heat shocks (10 min) at 38 °C or 40 °C, but not at 36 °C, were able to reduce the yield of chromatid aberrations induced by challenge treatment with MH or TEM. Reduction of heat shocking to one minute at 38 °C and 40 °C, respectively, triggered protection when the hs temperature was 40 °C; but only partial protection occurred when the hs temperature was 38 °C (Fig. 36 for MH challenging).
6) The question of how fast protective functions are expressed after a 10 min hs at 40 °C and how long they are effective have been tested for both MH and TEM challenging. The results (RIEGER and MICHAELIS, 1988b) document that hs protection is effective after less than 10 min and lasts for up to 210 min in the case of challenge with the nonalkylating agent MH (Fig. 37). Analogous protective responses against the alkylating agent TEM needed more than 10 min for expression and decayed more quickly. After an intertreatment time longer than 180 min, hs protection was absent (Fig. 38). Whatever the underlying processes may be, this clastogen-specific difference in hs response indicates that hs triggers at least two protective functions, one specific for MH, the

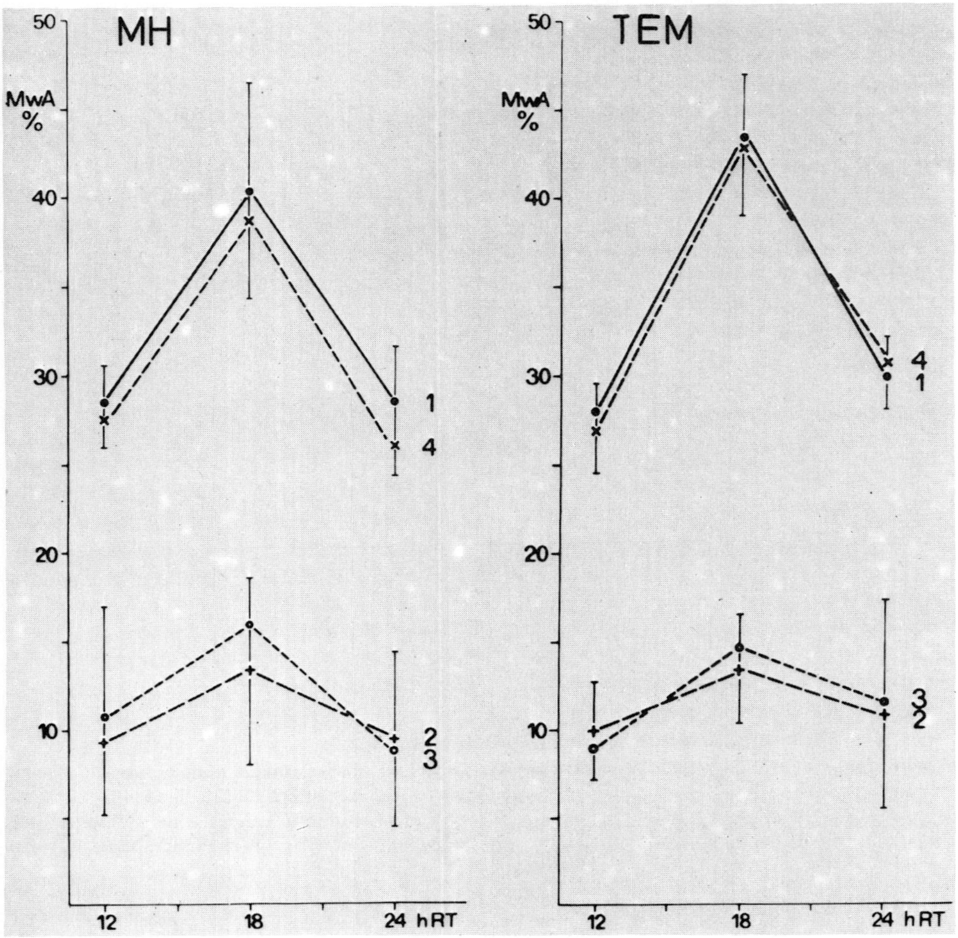

Fig. 37. Effects of short intertreatment times (IT) between heat shock (hs) and MH (left) resp. TEM challenge treatment (right) on the frequency of metaphases with chromatid aberrations (MwA %) at recovery times (RT) of 12, 18 and 24 h: hs protection is a very quick response (occurring with intertreatment times shorter than 10 min in the case of MH challenging; protective responses need more than 10 min intertreatment time in the case of TEM challenging. Left figure—curve 1: 0.5 h 5×10^{-4} M MH; curve 2: 10 min 40 °C hs – 1 h IT – 0.5 h 5×10^{-4} M MH; curve 3: 10 min 40 °C hs – 10 min IT – 0.5 h 5×10^{-4} M MH; curve 4: 10 min 40 °C hs – 2 min IT – 0.5 h 5×10^{-4} M MH (no longer protection).
Right figure—curve 1: 0.5 h 10^{-4} M TEM; curve 2: 10 min 40 °C hs – 1 h IT – 0.5 h 10^{-4} M TEM; curve 3: 10 min 40 °C hs – 0.5 h IT – 0.5 h 10^{-4} M TEM; curve 4: 10 min 40 °C hs – 10 min IT – 0.5 h 10^{-4} M TEM (**no** protection)

other for TEM. This inference is further substantiated by experiments in which benzamide (BA) was administered before hs and challenge treatment.

7) Benzamide applicated for one hour (5×10^{-3} M) before hs differentially affected hs protection against TEM and MH (Rieger and Michaelis, 1988a). BA is an inhibitor of ADP-ribosyltransferase (ADPRT) and ADP-ribosylation may have a role in the response of cells to DNA damaging agents (for reviews see Althaus and Richter, 1987; Shall, 1984; Gaal and Pearson, 1985). In the case of TEM challenging, BA completely prevented hs protection; no such effect occurred when MH was used for challenge treatment. The situation was similar, when, instead of hs, inor-

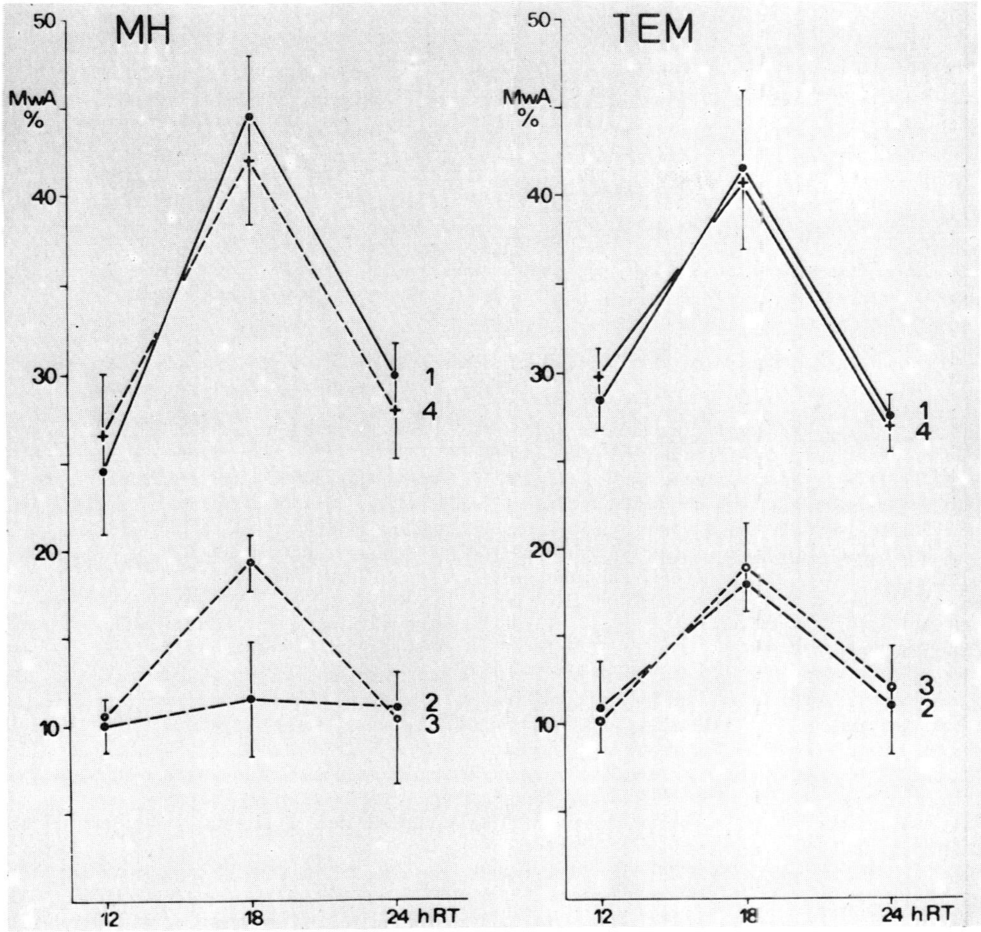

Fig. 38. Effects of long intertreatment times (IT) between heat shock (hs) and MH (left) resp. TEM challenge treatment (right) on the frequency of metaphases with chromatid aberrations (MwA %) at recovery times (RT) of 12, 18 and 24 h: hs protection is absent when IT is longer than 210 min in the case of MH challenging; protective functions disappear when IT is longer than 180 min in the case of TEM challenging.
Left figure—curve 1: 0.5 h 5×10^{-4} M MH; curve 2: 10 min 40 °C hs – 1 h IT – 0.5 h 5×10^{-4} M MH; curve 3: 10 min 40 °C hs – 210 min IT – 0.5 h 5×10^{-4} M MH; curve 4: 10 min 40 °C hs – 240 min IT – 0.5 h 5×10^{-4} M MH.
Right figure—curve 1: 0.5 h 10^{-4} M TEM; curve 2: 10 min 40 °C hs – 1 h IT – 0.5 h 10^{-4} M TEM; curve 3: 10 min 40 °C hs – 180 min IT – 0.5 h 10^{-4} M TEM; curve 4: 10 min 40 °C hs – 210 min IT – 0.5 h 10^{-4} M TEM

ganic salts were used for conditioning (Table 8). These observations provide additional evidence for different mechanisms underlying protection by hs (and inorganic salts) against TEM on the one hand, and MH on the other hand.

Taken together, the results show that conditioning pretreatments may, in many cases, render the *Vicia faba* root tip meristems less sensitive to DNA damage as evidenced by lower yields of induced chromatid aberrations. Analogous results have been obtained from experiments with other plants (Veleminsky et al., 1983; Nicoloff et al., 1985; Stepanyan et al., 1983; Dubinina et al., 1986) and also with animal cells cultivated *in vitro* (e.g., Kaina, 1982; Laval and Laval, 1984; Olivieri et al.,

1984; for review see FROSINA and ABBONDANDOLO, 1985). Together with the results reported for bacteria, these findings show that adaptive responses are a widespread phenomenon which enables cells to react on various stresses with transient protective reactions. The mechanisms triggered by the pretreatments and eventually resulting in partial protection against mutagen-induced chromatid structural changes and other genotoxic effects are presently unknown. The results now at hand pave the way for biochemical studies which might elucidate the possible role of stress proteins and/or inducible repair processes of DNA damage. Much more detailed data are available with respect to similar functions in bacteria (cf. TEO et al., 1984, 1986; WALKER, 1985) and *in vitro* cultivated animal cells (cf. YAROSH, 1985; WALKER et al., 1985). These may be a guide for future investigations in higher plants as well.

References

ABELES, F.B., R.P.BOSSHART, L.E.FORRENCE, W.H.HABIG, 1970, Preparation and purification of glucanase and chitinase from bean leaves. Plant Physiol. 47: 129–134.

ABERNETHY, R. H., D. S. THIEL, N. S. PETERSEN, K. HEIM, 1989, Thermotolerance is developmentally dependent in germinating wheat seed. Plant Physiol. 89: 569–576

AIST, J.R., 1976, Papillae and related wound plugs of plant cells. Annu. Rev. Phytopathol. 81: 107–121.

ALDEN, J., R.K.HERMANN, 1971, Aspects of the cold-hardiness mechanism in plants. Bot. Rev. 37: 37–142.

ALEXANDROV, V.YA., 1977, Cells, Molecules and Temperature. Springer Verlag, Berlin.

ALEXANDROV, V.YA., 1979, Cell reparation of non-DNA injury. Int. Rev. Cytol. 60: 223–269.

ALEXANDROV, V. YA., A. G. LOMAGIN, N. L. FELDMAN, 1970, The responsive increase in thermostability of plant cells. Protoplasma 69: 417–458.

ALLAN, E.F., A.J.TREWAVAS, 1987, The role of calcium in metabolic control. In: E.DAVIES (ed.), The Biochemistry of Plants, Vol. 12. Acad. Press, San Diego, pp. 117–149.

ALTHAUS, F.R., C.RICHTER, 1987, ADP-ribosylation of proteins. Enzymology and Biological Significance. Mol. Biol. Biochem. Biophys. Vol. 37. Springer Verlag, Berlin.

ALTMAN, A., R.FRIEDMAN, D.AMIR, N.LEVIN, 1982, Polyamine effects and metabolism in plants under stress conditions. In: P.F.WAREING (ed.), Plant Growth Substances. Acad. Press, London, pp. 483–494.

ALTON, T.H., H.F.LODISH, 1977, Translational control of protein synthesis during early stages of differentiation of the slime mould *Dictyostelium discoideum*. Cell 12: 301–310.

ALTSCHULER, M., J.P.MASCARENHAS, 1982, Heat shock proteins and effects of heat shock in plants. Plant Mol. Biol. 1: 103–115.

ALTSCHULER, M., J.P.MASCARENHAS, 1985, Transcription and translation of heat shock and normal proteins in seedlings and developing seeds of soybean exposed to a gradual temperature increase. Plant Mol. Biol. 5: 291–298.

AMARAL, M. D., L. GALEGO, C. RODRIGUES-POUSADA, 1988, Stress response of *Tetrahymena pyriformis* to arsenite and heat shock: Differences and similarities. Eur. J. Biochem. 171: 463–470.

AMIN, J., R. MESTRIL, P. SCHILLER, M. DREANO, R. VOELLMY, 1987, Organization of the *Drosophila melanogaster* hsp70 heat shock regulation unit. Mol. Cell. Biol. 7: 1055–1062.

AMIN, J., J. ANANTHAN, R. VOELLMY, 1988, Key features of heat shock regulatory elements. Mol. Cell. Biol. 8: 3761–3769.

AMINO, S., T.FUJIMARA, A.KOMAMINE, 1983, Synchrony induced by double phosphate starvation in a suspension culture of *Catharanthus roseus*. Physiol. Plant. 59: 393–396.

AN, G., A.MITRA, H.K.CHOI, M.A.COSTA, K.AN, R.W.THORNBURG, C.A.RYAN, 1989, Functional analysis of the 3′control region of the potato wound-inducible proteinase inhibitor II gene. Plant Cell 1: 115–122.

ANANTHAN, J., A.L.GOLDBERG, R.VOELLMY, 1986, Abnormal proteins serve as eukaryotic stress signals and trigger the activation of heat shock genes. Science 232: 522–524.

ANDREWS, C.J., 1987, Low temperature stress in field and forage crop production—An overview. Can. J. Plant Sci. 67: 1121–1133.

ANGHILERI, L.J., J.ROBERT (eds.), 1986, Hyperthermia in Cancer Treatment. CRC Press, Boca Raton, Fd.

ANTONIW, J.F., R.F.WHITE, 1986, Changes with time in the distribution of viruses and PR protein around single local lesions of TMV infected tobacco. Plant Mol. Biol. 6: 145–149.

APELBAUM, A., A.C.BURGOON, J.D.ANDERSON, M.LIEBERMAN, R.BEN ARIE, A.K.MATTOO, 1981, Polyamines inhibit biosynthesis of ethylene in higher plant tissue and fruit protoplasts. Plant Physiol. 68: 433–456.

APREES, T., 1969, Protein synthesis by disks of pumpkin mesocarp. Phytochemistry 8: 1879–1886.

Arcangeli, G., M. Benassi, A. Cividalli, G. A. Lovisolo, F. Mauro, 1987, Radiotherapy and hyperthermia: Analysis of clinical results and identification of prognostic variables. Cancer Res. 60: 950–956.

Arrigo, A.-P., 1987, Cellular localization of HSP23 during *Drosophila* development and following subsequent heat shock. Dev. Biol. 122: 39–48.

Arrigo, A.-P., C. Ahmad-Zadeh, 1981, Immunofluorescence localization of a small heat shock protein (hsp23) in salivary gland cells of *Drosophila melanogaster.* Mol. Gen. Genet. 184: 73–79.

Arrigo, A.-P., W. J. Welch, 1987, Characterization and purification of the small 28,000 Dalton mammalian heat shock protein. J. Biol. Chem. 262: 15359–15369.

Arrigo, A.-P., S. Fakan, A. Tissieres, 1980, Localization of the heat shock-induced proteins in *Drosophila melanogaster* tissue culture cells. Dev. Biol. 78: 86–103.

Arrigo, A.-P., J. P. Suhan, W. J. Welch, 1988, Dynamic changes in the structure and intracellular locale of the mammalian low-molecular-weight heat shock protein. Mol. Cell. Biol. 8: 5059–5071.

Asada, Y., I. Matsumoto, 1987, Induction of disease resistance in plants by a lignification-inducing factor. In: S. Nishimura, C. P. Vance, N. Dohe (eds.), Molecular Determinants of Plant Diseases. Springer-Verlag, Berlin, pp. 223–233.

Asare-Boamah, N. K., R. A. Fletcher, 1986, Protection of bean seedlings against heat and chilling injury by triadimefon. Physiol. Plant. 67: 353–358.

Asare-Boamah, N. K., G. Hofstra, R. A. Fletcher, E. B. Dumbroff, 1986, Triadimefon protects bean plants from water stress through its effects on abscisic acid. Plant Cell Physiol. 27: 383–390.

Ashburner, M., J. J. Bonner, 1979, The induction of gene activity in *Drosophila* by heat shock. Cell 17: 241–254.

Atkinson, B. G., D. B. Walden (eds.), 1985, Changes in Eukaryotic Gene Expression in Response to Environmental Stress. Acad. Press, Orlando.

Averyhart-Fullard, V., K. Datta, A. Marcus, 1988, A hydroxyproline-rich protein in the soybean cell wall. Proc. Natl. Acad. Sci. USA 85: 1082–1085.

Ayme, A., A. Tissieres, 1985, Locus 67B of *Drosophila melanogaster* contains seven, not four, closely related heat shock genes. EMBO J. 4: 2949–2954.

Ayme, A., R. Southgate, A. Tissieres, 1985, Nucleotide sequences responsible for the thermal inducibility of the *Drosophila* small heat-shock protein genes in monkey COS cells. J. Mol. Biol. 182: 469–475.

Ayres, P. G., 1984, The interaction between environmental stress injury and biotic disease physiology. Annu. Rev. Phytopathol. 22: 53–75.

Baker, E. A., 1974, The influence of environment on leaf wax development in *Brassica oleracea* var. *gemmifers.* New. Phytol. 73: 955–966.

Baker, J., C. Steele, L. Dure, 1988, Sequence and characterization of 6 Lea proteins and their genes from cotton. Plant Mol. Biol. 11: 277–291.

Bardwell, J. C. A., E. A. Craig, 1984, Major heat shock gene of *Drosophila* and *Escherichia coli* heat-inducible dnaK gene are homologous. Proc. Natl. Acad. Sci. USA 81: 848–852.

Bardwell, J. C. A., E. A. Craig, 1987, Eukaryotic Mr 83,000 heat shock protein has a homologue in *Escherichia coli.* Proc. Natl. Acad. Sci. USA 84: 5177–5181.

Barnett, T., M. Altschuler, C. N. McDaniel, J. P. Mascarenhas, 1980, Heat shock induced proteins in plant cells. Dev. Genet. 1: 331–340.

Barraclough, R., R. J. Ellis, 1980, Protein synthesis in chloroplasts. IX. Assembly of newly-synthesized large subunits into ribulosebisphosphate carboxylase in isolated intact pea chloroplasts. Biochim. Biophys. Acta 608: 19–31.

Barratt, D. H. P., 1986, Modulation by abscisic acid on storage protein accumulation in *Vicia faba* L. cotyledons cultured *in vitro.* Plant Sci. 46: 159–167.

Bartels, D., M. Singh, F. Salamini, 1988, Onset of desiccation tolerance during development of the barley embryo. Planta 175: 485–492.

Bass, H., W. T. Coakley, J. L. Moore, D. Tilley, 1982, Hyperthermia-induced changes in the morphology of CHO-K1 cells and their refractile inclusions. J. Therm. Biol. 7: 231–242.

Basso, L. C., T. A. Smith, 1974, Effect of mineral deficiency on amine formation in higher plants. Phytochemistry 13: 875–883.

Baszczynski, C. L., 1986, Immunochemical analysis of heat-shock protein synthesis in maize (*Zea mays* L.). Can. J. Genet. Cytol. 28: 1076–1087.

Baszczynski, C. L., D. B. Walden, B. G. Atkinson, 1982, Regulation of gene expression in corn (*Zea mays* L.) by heat shock. Can. J. Biochem. 60: 569–579.

Baszczynski, C. L., D. B. Walden, B. G. Atkinson, 1983, Regulation of gene expression in corn (*Zea mays* L.) by heat shock. II. *In vitro* analysis of RNAs from heat-shocked seedlings. Can. J. Cell Biol. 61: 395–403.

Bauer, H., 1972, CO_2-Gaswechsel nach Hitzestress bei *Abies alba* Mill. and *Acer pseudoplatanus* L. Phytosynthetica 6: 424–434.

BAUER, A., M. SENSER, 1979, Photosynthesis of ivy leaves (*Hedera helix*L.) after heat stress II. Activity of ribulose-bisphosphate carboxylase, Hill reaction, and chloroplast ultrastructure. Z. Pflanzenphysiol. 91: 359–369.

BAULCOMBE, D. C., 1987, Do plant hormones regulate gene expression during development? In: G. V. HOAD, J. R. LENTON, M. B. JACKSON, R. K. ATKIN (eds.), Hormone Action in Plant Development. Butterworth, London, pp. 63–70.

BAULIEU, E. E., 1987, Steroid hormone antagonists at the receptor level: A role for the heat-shock protein MW 90,000 (hsp90). J. Cell. Biochem. 35: 161–174.

BAULIEU, E.-E., M.-G. CATELLI, 1989, Steroid hormone receptors and heat shock protein Mr 90,000 (HSP90): A functional interaction? In: PARDUE et al. (eds.) 1989: pp. 203–219.

BAUMANN, G., E. RASCHKE, M. BEVAN, F. SCHOEFFL, 1987, Functional analysis of sequences required for transcriptional activation of a soybean heat shock gene in transgenic tobacco plants. EMBO J. 6: 1161–1166.

BEAULIEU, J. F., R. M. TANGUAY, 1988, Members of the *Drosophila* HSP70 family share ATP-binding properties. Eur. J. Biochem. 172: 341–347.

BEAULIEU, J. F., A.-P. ARRIGO, R. M. TANGUAY, 1989, Interaction of *Drosophila* 27,000 Mr heat-shock protein with the nucleus of heat-shocked and ecdysone-stimulated culture cells. J. Cell Sci. 92: 29–36.

BELANGER, F. C., M. R. BRODL, T.-H. D. HO, 1986, Heat shock causes destabilization of specific mRNAs and destruction of endoplasmic reticulum in barley aleurone cells. Proc. Natl. Acad. Sci. USA 83: 1354–1358.

BELL, A. A., 1981, Biochemical mechanisms of disease resistance. Annu. Rev. Plant Physiol. 32: 21–81.

BELL, J., L. NEILSON, M. PELLEGRINI, 1988, Effect of heat shock on ribosome synthesis in *Drosophila melanogaster.* Mol. Cell. Biol. 8: 91–95.

BENNETZEN, J. L., T. L. ADAMS, 1984, Selection and characterization of cadmium-resistant suspension cultures of the wild tomato *Lycopersicon peruvianum.* Plant Cell Rep. 3: 258–261.

BENZIONI, A., C. ITAI, 1973, Short- and long-term effects of higher temperature (47-49 C) on tobacco leaves. III. Efflux and 32-P incorporation into phospholipids. Physiol. Plant. 28: 493–497.

BENZIONI, A., C. ITAI, 1975, Preconditioning of tobacco and bean leaves to heat shock by high temperature or NaCl. Physiol. Plant. 35: 80–84.

BERNSTEIN, C., V. JOHNS, 1989, Sexual reproduction as a response to H_2O_2 damage in *Schizosaccharomyces pombe.* J. Bacteriol. 171: 1893–1897.

BERRIOS, S., P. A. FISHER, 1988, Thermal stabilization of putative karyoskeletal protein-enriched fractions from *Saccharomyces cerevisiae.* Mol. Cell. Biol. 8: 4573–4575.

BERRY, J. A., O. BJÖRKMAN, 1980, Photosynthetic response and adaptation to temperature in higher plants. Annu. Rev. Plant Physiol. 31: 491–543.

BERTANI, A., F. MENEGUS, R. BOLLINI, 1981, Some effects of anaerobiosis on protein metabolism in rice roots. Z. Pflanzenphysiol. 103: 37–43.

BEWLEY, J. D., K. M. LARSEN, J. E. T. PAPP, 1983, Water stress induced changes in the pattern of protein synthesis in maize seedling mesocotyls. A comparison with the effects of heat shock. J. Exp. Bot. 34: 1126–1133.

BIELKA, H. (ed.), 1982, The Eukaryotic Ribosome. Akademie-Verlag, Berlin.

BIENZ, M., 1984, *Xenopus* hsp70 genes are constitutively expressed in injected oocytes. EMBO J. 3: 2477–2483.

BIENZ, M., 1985, Transient and developmental activation of heat-shock genes. Trends Biochem. Sci. 10: 157–161.

BIENZ, M., 1986, A CCAAT box confers cell-type-specific regulation on the *Xenopus* hsp70 gene in oocytes. Cell 46: 1037–1042.

BIENZ, M., H. R. B. PELHAM, 1982, Expression of a *Drosophila* heat-shock protein in *Xenopus* oocytes: Conserved and divergent regulatory signals. EMBO J. 1: 1583–1588.

BIENZ, M., H. R. B. PELHAM, 1986, Heat shock regulatory elements function as an inducible enhancer in the *Xenopus* hsp70 gene and when linked to a heterologous promoter. Cell 45: 753–760.

BIENZ, M., H. R. B. PELHAM, 1987, Mechanisms of heat-shock gene activation in higher eukaryotes. Adv. Genet. 24: 31–72.

BINZEL, M. L., F. D. HESS, R. A. BRESSAN, P. M. HASEGAWA, 1988, Intracellular compartmentation of ions in salt adapted tobacco cells. Plant Physiol. 86: 607–614.

BISHOP, P. D., D. J. MAKUS, G. PEARCE, C. A. RYAN, 1981, Proteinase inhibitor-inducing factor activity in tomato leaves resides in oligosaccharides enzymically released from cell walls. Proc. Natl. Acad. Sci. USA 78: 3536–3540.

BISHOP, D., G. PEARCE, J. E. BRYANT, C. A. RYAN, 1984, Isolation and characterization of the proteinase inhibitor-inducing factor from tomato leaves: Indentity and activity of poly- and oligogalacturonide fragments. J. Biol. Chem. 259: 13172–13177.

BLACKMAN, R. K., M. MESELSON, 1986, Interspecific nucleotide sequence comparisons used to identify regulatory and structural features of the *Drosophila* hsp82 gene. J. Mol. Biol. 188: 499–516.

BLUM, A., 1975, Effect of the Bm gene on epicuticular wax and the water relations of *Sorghum bicolor* L. (Moench). Israel J. Bot. 24: 50–51.

Bochner, B.R., P.C.Lee, S.W.Wilson, Ch.W.Cutler, B.N.Ames, 1984, ApppppA and related adenylated nucleotides are synthesized as a consequence of oxidation stress. Cell 37: 225–232.

Bochner, B.R., M.Zylicz, C.Georgopoulos, 1986, *Escherichia coli* DnaK protein possesses a 5′-nucleotidase activity that is inhibited by AppppA. J. Bacteriol. 168: 931–935.

Bodner, M., J.-L.Castrillo, L.E.Theill, T.Deerinck, M.Ellisman, M.Karin, 1988, The pituitary-specific transcription factor GHF-1 is a homeobox-containing protein. Cell 55: 505–518.

Bohlmann, H., S. Clausen, S. Behnke, H. Giese, C. Hiller, U. Reimann-Philipp, G. Schrader, V. Barkholt, K. Apel, 1988, Leaf-specific thionins of barley—a novel class of cell wall proteins toxic to plant-pathogenic fungi and possibly involved in the defense mechanism of plants. EMBO J. 7: 1559–1566.

Boller, T., 1985, Induction of hydrolases as a defense reaction against pathogens. In: Key and Kosuge (eds.), 1985, pp. 247–262.

Boller, T., 1987, Hydrolytic enzymes in plant disease resistance. In: Kosuge and Nester (eds.), 1987, pp. 385–413.

Boller, T., 1988, Ethylene and the regulation of antifungal hydrolases in plants. Oxford Surv. Plant Mol. Cell Biol. 5: 145–174.

Boller, T., 1989, Primary signals and second messenger in the reaction of plants to pathogenes. In: W. F. Boss, I.J. Morré (eds.), 1989, Second Messengers in Plant Growth and Development. Alan R. Liss, Inc. New York, pp. 227–255.

Boller, T., J. P. Metraux, 1988, Extracellular localization of chitinase in cucumber. Physiol. Mol. Plant Pathol. 33: 11–16.

Boller, T., A. Gehri, F. Mauch, U. Voegeli, 1983, Chitinase in bean leaves: Induction by ethylene, purification, properties and possible function. Planta 157: 22–31.

Bonato, M. C. M., A. M. Silva, S. L. Gomes, J. C. C. Maia, M. H. Juliani, 1987, Differential expression of heat-shock proteins and spontaneous synthesis of HSP70 during the life cycle of *Blastocladiella emersonii.* Eur. J. Biochem. 163: 211–220.

Bond, U., 1988, Heat shock but not other stress inducers leads to the disruption of a sub-set of snRNPs and inhibition of *in vitro* splicing in HeLa cells. EMBO J. 7: 3509–3518.

Bond, U., M.J. Schlesinger, 1986, The chicken ubiquitin gene contains a heat shock promoter and expresses an unstable mRNA in heat-shocked cells. Mol. Cell. Biol. 6: 4602–4610.

Bond, U., M.J. Schlesinger, 1987, Heat-shock proteins and development. Adv. Genet. 24: 1–29.

Bond, U., N. Agell, A.L. Haas, K. Redman, M.J. Schlesinger, 1988, Ubiquitin in stressed chicken embryo fibroblasts. J. Biol. Chem. 263: 2384–2388.

Bonham-Smith, P.C., M.Kapoor, J.D.Bewley, 1987, Establishment of thermotolerance in maize by exposure to stresses other than a heat shock does not require heat shock protein synthesis. Plant Physiol. 85: 575–580.

Bonham-Smith, P.C., M.Kapoor, J.D.Bewley, 1988a, A comparison of the stress response of *Zea mays* seedlings as shown by qualitative changes in protein synthesis. Can. J. Bot. 66: 1883–1890.

Bonham-Smith, P.C., M.Kapoor, J.D.Bewley, 1988b, Exogenous application of abscisic acid or triadimefon affects the recovery of *Zea mays* from heat shock. Physiol. Plant. 73: 27–30.

Bonhoff, A., B.Rieth, J.Golecki, H.Grisebach, 1987, Race cultivar-specific differences in callose deposition in soybean roots following infection with *Phytophthora megasperma* f. sp. *glycinea.* Planta 172: 101–105.

Boon-Niermeijer, E.K., M.Tuyl, H. van de Scheur, 1986, Evidence for two states of thermotolerance. Int. J. Hyperthermia 2: 93–105.

Borelli, M.J., R.S.L.Wong, W.C.Dewey, 1986, A direct correlation between hyperthermia induced membrane blebbing and survival in synchronous G1 CHO cells. J. Cell. Physiol. 126: 181–190.

Bostock, R.M., R.A.Laine, J.A.Kuc, 1982, Factors affecting the elicitation of sesquiterpenoid phytoalexin accumulation by eicosapentaenoic and arachidonic acids in potato. Plant Physiol. 70: 1417–1427.

Bournias-Vardiabasis, N., C.H.Buzin, 1986, Developmental effects of chemicals and the heat shock response in *Drosophila* cells. Teratog. Carcinog. Mutag. 6: 523–537.

Boussiba, S., A. Rikin, A. E. Richmond, 1975, The role of abscisic acid in cross-adaptation of tobacco plants. Plant Physiol. 56: 337–339.

Boyer, J. S., 1982, Plant productivity and environment. Science 218: 443–448.

Brady, F. O., 1982, The physiological function of metallothionein. Trends Biochem. Sci. 7: 143–145.

Bray, E. A., R. N. Beachy, 1985, Regulation by ABA of betaconglycinin expression in cultured developing soybean cotyledons. Plant Physiol. 79: 746–750.

Brevet, A., P. Plateau, M. Best-Belpomme, S. Blanquet, 1985, Variation of $A_{P_4}A$ and other dinucleotide polyphosphates in stressed *Drosophila* cells. J. Biol. Chem. 260: 15566–15570.

Brock, Th. D., 1978, Thermophilic Microorganisms and Life at High Temperatures. Springer Verlag, New York.

Brock, Th. D., 1985, Life at high temperatures. Science 230: 132–138.

Broertjes, C., 1972, Use in plant breeding of acute, chronic or fractionated doses of X-rays or fast neutrons as illustrated with leaves of *Saintpaulia.* Agric. Res. Rep. (Centre Agric. Publ. Document.) 776: 74.

Broglie, K. E., J. J. Gaynor, R. M. Broglie, 1986, Ethylene-regulated gene expression: Molecular cloning of the genes encoding an endochitinase from *Phaseolus vulgaris*. Proc. Natl. Acad. Sci. USA 83: 6820–6824.

Brugge, J., W. Yonemoto, D. Darrow, 1983, Interaction between the *Rous sarcoma* virus transforming protein and two cellular phosphoproteins: Analysis of the turnover and distribution of this complex. Mol. Cell. Biol. 3: 9–19.

Bruns, P., 1887, Die Heilwirkung des Erysipels auf Geschwulste. Beitraege Klin. Chir. 3: 443–466.

Bünger, W. and J. Feierabend, 1980, Capacity for RNA synthesis in 70S ribosome-deficient plastids of heat-bleached rye leaves. Planta 149: 163–169.

Burke, M. J., L. V. Gusta, H. A. Ouamme, C. J. Weiser and P. H. Li, 1976, Freezing and injury in plants. Annu. Rev. Plant Physiol. 27: 507–528.

Burke, J. J., J. L. Hatfield, R. P. Klein and J. E. Mullet, 1985, Accumulation of heat shock proteins in field grown cotton. Plant Physiol. 78: 394–398.

Busch, W., 1866, Über den Einfluss welchen heftigere Erysipeln zuweilen auf organisierte Neubildungen ausüben. Verhandl. Naturh. Preuss. Rhein. Westphal. 23: 28–30.

Byfield, J. E. and O. H. Scherbaum, 1967, Temperature-dependent decay of RNA and of protein synthesis in a heat-synchronized protozoan. Proc. Natl. Acad. Sci. USA 57: 602–606.

Caldecott, R. S. and L. Smith, 1952, The influence of heat treatments on the injury and cytogenetic effects of X-rays on barley. Genetics 37: 136–157.

Cadrin, M., M. Ashraf and J. G. Lafontaine, 1984, Effects of heat shock on nuclear organization in plasmodia of *Physarum polycephalum*. Can. J. Bot. 62: 1550–1560.

Cardelli, J. A. and R. L. Dimond, 1981,, Regulation of protein synthesis in *Dictyostelium discoideum:* Effects of starvation and anoxia on initiation. Biochemistry 20: 7391–7398.

Carr, J. P., D. C. Dixon and D. F. Klessig, 1985, Synthesis of pathogenesis-related proteins in tobacco is regulated at the level of mRNA accumulation and occurs on membrane-bound polysomes. Proc. Natl. Acad. Sci. USA 82: 7999–8003.

Carr, J. P., D. C. Dixon, B. J. Nikolau, K. V. Voelkerding, D. F. Klessig, 1987, Synthesis and localization of pathogenesis-related proteins in tobacco. Mol. Cell. Biol. 7: 1580–1583.

Cathcart, R., E. Schwiers, R. L. Saul, B. N. Ames, 1984, Thymine glycol in human and rat urine: A possible assay for oxidative DNA damage. Proc. Natl. Acad. Sci. USA 81: 5633–5637.

Cattivelli, L., D. Bartels, 1989, Cold induced mRNAs accumulate with different kinetics in barley coleoptiles. Planta 178: 184–188

Chang, C., M. Meyerowitz, 1986, Molecular cloning and DNA sequence of the *Arabidopsis thaliana* alcohol dehydrogenase gene. Proc. Natl. Acad. Sci. USA 83: 1408–1412.

Chang, S. C., S. K. Wooden, T. Nakaki, Y. K. Kim, A. Y. Lin, L. K. Kung, J. W. Attenello, A. S. Lee, 1987, Rat gene encoding the 78-kDa glucose-regulated protein GRP78: Its regulatory sequences and the effect of protein glycosylation on its expression. Proc. Natl. Acad. Sci. USA 84: 680–684.

Chappell, J., K. Hahlbrock, 1984, Transcription of plant defence genes in response to UV light or fungal elicitor. Nature 311: 76–78.

Chappell, J., K. Hahlbrock, T. Boller, 1984, Rapid induction of ethylene biosynthesis in cultured parsley cells by fungal elicitor and its relationship to the induction of phenylalanine ammonia-lyase. Planta 161: 475–480.

Chappell, T. G., W. J. Welch, D. M. Schlossman, K. B. Palter, M. J. Schlesinger, J. E. Rothman, 1986, Uncoating ATPase is a member of the 70 kilodalton family of stress proteins. Cell 45: 3–13.

Chaturvedi, S. N., G. Zabka, 1977, Studies on dark fixation of carbon dioxide in *Kalanchoe:* I Effect of water stress and growth retardants. Ann. Bot. 41: 493–500.

Cheeseman, J. M., 1988, Mechanisms of salinity tolerance in plants. Plant Physiol. 87: 547–550.

Chen, T. H. H., L. V. Gusta, 1983, Abscisic acid-induced freezing resistance in cultured plant cells. Plant Physiol. 73: 71–75.

Chen, J., J. E. Varner, 1985a, Isolation and characterization of cDNA clones for carrot extensin and a proline-rich 33-kDa protein. Proc. Natl. Acad. Sci. USA 82: 4399–4403.

Chen, J., J. E. Varner, 1985b, An extracellular matrix protein in plants: Characterization of a genomic clone for carrot extensin. EMBO J. 4: 2145–2151.

Chen, Y.-M., S. Kamisaka, Y. Masuda, 1986, Enhancing effects of heat shock and gibberellic acid on the thermotolerance in etiolated *Vigna radiata*. I. Physiological aspects on thermotolerance. Physiol. Plant 66: 595–601.

Cheng, M. Y., F.-U. Hartl, J. Martin, R. A. Pollock, F. Kalonsek, W. Neupert, E. M. Hallberg, R. L. Hallberg, A. L. Horwich, 1989, Mitochondrial heat-shock protein hsp60 is essential for assembly of proteins imported into yeast mitochondria. Nature 337: 620–625.

Chirico, W. J., M. G. Waters, G. Blobel, 1988, 70 K heat shock related proteins stimulate protein translocation into microsomes. Nature 332: 805–810.

CHON, M., Y.-M. CHEN, C.-Y. LIN, 1989, Thermotolerance of isolated mitochondria associated with heat shock proteins. Plant Physiol. 89: 617–621.

CHRISPEELS, M. J., 1969, Synthesis and secretion of hydroxyproline containing macromolecules in carrots. I. Kinetic analysis. Plant Physiol. 44: 1187–1193.

CHRISPEELS, M., J. S. GREENWOOD, 1987, Heat stress enhances phytohemaglutinin synthesis but inhibits its transport out of the endoplasmic reticulum. Plant Physiol. 83: 778–784.

CHRISTMAN, M. F., R. W. MORGAN, F. S. JACOBSON, B. N. AMES, 1985, Positive control of a regulon for defenses against oxidative stress and some heat-shock proteins in *Salmonella typhimurium*. Cell 41: 753–762.

CLARKE, C. F., K. CHENG, A. B. FREY, R. STEIN, P. W. HINDS, A. J. LEVINE, 1988, Purification of complexes of nuclear oncogene p53 with rat and *Escherichia coli* heat shock proteins. *In vitro* dissociation of HSC70 and DnaK from murine p53 by ATP. Mol. Cell. Biol. 8: 1206–1215.

CLEVELAND, D. W., 1988, Autoregulated instability of tubulin mRNAs: a novel eukaryotic regulatory mechanism. Trends Biochem. Sci. 13: 339–343.

CLEVELAND, T. E., R. W. THORNBURG, C. A. RYAN, 1987, Molecular characterization of a wound-inducible inhibitor I gene from potato and the processing of its mRNA and protein. Plant Mol. Biol. 8: 199–207.

CLOUTIER, Y., 1984, Changes of protein patterns in winter rye following cold acclimation and dessication stress. Can. J. Bot. 62: 366–371.

COIAS, R., L. GALEGO, I. BARCHONA, C. RODRIGUES-POUSADA, 1988, Destabilization of tubulin mRNA during heat-shock in *Tetrahymena pyriformis*. Eur. J. Biochem. 175: 467–474.

COLEY, W. B., 1893, The treatment of malignant tumors by repeated inoculations of erysipels with a report of ten original cases. Amer. J. Med. Sci. 105: 487–511.

COLLIER, N. C., M. J. SCHLESINGER, 1986, The dynamic state of heat shock proteins in chicken embryo fibroblasts. J. Cell Biol. 103: 1495–1507.

COLLIER, N. C., J. HEUSER, M. A. LEVY, M. J. SCHLESINGER, 1988, Ultrastructural and biochemical analysis of the stress granule in chicken embryo fibroblasts. J. Cell Biol. 106: 1131–1139.

COLLINGE, D. B., A. J. SLUSARENKO, 1987, Plant gene expression in response to pathogens. Plant Mol. Biol. 9: 389–410.

COLLINGE, D. B., D. E. MILLIGAN, J. M. DOW, G. SCOFIELD, M. J. DANIELS, 1987, Gene expression in *Brassica campestris* showing a hypersensitive response to the incompatible pathogen *Xanthomonas campestris* pv. *vitians*. Plant Mol. Biol. 8: 405–414.

CONDIT, C. M., R. B. MEAGHER, 1987, Expression of a gene encoding a glycine-rich protein in *Petunia*. Mol. Cell. Biol. 7: 4273–4279.

CONEJERO, V., I. PICAZO, P. SEGADO, 1979, *Citrus exocortis* viroid (CEV): protein alterations in different hosts following viroid infection. Virology 97: 454–456.

CONROY, J. P., J. M. VIRGONA, R. M. SMILLIE, E. W. BARLOW, 1988, Influence of drought acclimation and CO_2 enrichment on osmotic adjustment and chlorophyll a fluorescence of sunflower during drought. Plant Physiol. 86: 1108–1115.

COOPER, P., T.-H. D. HO, 1983, Heat shock proteins in maize. Plant Physiol. 71: 215–222.

COOPER, P., T.-H. D. HO, 1987, Intracellular localization of heat shock proteins in maize. Plant Physiol. 84: 1197–1203.

COOPER, J. B., J. E. VARNER, 1983, Insolubilization of hydroxyproline-rich cell wall glycoprotein in aerated carrot root slices. Biochem. Biophys. Res. Commun. 112: 161–167.

COOPER, J. B., J. E. VARNER, 1984, Cross-linking of soluble extensin in isolated cell walls. Plant Physiol. 76: 414–417.

COOPER, P., T.-H. D. HO, R. M. HAUPTMANN, 1984, Tissue specificity of the heat-shock response in maize. Plant Physiol. 75: 431–441.

COOPER, J. B., J. A. CHEN, G. J. VAN HOLST, J. E. VARNER, 1987, Hydroxyproline-rich glycoproteins of plant cell walls. Trends Biochem. Sci. 12: 24–27.

CORBIN, D. R., N. SAUER, C. J. LAMB, 1987, Differential regulation of a hydroxyproline-rich glycoprotein gene family in wounded and infected plants. Mol. Cell. Biol. 7: 4337–4344.

CORNELISSEN, B. J. C., R. A. M. HOOFT VAN HUJSCLUIJNEN, J. F. BOL, 1986, A tobacco mosaic virus-induced tobacco protein is homologous to the sweet-tasting protein thaumatin. Nature 321: 531–532.

CORNELISSEN, B. J. C., J. HOROWITZ, J. A. L. VAN KAN, R. B. GOLDBERG, J. F. BOL, 1987, Structure of tobacco genes encoding pathogenesis-related proteins from the PR1 group. Nucl. Acids Res. 15: 6799–6811.

CORNISH, K., J. A. D. ZEEVAART, 1984, Abscisic acid metabolism in relation to water stress and leaf age in *Xanthium strumarium*. Plant Physiol. 76: 1029–1035.

CRAIG, E. A., 1985, The heat shock response. CRC Crit. Rev. Biochem. 18: 239–280.

CRAIG, E. A., T. KRAMER, J. KOSIC-SMITHERS, 1987, SSC1, a member of the 70-kDa heat shock protein multigene family of *Saccharomyces cerevisiae*, is essential for growth. Proc. Natl. Acad. Sci. USA 84: 4156–4160.

CRAIG, E., W. BOORSTEIN, H.-O-PARK, D. STONE, C. NICOLET, 1989, Complex regulation of three heat inducible HSP70 related genes in *Saccharomyces cerevisiae.* In: PARDUE et al. (eds.), 1989, pp. 51–61.

CRAMER, C. L., J. N. BELL, T. B. RYDER, J. A. BAILEY, W. SCHUCH, G. P. BOLWELL, M. P. ROBBINS, R. A. DIXON, C. J. LAMB, 1985, Co-ordinated synthesis of phytoalexin biosynthetic enzymes in biologically-stressed cells of bean (*Phaseolus vulgaris* L.). EMBO J. 4: 285–289.

CURRIE, R. W., F. P. WHITE, 1981, Trauma-induced protein in rat tissues: A physiological role for a "heat shock" protein?. Science 214: 72–73.

CUTT, J. R., D. C. DIXON, J. P. CARR, D. F. KLESSIG, 1988, Isolation and nucleotide sequence of cDNA clones for the pathogenesis-related proteins PR1a, PR1b and PR1c of *Nicotiana tabacum* cultivar Xanthi NC induced by TMV infection. Nucl. Acids. Res. 16: 9861.

CUYPERS, B., E. SCHMELZER, K. HAHLBROCK, 1988, *In situ* localization of rapidly accumulated phenylalanine ammonia-lyase mRNA around penetration sites of *Phytophthora infestans* in potato leaves. Mol. Plant-Microbe Interact. 1: 157–160.

CZARNECKA, E., L. EDELMAN, F. SCHOEFFL, J. L. KEY, 1984, Comparative analysis of physical stress responses in soybean seedlings using cloned heat shock cDNAs. Plant Mol. Biol. 3: 45–58.

CZARNECKA, E., W. B. GURLEY, R. T. NAGAO, L. A. MOSQUERA, J. L. KEY, 1985, DNA sequence and transcript mapping of a soybean gene encoding a small heat shock protein. Proc. Natl. Acad. Sci. USA 82: 3726–3730.

CZARNECKA, E., R. T. NAGAO, J. L. KEY, W. B. GURLEY, 1988, Characterization of Gm hsp26-A, a stress gene encoding a divergent heat shock protein of soybean: Heavy-metal-induced inhibition of intron processing. Mol. Cell. Biol. 8: 1113–1122.

DAIE, J. W. F. CAMPBELL, 1981, Response of tomato plants to stressful temperatures. Increase in abscisic acid concentrations. Plant Physiol. 67: 26–29.

DAMERON, C. T., R. N. REESE, R. K. MEHRA, A. R. KORTAN, P. J. CARROLL, M. L. STEIGERWALD, L. E. BRUS, D. R. WINGE, 1989, Biosynthesis of cadmium sulphide quantum semiconductor crystallites. Nature 338: 596–597.

DANGL, J. L., K. D. HAUFFE, S. LIPPHARDT, K. HAHLBROCK, D. SCHEEL, 1987, Parsley protoplasts retain differential responsiveness to u. v. light and fungal elicitor. EMBO J. 6: 2551–2556.

DANIELL, J. W., W. E. CHAPPELL, H. B. COUCH, 1969, Effect of sublethal and lethal temperatures on plant cells. Plant Physiol. 44: 1684–1689.

DANIELS, C. H., L. A. HADWIGER, Y. S. CODY, D. F. KENDRA, 1987a, Disease resistance response genes: Induction in peas by fungal and bacterial wall components and blockage by heat shock. In: Plant Gene Systems and Their Biology. Alan R. Liss Inc., New York, pp. 161–170.

DANIELS, C. H., B. FRISTENSKY, W. WAGONER, L. A. HADWIGER, 1987b, Pea genes associated with non-host disease resistance to *Fusarium* are also active in race-specific disease resistance to *Pseudomonas.* Plant Mol. Biol. 8: 309–316.

DARASCH, S., D. D. MOSSER, N. C. BOLS, J. J. HEIKKILA, 1988, Heat shock gene expression in *Xenopus laevis* A6 cells in response to heat shock and sodium arsenite. Biochem. Cell Biol. 66: 862–870.

DARVILL, A. G., P. ALBERSHEIM, 1984, Phytoalexins and their elicitors—A defense against microbial infection in plants. Annu. Rev. Plant Physiol. 35: 243–275.

DASHEK, W. V., S. S. ERICKSON, 1981, Isolation, assay, biosynthesis, metabolism, uptake and translocation, and function of proline in plant cells and tissues. Bot. Rev. 47: 349–385.

DAVIDSON, A. D., J. M. MANNERS, R. S. SIMPSON, K. J. SCOTT, 1987, cDNA cloning of mRNAs induced in resistant barley during infection by *Erysiphe graminis* f. sp. *hordei.* Plant Mol. Biol. 8: 77–85.

DAVIES, E., 1987, Plant responses in wounding. In: E. DAVIES (ed.), The Biochemistry of Plants, Vol. 12. Acad. Press, San Diego, pp. 243–264.

DAVIES, E., A. SCHUSTER, 1981, Intercellular communication in plants: Evidence for a rapidly generated, bidirectionally transmitted wound signal. Proc. Natl. Acad. Sci. USA 78: 2422–2426.

DAVIES, E., K. V. A. RAMAIAH, S. ABE, 1986, Wounding inhibits protein synthesis yet stimulates polysome formation in aged, excised pea epicotyls. Plant Cell Physiol. 27: 1377–1386.

DAWSON, W. O., 1976, Synthesis of TMV RNA at restrictive high temperatures. Virology 73: 319–326.

DAWSON, W. O., 1983, Tobacco mosaic virus protein synthesis is correlated with double-stranded RNA synthesis and not single-stranded RNA synthesis. Virology 125: 314–323.

DAWSON, W. O., G. L. GRANTHAM, 1981, Inhibition of stable RNA synthesis and production of a noval RNA in heat-stressed plants. Biochem. Biophys. Res. Commun. 100: 23–30.

DAWSON, W. O., C. BOYD, 1987, TMV protein synthesis is not translationally regulated by heat shock. Plant Mol. Biol. 8: 145–149.

DAY, P. R., G. J. JELLIS, 1987, Genetics and Plant Pathogenesis. Blackwell Sci. Publ., Oxford.

DEBENEDETTI, A., C. BAGLIONI, 1986a, Activation of hemin-regulated initiation factor-2 kinase in heat-shocked HeLa cells. J. Biol. Chem. 261: 338–342.

DeBenedetti, A., C. Baglioni, 1986b, Translational regulation of the synthesis of a major heat shock protein in HeLa cells. J. Biol. Chem. 261: 15800–15805.

Degani, N., C. Itai, 1978, The effect of radiation on growth and abscisic acid in wheat seedlings. Env. Exp. Bot. 18: 113–115.

Delhaize, E., P. J. Jackson, L. D. Lujan, N. J. Robinson, 1989, Poly(gamma-glutamylcysteinyl)glycine synthesis in *Datura innoxia* and binding with cadmium. Plant Physiol. 89: 700–706.

Denis, M., S. Cuthill, A.-C. Wikstrom, L. Poellinger, J.-A. Gustafsson, 1988, Association of the dioxin receptor with the Mr 90,000 heat shock protein. A structural kinship with the glucocorticoid receptor. Biochem. Biophys. Res. Commun. 155: 801–807.

Dennis, E. S., W. L. Gerlach, A. J. Pryor, J. L. Bennetzen, A. Inglis, D. L. Jewellyn, M. M. Sachs, R. J. Ferl, W. J. Peacock, 1984, Molecular analysis of the alcohol dehydrogenase (Adh1) gene of maize. Nucl. Acids Res. 12: 3983–4000.

Dennis, E. S., M. M. Sachs, W. L. Gerlach, E. J. Finnegan, W. J. Peacock, 1985, Molecular analysis of the alcohol dehydrogenase 2 (Adh2) gene of maize. Nucl. Acids Res. 13: 727–743.

Dennis, E. S., J. C. Walker, D. J. Llewellyn, J. G. Ellis, K. Singh, J. G. Tokuhisa, D. R. Wolsterholme, W. J. Peacock, 1987, The response to anaerobic stress: transcriptional regulation of genes for anaerobically induced proteins. In: D. v. Wettstein, N. H. Chua (eds.), Plant Molecular Biology. Plenum Press, New York, pp. 407–417.

Deshaies, R. J., B. D. Koch, M. Werner-Washburne, E. A. Craig, R. Schekman, 1988a, A subfamily of stress proteins facilitates translocation of secretory and mitochondrial precursor polypeptides. Nature 332: 800–805.

Deshaies, R. J., B. D. Koch, R. Schekman, 1988b, The role of stress proteins in membrane biogenesis. Trends Biochem. Sci. 13: 384–388.

DeWit, P. J. G. M., J. Bakker, 1980, Differential changes in soluble tomato leaf proteins after inoculation with virulent and avirulent races of *Cladosporium fulvum* (syn. *Fulvia fulva*). Physiol. Plant Pathol. 17: 121–130.

DeWit, P. J. G. M., M. B. Buurlage, 1986, The occurrence of host-, pathogen- and interaction-specific proteins in the apoplast of *Cladosporium fulvum* (syn. *Fulvia fulva*) infected tomato leaves. Physiol. Mol. Plant Pathol. 29: 159–172.

DeWit, P. J. G. M., F. E. van der Meer, 1986, Accumulation of the pathogenesis related tomato leaf protein P14 as an early indicator of incompatibility in the interaction between *Cladosporium fulvum* (syn. *Fulvia fulva*) and tomato. Physiol. Mol. Plant Pathol. 28: 203–214.

Dhindsa, R. S., R. E. Cleland, 1975, Water stress and protein synthesis I. Differential inhibition of protein synthesis. Plant Physiol. 55: 778–781.

DiCosmo, F., G. H. N. Towers, 1984, Stress and secondary metabolism in cultured plant cells. Rec. Advanc. Phytochem. 18: 97–175.

Didden-Zopfy, B., P. S. Nobel, 1982, High temperature tolerance and heat acclimation of *Opuntia bigelovii.* Oecologia 52: 176–180.

DiDomenico, B. J., G. E. Bugaisky, S. Lindquist, 1982a, The heat shock response is self-regulated at both the transcriptional and posttranscriptional levels. Cell 31: 593–603.

DiDomenico, B. J., G. E. Bugaisky, S. Lindquist, 1982b, Heat shock and recovery are mediated by different translational mechanisms. Proc. Natl. Acad. Sci. USA 79: 6181–6185.

Dinar, M., J. Rudich, E. Zamski, 1983, Effects of heat stress on carbon transport from tomato leaves. Ann. Bot. 51: 97–103.

Dixon, R. A., 1986, The phytoalexin response: Elicitation, signalling and control of host gene expression. Biol. Rev. 61: 239–291.

Dixon, R. A., P. M. Dey, M. A. Lawton, C. J. Lamb, 1983, Phytoalexin induction in French bean. Plant Physiol. 71: 251–256.

Dixon, R. A., J. A. Bailey, J. N. Bell, G. P. Bolwell, C. L. Cramer, K. Edwards, M. A. M. Hamdan, C. J. Lamb, M. P. Robbins, T. B. Ryder, W. Schuch, 1986, Rapid changes in gene expression in response to microbial elicitation. Phil. Trans. R. Soc. Lnd. B 314: 411–426.

Dodson, M., H. Echols, S. Wickner, C. Alfano, K. Mensa-Wilmot, B. Gomes, J. Lebowitz, J. D. Roberts, R. McMacken, 1986, Specialized nucleoprotein structures at the origin of replication of bacteriophage lambda: Localized unwinding of duplex DNA by a six-protein complex Proc. Natl. Acad. Sci. USA 83: 7638–7642.

Dolferus, R., G. Marbaix, M. Jacobs, 1985, Alcohol dehydrogenase in *Arabidopsis:* Analysis of the induction phenomenon in plantlets and tissue cultures. Mol. Gen. Genet. 199: 256–264.

Dorner, A. J., M. G. Krane, R. J. Kaufmann, 1988, Reduction of endogenous GRP78 levels improves secretion of a heterologous protein in CHO cells. Mol. Cell. Biol. 8: 4063–4070.

Douglas, C., H. Hoffmann, W. Schulz, K. Hahlbrock, 1987, Structure and elicitor or u.v.-light-stimulated expression of two 4-coumarate: CoA ligase genes in parsley. EMBO J. 6: 1189–1195.

DOVER, L. D., 1968, Nitrogen distribution in tomato plants during drought (*Lycopersicon esculentum* Marglobe). Phyton 25: 49–52.

DOWGERT, M. F., P. L. STEPONKUS, 1984, Behaviour of the plasma membrane of isolated protoplasts during freeze-thaw cycle. Plant Physiol. 75: 1139–1151.

DRAGON, E. A., S. R. SIAS, E. A. KATO, J. D. GABE, 1987, The genome of *Trypanosoma cruzi* contains a constitutively expressed, tandemly arranged multicopy gene homologous to a major heat shock protein. Mol. Cell. Biol. 7: 1271–1275.

DRON, M., S. D. CLOUSE, R. A. DIXON, M. A. LAWTON, C. J. LAMB, 1988, Glutathione and fungal elicitor regulation of a plant defense gene promoter in electroporated protoplasts. Proc. Natl. Acad. Sci. USA 85: 6738–6742.

DUBININA, L. G., Z. I. KURASHOVA, 1986, Mutagenesis induced by mitomycin C (russ.). Genetika (Moskau) 10: 2444–2451.

DUBININA, L. G., Z. I. KURASHOVA, S. P. SERGIEVKAYA, 1986, The influence of pretreatment with nonmutagenic doses of ethyleneimine on mutagenesis in *Crepis cappilaris* cells (russ.). Genetika (Moskau) 12: 2805–2812.

DUDLER, R., A. A. TRAVERS, 1984, Upstream elements necessary for optimal function of the hsp70 promoter transformed flies. Cell 38: 391–398.

DULSON, J., J. D. BEWLEY, R. N. JOHNSTON, 1988, Abscisic acid is an endogenous inhibitor in the regulation of mannase production by isolated lettuce endosperms. Plant Physiol. 87: 660–665.

DUNCAN, R., J. W. B. HERSHEY, 1984, Heat shock-induced translational alterations in HeLa cells. J. Biol. Chem. 259: 11882–11889.

DUNKLE, L. D., R. MAHESHWARI, P. J. ALLEN, 1969, Infection structures from rust urediospores: Effect of RNA and protein synthesis inhibitors. Science 163: 481–482.

DURA, J. M., 1981, Stage dependent synthesis of heat shock induced proteins in early embryos of *Drosophila melanogaster.* Mol. Gen. Genet. 184: 381–385.

DWORNICZAK, B., M.-E. MIRAULT, 1987, Structure and expression of a human gene coding for a 71 kd heat shock 'cognate' protein. Nucl. Acids Res. 15: 5161–5198.

EBEL, J., 1986, Phytoalexin synthesis: The biochemical analysis of the induction process. Annu. Rev. Phytopathol. 24: 235–264.

EBEL, J., H. GRISEBACH, 1988, Defense strategies of soybean against the fungus *Phytophthora megasperma* f. sp. *glycinea:* A molecular analysis. Trends Biochem. Sci. 13: 23–27.

ECKER, J. R., R. W. DAVIS, 1987, Ethylene regulation of plant defense genes. In: J. E. FOX, and M. JACOBS (eds.) Molecular Biology of Plant Growth Control. Alan R. Liss, Inc., New York, pp. 133–143.

EDELMAN, J., M. A. HALL, 1965, Enzyme formation in higher plant tissues. Development of invertase and ascorbate oxidase activities in mature storage tissue of *Helianthus tuberosus* L. Biochem. J. 95: 403–410.

EDELMAN, L., E. CZARNECKA, J. L. KEY, 1988, Induction and accumulation of heat shock-specific poly(A+)RNAs and proteins in soybean seedlings during arsenite and cadmium treatments. Plant Physiol. 86: 1048–1056.

EDERY, I., R. PETRYSHYN, A. SONENBERG, 1989, Activation of double-stranded RNA-dependent kinase (dsI) by the TAR region of HIV-1 mRNA: A novel translational control mechanism. Cell 56: 303–312.

EICHHOLZ, R., J. HARPER, G. FELIX, F. MEINS, 1983, Evidence for an abundant 33,000-dalton polypeptide regulated by cytokinins in cultured tobacco tissues. Planta 158: 410–415.

EILERT, U., 1987, Elicitation: Methodology and aspects of application. In: F. CONSTABEL and I. K. VASIL, (eds.), 1987, Cell Culture and Somatic Cell Genetics of Plants, Vol. 4. Acad. Press, San Diego 153–196.

ELLGAARD, E. G., U. CLEVER, 1971, RNA metabolism during puff induction in *Drosophila melanogaster.* Chromosoma 36: 60–78.

ELLIS, J. G., D. J. LLEWELLYN, E. S. DENNIS, W. J. PEACOCK, 1987a, Maize Adh-1 promoter sequences control anaerobic regulation: addition of upstream promoter elements from constitutive genes is necessary for expression in tobacco. EMBO J. 6: 11–16.

ELLIS, J. G., D. J. LLEWELLYN, J. C. WALKER, E. S. DENNIS, W. J. PEACOCK, 1987b, The ocs element: a 16 base pair palindrome essential for activity of the octopine synthase enhancer. EMBO J. 6: 3203–3208.

ENGELBRECHT, L., K. MOTHES, 1964, Weitere Untersuchungen zur experimentellen Beeinflussung der Hitzewirkung bei Blättern von *Nicotiana rustica.* Flora 154: 279–298.

ERICSON, M. C., S. H. ALFINITO, 1984, Proteins produced during salt stress in tobacco cell culture. Plant Physiol. 74: 506–509.

ERICKSON, S. S., W. V. DASHEK, 1982, Accumulation of foliar soluble proline in sulphur dioxide-stressed *Glycine max* cvs. "Proctor" and "Excelsior" seedlings. Environ. Pollut. Ser. A 28: 89–108.

ESAKA, M., T. TAKAHASHI, T. ASAHI, 1983, Effect of ethylene on the increase in catalase activity through microbody development in wounded sweet potato root tissue. Plant Cell Physiol. 24: 625–633.

ESQUERRÉ-TUGAYÉ, M. T., C. LAFITTE, D. MAZAU, A. TOPPAN, A. TOUZÉ, 1979, Cell surfaces in plant-microorganism interactions II. Evidence for the accumulation of hydroxyproline-rich glycoproteins in the cell wall of diseased plants as a defense mechanism. Plant Physiol. 64: 320–326.

Esquerré-Tugayé, M.T., D. Mazau, B. Pelissier, D. Roby, D. Rumean, A. Toppan, 1985, Induction by elicitors and ethylene of proteins associated to the defense of plants. In: Key and Kosuge (eds.), 1985, pp. 459–466.

Evans, D.A., W.R. Sharp, P.V. Ammirato, Y. Yamada (eds.), 1983, Handbook of Plant Cell Culture, Vol. 1. Macmillan Publ. Comp., New York.

Fabijanski, S., I. Altosaar, P.G. Arnison, 1987, Heat shock response of *Brassica oleracea* L. broccoli. J. Plant Physiol. 128: 29–38.

Fakan, S., E. Puvion, 1980, The ultrastructural visualization of nucleolar and extranucleolar RNA synthesis and distribution. Int. Rev. Cytol. 65: 255–299.

Falkner, F.G., H. Saumweber, H. Biessmann, 1981, Two *Drosophila melanogaster* proteins related to intermediate filament proteins of vertebrate cells. J. Cell Biol. 91: 175–183.

Farmer, E.E., G. Pearce, C.A. Ryan, 1989, *In vitro* phosphorylation of plant plasma membrane proteins in response to the proteinase inhibitor inducing factor. Proc. Natl. Acad. Sci. USA 86: 1539–1542.

Farrell-Towt, J., M.M. Sanders, 1984, Noncoordinate histone synthesis in heat-shocked *Drosophila* cells is regulated at multiple levels. Mol. Cell. Biol. 4: 2676–2685.

Farrelly, F.W., D.B. Finkelstein, 1984, Complete sequence of the heat shock-inducible HSP90 gene of *Saccharomyces cerevisiae*. J. Biol. Chem. 259: 5745–5751.

Feierabend, J., 1979, Role of cytoplasmic protein synthesis and its coordination with the plastidic protein synthesis in the biogenesis of chloroplasts. Ber. Dtsch. Bot. Ges. 92: 553–574.

Feierabend, J., M. Mikus, 1977, Occurrence of a high temperature sensitivity of chloroplast ribosome formation in several higher plants. Plant Physiol. 59: 863–867.

Feierabend, J., W. Schlueter, K. Tebartz, 1988, Unassembled polypeptides of the plastidic ribosomes in heat-treated 70S ribosome-deficient rye leaves. Planta 174: 542–550.

Felix, G., F. Meins, 1985, Purification, immunoassay and characterization of an abundant, cytokinin-regulated polypeptide in cultured tobacco tissues. Evidence that the protein is a β-1,3-glucanase. Planta 164: 423–428.

Felix, G., F. Meins, 1987, Ethylene regulation of beta-1,3-glucanase in tobacco. Planta 172: 386–392.

Ferl, R.J., H. Nick, 1987, *In vivo* detection of regulatory factor binding sites in the 5′ flanking region of maize Adh1. J. Biol. Chem. 262: 7947–7950.

Ferl, R.J., S.R. Dlonhy, D. Schwartz, 1979, Analysis of maize alcohol dehydrogenase by native-SDS two dimensional electrophoresis and autoradiography. Molec. & Gener. Genet. 169: 7–12.

Ferl, R.J., M.D. Brennan, D. Schwartz, 1980, *In vitro* translation of maize alcohol dehydrogenase (EC 1.1.1.1.): Evidence for the anaerobic induction of messenger RNA. Biochem. Genet. 18: 681–692.

Finkelstein, R.R., K.M. Tenbarge, J.E. Shumway, M.L. Crouch, 1985, Role of ABA in maturation of rape seed embryos. Plant Physiol. 78: 630–636.

Finkelstein, R.R., A.J. DeLisle, A.E. Simon, M. Crouch, 1987, Role of abscisic acid and restricted water uptake during embryogeny in *Brassica*. In: J.E. Fox and M. Jacobs (eds.), Molecular Biology of Plant Growth Control. Alan R. Liss Inc., New York, pp. 73–84.

Finley, D., E. Oezkaynak, A. Varshavsky, 1987, The yeast polyubiquitin gene is essential for resistance to high temperatures, starvation and other stresses Cell 48: 1035–1946.

Fischer, W., R. Bergfeld, P. Schopfer, 1987, Induction of storage protein synthesis in embryos of mature plant seeds. Naturwissenschaften 74: 86–88.

Fleck, J., A. Durr, C. Fritsch, T. Vernet, L. Hirth, 1982, Osmotic-shock 'stress proteins' in protoplasts of *Nicotiana sylvestris*. Plant Sci. Letters 26: 159–165.

Fletcher, R.A., G. Hofstra, 1985, Triadimefon, a plant multiprotectant. Plant Cell Physiol. 26: 821–827.

Flores, H.E., A.W. Galston, 1982, Polyamines and plant stress: Activation of putrescine biosynthesis by osmotic shock. Science 217: 1259–1261.

Flores, H.E., A.W. Galston, 1984a, Osmotic stress-induced polyamine accumulation in cereal leaves. Plant Physiol. 75: 102–109.

Flores, H.E., A.W. Galston, 1984b, Osmotic stress-induced polyamine accumulation in cereal leaves. II. Relation to amino acid pools. Plant Physiol. 75: 110–113.

Flores, H.E., N.D. Young, A.W. Galston, 1985, Polyamine metabolism and plant stress. In: Key and Kosuge (eds.), 1985, pp. 93–114.

Foy, C.D., R.L. Chaney, M.C. White, 1978, The physiology of metal toxicity in plants. Annu. Rev. Plant Physiol. 29: 511–566.

Fransolet, S., R. Deltour, R. Bronchart, C. van de Walle, 1979, Changes in ultrastructure and transcription by elevated temperatures in *Zea mays* embryonic root cells. Planta 146: 7–18.

Freeling, M., 1973, Simultaneous induction by anaerobiosis or, 2,4-D of multiple enzymes specified by two unlinked genes: Differential Adh1—Adh2 expression in maize. Mol. Gen. Genet. 127: 215–227.

Freeling, M., D.C. Bennett, 1985, Maize Adh1. Annu. Rev. Genetics 19: 297–323.

Frosina, G., A. Abbondandolo, 1985, The current evidence for an adaptive response to alkylating agent in mammalian cells, with special reference to experiments with *in vitro* cell cultures. Mutat. Res. 154: 85–100.

FUHRER, J., 1982, Ethylene biosynthesis and cadmium toxicity in leaf tissue of beans (*Phaseolus vulgaris* L.). Plant Physiol. 70: 162–167.

GAAL, I. C., C. K. PEARSON, 1985, Eukaryotic nuclear ADP-ribosylation reactions. Biochem. J. 230: 1–18.

GAESTEL, M., B. GROSS, R. BENNDORF, M. STRAUSS, W.-H. SCHUNK, R. KRAFT, A. OTTO, H. BOEHM, J. STAHL, H. DRABASCH, H. BIELKA, 1989, Molecular cloning, sequencing and expression in *Escherichia coli* of the 25-kDa growth-related protein of Ehrlich ascites tumor and its homology to mammalian stress proteins. Eur. J. Biochem. 179: 209–213.

GALAU, G. A., D. W. HUGHES, L. DURE, 1986, Abscisic acid induction of cloned cotton late embryogenesis-abundant (Lea) mRNAs. Plant Mol. Biol. 7: 155–170.

GASSER, S. M., U. K. LAEMMLI, 1987, A glimpse at chromosmal order. Trends Genet. 3: 16–22.

GATES D. M., R. ALDERFER, E. TAYLOR, 1968, Leaf temperature of desert plants. Science 159: 994–995.

GAUSING, K., R. BARKARDOTTIR, 1986, Structure and expression of ubiquitin genes in higher plants. Eur. J. Biochem. 158: 57–62.

GEORGOPOULOS, C., V. TILLY, D. ANG, G. N. CHANDRASEKHAR, O. FAYET, J. SPENCE, T. ZIEGELHOFFER, K. LIBEREK, M. ZYLICZ, 1989, The role of the *Escherichia coli* heat shock proteins in bacteriophage lambda growth. In: PARDUE et al. (eds.), 1989, pp. 37–47.

GERLACH, W. L., A. J. PRYOR, E. S. DENNIS, R. J. FERL, M. M. SACHS, W. J. PEACOCK, 1982, cDNA cloning and induction of the alcohol dehydrogenase gene (Adh1) of maize. Proc. Natl. Acad. Sci. USA 79: 2981–2985.

GERNER, E. W., D. K. HOLMES, D. G. STICKNEY, J. A. NOTERMAN, D. J. M. FULLER, 1980, Enhancement of hyperthermia-induced cytotoxicity by polyamines. Cancer Res. 40: 432–438.

GERSHENZON, J., 1984, Changes in the levels of plant secondary metabolites under water and nutrient stress. Rec. Advanc. Phytochem. 18: 273–320.

GIEBEL L. B., B. P. DWORNICZAK, E. K. F. BAUTZ, 1988, Developmental regulation of a constitutively expressed mouse mRNA encoding a 72-kDa heat shock-like protein. Dev. Biol. 125: 200–207.

GILMOUR, S. J., R. K. HAJELA, M. F. THOMASHOW, 1988, Cold acclimation in *Arabidopsis thaliana.* Plant Physiol. 87: 745–750.

GINZBURG, C., R. SALOMON, 1986, The effect of dormancy on the heat shock response in *Gladiolus gandavensis* cormels. Plant Physiol. 81: 259–267.

GLACZINSKI, H., K. KLOPPSTECH, 1988, Temperature-dependent binding to the thylakoid membranes of nuclear coded chloroplast heat-shock proteins. Eur. J. Biochem. 173: 579–583.

GLAEVER, H. M., O. B. STYRVOLD, I. KAASEN, A. R. STROM, 1988, Biochemical and genetic characterization of osmoregulatory trehalose synthesis in *Escherichia coli.* J. Bacteriol. 170: 2841–2849.

GLASS, J. R., R. G. DEWITT, A. E. CRESS, 1985, Rapid loss of stress fibers in Chinese hamster ovary cells after hyperthermia. Cancer Res. 45: 258–262.

GLASS, D. J., R. J. POLVERE, L. H. T. VAN DER PLOEG, 1986, Conserved sequences and transcription of the hsp 70 gene family in *Trypanosoma brucei.* Mol. Cell. Biol. 6: 4657–4666.

GLOVER, C. V. C., 1982, Heat shock induces rapid dephosphorylation of a ribosomal protein in *Drosophila.* Proc. Natl. Acad. Sci. USA 79: 1781–1785.

GOFF , S. A., A. L. GOLDBERG, 1985, Production of abnormal proteins in *E. coli* stimulates transcription of lon and other heat shock genes. Cell 41: 587–595.

GOLDBERG, R. B., 1986, Regulation of plant gene expression. Phil. Trans. Royal Soc. Lnd. B 313: 343–353.

GOLOUBINOFF, P., A. A. GATENBY, G. H. LORIMER, 1989, GroE heat shock proteins promote assembly of foreign prokaryotic ribulose bisphosphate carboxylase oligomers in *Escherichia coli.* Nature 337: 44–47.

GOMEZ, J., D. SANCHEZ-MARTINEZ, V. STIEFEL, J. RIGAN, P. PUIG-DOMENECH, M. PAGES, 1988, A gene induced by the plant hormone abscisic acid in response to water stress encodes a glycine-rich protein. Nature 334: 262–264.

GOULD, A. R., N. P. EVERETT, T. L. WANG, H. E. STREET, 1981, Studies on the control of cell cycle in cultured plant cells. I. Effects of nutrient limitation and nutrient starvation. Protoplasma 106: 1–13.

GRAHAM, D., B. D. PATTERSON, 1982, Responses of plants to low, non-freezing temperatures: protein, metabolism and acclimation. Annu. Rev. Plant Physiol. 33: 347–372.

GRAHAM, J. S., G. PEARCE, J. MERRYWEATHER, K. TITANI, L. ERICSSON, C. A. RYAN, 1985a, Wound-induced proteinase inhibitors from tomato leaves. I. The cDNA-deduced primary structure of pre-inhibitor I and its post-translational processing. J. Biol. Chem. 260: 6555–6560.

GRAHAM, J., G. PEARCE, J. MERRYWEATHER, K. TITANI, L. ERICSSON, C. A. RYAN, 1985b, Wound-induced proteinase inhibitors from tomato leaves. II. The cDNA-deduced primary structure of pre-inhibitor II. J. Biol. Chem. 260: 6560–6564.

GRAHAM, R. W., D. JONES, E. P. M. CANDIDO, 1989, UbiA, the major polyubiquitin locus in *Caenorhabditis elegans,* has unusual structural features and is constitutively expressed. Mol. Cell. Biol. 9: 268–277.

GRANELL, A., J. M. BELLES, V. CONEJERO, 1987, Induction of pathogenesis-related proteins in tomato by citrus exocortis viroid, silver ion, and etephon. Physiol. Mol. Plant Pathol. 31: 83–89.

GREEN, T.R., C.A. RYAN, 1972, Wound-induced proteinase inhibitor in plant leaves: A Possible defense mechanism against insects. Science 175: 776–777.

GRIERSON, D., 1982, RNA processing and other posttranscriptional modifications. Encycl. Plant Physiol. New Series, Vol. 14B, Springer Verlag, Berlin, 1982, pp. 192–223.

GRILL, E., E.-L. WINNACKER, M. H. ZENK, 1985, Phytochelatins: The principal heavy-metal complexing peptides of higher plants. Science 230: 674–676.

GRILL, E., E.-L. WINNACKER, M. H. ZENK, 1987, Phytochelatins, a class of heavy-metal-binding peptides from plants, are functionally analogous to metallothioneins. Proc. Natl. Acad. Sci. USA 84: 439–443.

GRILL, E., J. THUMANN, E.-L. WINNACKER, M.H. ZENK, 1988, Induction of heavy-metal binding phytochelatins by inoculation of cell cultures in standard media. Plant Cell. Rep. 7: 375–378.

GRIMM, B., D. ISH-SHALOM, D. EVEN, H. GLACZINSKI, P. OTTERSBACH, I. OHAD, K. KLOPPSTECH, 1989, The nuclear-coded chloroplast 22-kDa heat-shock protein of *Chlamydomonas*. Evidence for translocation into the organelle without a processing step. Europ. J. Biochem. 182: 539–546.

GROSS, D., 1987, Chemische Abwehrstoffe der Pflanze. Biol. Rundschau 25: 225–237.

GROSSMAN, A. D., D. B. STRAUS, W. A. WALTER, C. A. GROSS, 1987, Sigma 32 synthesis can regulate the synthesis of heat shock proteins in *Escherichia coli*. Genes Devel. 1: 179–184.

GULIK, P., J. DVORAK, 1987, Gene induction and repression by salt treatment in roots of the salinity-sensitive Chinese spring wheat and salinity-tolerant Chinese spring x *Elytrigia elongata* amphiploid. Proc. Natl. Acad. Sci. USA 84: 99–103.

GUPTA, A. S., G. A. BERKOWITZ, 1988, Chloroplast osmotic adjustment and water stress effects on photosynthesis. Plant Physiol. 88: 200–206.

GURLEY W. B., E. CZARNECKA, R. T. NAGAO, J. L. KEY, 1986, Upstream sequences required for efficient expression of a soybean heat shock gene. Mol. Cell. Biol. 6: 559–565.

GUTTMAN, S. D., C. V. C. GLOVER, C. D. ALLIS, M. A. GOROVSKY, 1980, Heat shock, deciliation and release from anoxia induce the synthesis of the same set of polypetides in starved *T. pyriformis*. Cell 22: 299–307.

GUY, C. L., K. J. NIEMI, R. BRAMBL, 1985, Altered gene expression during cold acclimation of spinach. Proc. Natl. Acad. Sci. USA 82: 3673–3677.

GUY, C. L., N. PLESOFSKY-VIG, R. BRAMBL, 1986, Heat shock protects germinating conidiospores of *Neurospora crassa* against freezing injury. J. Bacteriol. 167: 124–129.

GWYNNE, D. I., B. P. BRANDHORST, 1982, Alterations in gene expression during heat shock of *Achlya ambisexualis*. J. Bacteriol. 149: 488–493.

HABEREDER, H., G. SCHROEDER, J. EBEL, 1989, Rapid induction of phenylalanine ammonia lyase and chalcone synthase mRNAs during fungus infection of soybean (*Glycine max* L.) roots or elicitor treatment of soybean cell cultures at the onset of phytoalexin synthesis. Planta 177: 58–65.

HACKET, R. W., J. T. LIS, 1983, Localization of the hsp83 transcript within a 3292 nucleotide-sequence from the 63B heat shock locus of *Drosophila melanogaster*. Nucl. Acids Res. 11: 7011–7030.

HADJIOLOV, A. A., 1985, The Nucleolus and Ribosome Biogenesis. Cell Biol. Monographs Vol. 12, Springer Verlag, Wien.

HADWIGER, L. A., W. WAGONER, 1983, Effect of heat shock on the mRNA-directed disease resistance response of peas. Plant Physiol. 72: 553–556.

HAGEMAN, R. H., D. FLESHER, 1960, The effect of anaerobic environment on the activity of alcohol dehydrogenase and other enzymes of corn seedlings. Arch. Biochem. Biophys. 87: 203–209.

HAGEN, G., N. UHRHAMMER, T. J. GUILFOYLE, 1988, Regulation of expression of an auxin-induced soybean sequence by cadmium. J. Biol. Chem. 263: 6442–6446.

HAGMANN, M. L., W. HELLER, H. GRISEBACH, 1984, Induction of phytoalexin synthesis in soybean. Stereospecific 3,9-dihydroxypterocarpan-6a-hydrolase from elicitor-induced soybean cell cultures. Eur. J. Biochem. 142: 127–131.

HAHN, G. M., G. C. LI, 1982, Thermotolerance and heat shock proteins in mammalian cells. Radiat. Res. 92: 452–457.

HAKE, S., P. M. KELLEY, W. C. TAYLOR, M. FREELING, 1985, Coordinate induction of alcohol dehydrogenase 1, aldolase, and other anaerobic RNAs in maize. J. Biol. Chem. 260: 5050–5054.

HALLBERG, R. L., K. W. KRAUS, R. C. FINDLY, 1984, Starved *Tetrahymena thermophila* cells that are unable to mount an effective heat shock response selectively degrade their rRNA. Mol. Cell. Biol. 4: 2170–2179.

HAMER, D. H., D. J. THIELE, J. E. LEMONTT, 1985, Function and autoregulation of yeast copperthionein. Science 228: 685–690.

HANDA, S., R. A. BRESSAN, A. K. HANDA, N. C. CARPITA, P. M. HASEGAWA, 1983, Solutes contributing to osmotic adjustment in cultured plant cells adapted to water stress. Plant Physiol. 73: 834–843.

HANDA, S., A. K. HANDA, P. M. HASEGAWA, R. A. BRESSAN, 1986, Proline accumulation and the adaptation of cultured plant cells to water stress. Plant Physiol. 80: 938–945.

HANNINK, M., D. J. DONOGHUE, 1985, Lysine residue 121 in the proposed ATP-binding site of the v-mos protein is required for transformation. Proc. Natl. Acad. Sci. USA 82: 7894–7898.

HANSON, A. D., W. D. HITZ, 1982, Metabolic responses of mesophytes to plant water deficits. Annu. Rev. Plant Physiol. 33: 163–203.

HANSON, A. D., J. V. JACOBSEN, 1984, Control of lactate dehydrogenase, lactate glycolysis, and α-amylase by O_2 deficit in barley aleurone layers. Plant Physiol. 75: 566–572.

HANSON, A. D., J. V. JACOBSEN, J. A. ZWAR, 1984, Regulated expression of three acohol dehydrogenase genes in barley aleurone layers. Plant Physiol. 75: 573–581.

HANSON, A. D., A. M. MAY, R. GRUMET, J. BODE, G. C. JAMIESON, D. RHODES, 1985, Betaine synthesis in chenopods: Localization in chloroplasts. Proc. Natl. Acad. Sci. USA 82: 3678–3682.

HARARI, A. D., M. E. TOME, E. W. GERNER, 1986, Heat shock induced polyamine oxidation in mammalian cells. J. Cell Biol. 103: 175a.

HARARI, P. M., D. J. FULLER, E. W. GERNER, 1989, Heat shock stimulates polyamine oxidation by two distinct mechanisms in mammalian cell cultures. Int. J. Radiat. Oncol. 16: 451–457.

HARGREAVES, J. A., 1981, Accumulation of phytoalexins in cotyledons of French bean (*Phaseolus vulgaris* L.) following treatment with triton (T-octylphenol polyethoxyethanol) surfactants. New Phytol. 87: 733–741.

HARRIS, M. J., W. H. OUTLAW, R. MERTENS, E. W. WEILER, 1988, Waterstress-induced changes in the abscisic acic content of guard cells and other cells of *Vicia faba* L. leaves as determined by enzyme-amplified immunoassay. Proc. Natl. Acad. Sci. USA 85: 2584–2588.

HARRY, D. E., C. S. KINLAW, R. R. SEDEROFF, 1988, The anaerobic stress response and its use for studying gene expression in conifers. In: J. W. HANOVER and D. E. KEATHLEY (eds.), Genetic Manipulation of Woody Plants. Plenum Press, New York, pp. 275–290

HASSAN, N. S., D. A. WILKINS, 1988, *In vitro* selection for salt tolerant lines in *Lydopersicon peruvianum.* Plant Cell Rep. 7: 463–466

HAVAUX, M., O. CANAANI, S. MALKIN, 1987a, Rapid screening for heat tolerance in *Phaseolus* species using the photoacoustic technique. Plant Sci. 48: 143–149.

HAVAUX, M., O. CANAANI, S. MALKIN, 1987b, Oxygen uptake by tobacco leaves after heat shock. Plant Cell Environ. 10: 677–684.

HAZEN, B. E., W. R. BUSHNELL, 1983, Inhibition of the hypersensitive reaction in barley *Hordeum vulgare* to powdery mildew by heat shock and cytochalasin B. Physiol. Plant Pathol. 23: 421–438.

HEATH, M. C., 1979, Effects of heat shock, actinomycin D, cycloheximide and blasticidin S on nonhost interactions with rust fungi. Physiol. Plant Pathol. 15: 211–218.

HEDRICK, S. A., J. N. BELL, T. BOLLER, C. J. LAMB, 1988, Chitinase cDNA cloning and mRNA induction by fungal elicitor, wounding and infection. Plant Physiol. 182–186.

HEIKKILA, J. J., J. E. T. PAPP, G. A. SCHULTZ, D. BEWLEY, 1984, Induction of heat shock protein messenger RNA in maize mesocotyls by water stress, abscisic acid and wounding. Plant Physiol. 76: 270–274.

HEINDORFF, K., R. RIEGER, I. SCHUBERT, A. MICHAELIS, O. AURICH, 1987a, Clastogenic adaptation of plant cells—Reduction of the yield of clastogen-induced chromatid aberrations by various pretreatment procedures. Mutat. Res. 181: 157–171.

HEINDORFF, K. R., R. RIEGER, A. MICHAELIS, S. TAKEHISA, 1987b, Clastogenic adaptation triggered by S-phase-independent clastogens in *Vicia faba.* Mutat. Res. 190: 131–135.

HEINDORF, K., I. SCHUBERT, R. RIEGER, A. MICHAELIS, 1987c, Clastogenic adaptation of *Vicia faba* root-tip meristem cells after consecutive treatments with S-phase dependent and S-phase independent agents. Biol. Zent. bl. 106: 439–448.

HEINE, U., L. SVERAK, J. KONDRATICK, R. A. BONAR, 1971, The behaviour of HeLa-S3 cells under the influence of supranormal temperatures. J. Ultrastruct. Res. 34: 375–396.

HEMMINGSEN, S. M., R. J. ELLIS, 1986, Purification and properties of ribulosebisphosphate carboxylase large subunit binding protein. Plant Physiol. 80: 269–276.

HEMMINGSEN, S. M., C. WOOLFORD, S. M. VAN DER VIES, K. TILLY, D. T. DENNIS, C. P. GEORGOPOULOS, R. W. HENDRIX, R. J. ELLIS, 1988, Homologous plant and bacterial proteins chaperone oligomeric protein assembly. Nature 333: 330–335.

HENDERSHOT, L. M., J. TING, A. S. LEE, 1988, Identity of the immunoglobulin heavy-chain-binding protein with the 78,000 dalton glucose-regulated protein and the role of post-translational modifications in its binding function. Mol. Cell. Biol. 8: 4250–4256. [1844–1851.

HENLE, K. J., L. A. DETHLEFSEN, 1978, Heat fractionation and thermotolerance. A review. Cancer Res. 38:

HERMOLIN, J., A. M. ZIMMERMAN, 1976, RNA synthesis in division synchronized *Tetrahymena:* Macronuclear and cytoplasmic RNA. J. Protozool. 23: 594–600.

HERR, W., R. A. STURM, R. G. CLERC, L. M. CORCORAN, D. BALTIMORE, P. A. SHARP, H. A. INGRAHAM, M. G. ROSENFELD, M. FINNEY, G. RUVKUN, H. R. HORVITZ, 1988, The POU domain: a large conserved region in the mammalian pit-1, oct-1, oct-2, and *Caenorhabditis elegans* nuc-86 gene products. Genes Devel. 2: 1513–1516.

HERRMANN, R. G., J. FEIERABEND, 1980, The presence of DNA in ribosome-deficient plastids of heat-bleached rye leaves. Eur. J. Biochem. 104: 603–609.
HERRMANN, A., W. SCHULZ, K. HAHLBROCK, 1988, Two alleles of the single-copy chalcone synthase gene in parsley differ by a transposon-like element. Mol. Gen. Genet. 212: 93–98.
HERRMANN, G., J. LEHMANN, A. PETERSON, G. SEMBDNER, R. A. WEIDHASE, B. PARTHIER, 1989, Species and tissue specificity of jasmonate-induced abundant proteins. J. Plant Physiol. 134: 703–709.
HEUSS-LAROSA, K., R. R. MAYER, J. H. CHERRY, 1987, Synthesis of only two heat shock proteins is required for thermoadaptation in cultured cowpea cells. Plant Physiol. 85: 4–7.
HICKEY, E., S. E. BRANDON, S. SADIS, G. SMALE, L. A. WEBER, 1986, Molecular cloning of sequences encoding the human heat-shock proteins and their expression during hyperthermia. Gene 43: 147–154.
HIGHTOWER, L. E., F. P. WHITE, 1981, Cellular response to stress: Comparison of a family of 71–73 kilodalton proteins rapidly synthesized in rat tissue slices and canavanine-treated cells in culture. J. Cell. Physiol. 108: 261–275.
HIGO, H., K. HIGO, Y. SATOW, J.-Y. LEE, 1987, Induction of an hsp 70-like protein during organ culture of rat embryonic heart. Biochem. Int. 15: 727–734.
HÖFNER, R., L. VAZQUEZ-MORENO, K. WINTER, H. J. BOHNERT, J. M. SCHMITT, 1987, Induction of Crassulacean acid metabolism in *Mesembryanthemum crystallinum* by high salinity: Mass increase and de novo synthesis of PEP-carboxylase. Plant Physiol. 83: 915–919.
HOFFMAN, E. P., V. G. CORCES, 1984, Correct temperature induction and developmental regulation of a cloned heat shock gene transformed into *Drosophila* germ line. Mol. Cell. Biol. 4: 2883–2889.
HOGSETT, W. E., R. M. RABA, D. TINGEY, 1981, Biosynthesis of stress ethylene in soybean seedlings: similarities to endogenous ethylene biosynthesis. Physiol. Plant. 53: 307–314.
HOLTUM, J. A. M., K. WINTER, 1982, Activity of enzymes of carbon metabolism during the induction of Crassulacean acid metabolism in *Mesembryanthemum crystallinum* L. Planta 155: 8–16.
HONG, B., S. J. UKNES, T. D. HO, 1988, Cloning and characterization of a cDNA encoding a mRNA rapidly induced by ABA in barley aleurone layers. Plant Mol. Biol. 11: 495–506.
HONG-QI, Z., A. F. CROES, H. F. LINSKENS, 1984, Qualitative changes in protein synthesis in germinating pollen of *Lilium longiflorum* after heat shock. Plant Cell Environm. 7: 689–691.
HORSFALL, J. C., E. B. COWLING (eds.), 1977–1980, Plant Disease: An Advanced Treatise. 5 volumes. Acad. Press, Orlando.
HOSOKAWA, D., Y. OHASHI, 1988, Immunochemical localization of pathogenesis-related proteins secreted into the intercellular spaces of salicylate-treated tobacco leaves. Plant Cell Physiol. 29: 1035–1040.
HOWARD, E. A., J. C. WALKER, E. S. DENNIS, W. J. PEACOCK, 1987, Regulated expression of an alcohol dehydrogenase 1 chimeric gene introduced into maize protoplasts. Planta 170: 535–540.
HOWE, J. G., J. W. B. HERSHEY, 1984, Translation initiation factor and ribosome association with the cytoskeletal frame work fraction from HeLa cells. Cell 37: 85–93.
HSIAO, T. C., 1970, Papid changes in levels of polyribosomes in *Zea mays* in response to water stress. Plant Physiol. 46: 281–285
HSIAO, T. C., 1973, Plant responses to water stress. Annu. Rev. Plant Physiol. 24: 519–570.
HUDSPETH, R. L., C. A. GLACKIN, J. BONNER, J. W. GRULA, 1986, Genomic and cDNA clones for maize phosphoenolpyruvate carboxylase and pyruvate-orthophosphate dikinase: Expression of different gene-family members in leaves and roots. Proc. Natl. Acad. Sci. USA 83: 2884–2888.
HUGHES, E. N., J. T. AUGUST, 1982, Coprecipitation of heat shock proteins with a cell surface glycoprotein. Proc. Natl. Acad. Sci. USA 79: 2305–2309.
HUGHES, D. W., G. A. GALAU, 1987, Translation efficiency of Lea mRNAs in cotton embryos: minor changes during embryogenesis and germination. Plant Mol. Biol. 9: 301–313.
HULTMARK, D., R. KLEMENZ, W. J. GEHRING, 1986, Translational and transcriptional control elements in the untranslated leader of the heat-shock gene hsp. 22. Cell 44: 429–438.
HUNT, C., R. J. MORIMOTO, 1985, Conserved features of eukaryotic hsp 70 genes revealed by comparison with the nucleotide sequence of human hsp 70. Proc. Natl. Acad. Sci. USA 82: 6455–6459.
HURKMAN, W. J., C. K. TANAKA, 1987, The effect of salt on the pattern of protein synthesis in barley roots. Plant Physiol. 83: 517–524.
IIDA, H., I. YAHARA, 1984, Durable synthesis of high molecular weight heat shock proteins in G_O cells of the yeast and other eukaryotes. J. Cell. Biol. 99: 199–207.
IIDA, K., H. IIDA, I. YAHARA, 1986, Heat shock induction of intranuclear actin rods in cultured mammalian cells. Exp. Cell Res. 165: 207–215.
IMAZEKI, H., 1985, Hormonal control of wound-induced responses. Encycl. Plant Physiol., New Ser. Vol 11 (III), Springer Verlag, Berlin, pp. 485–517.
IMPERIALE, M. J., H.-T. KAO, L. T. FELDMAN, J. R. NEVINS, S. STRICKLAND, 1984, Common control of the heat shock gene and early adenovirus genes: Evidence for a cellular E1A-like activity. Mol. Cell. Biol. 4: 867–874.

INGOLIA, T. D., E. A. CRAIG, B. J. MCCARTHY, 1980, Sequence of three copies of the gene for the major *Drosophila* heat shock induced protein and their flanking regions. Cell 21: 669–679.

ISHIZUKA, M., T. SATO, A. WATANABE, H. IMAZEKI, 1981, Alteration of coding properties of polysome-associated messenger RNA in potato tuber slices during ageing. Plant Physiol. 68: 154–157.

JACKSON, R. J., 1986, The heat-shock response in *Drosophila* KC 161 cells. mRNA competition is the main explanation for reduction of normal protein synthesis. Eur. J. Biochem. 158: 623–634.

JACKSON, P. J., C. M. NARANJO, P. R. MC CLURE, E. J. ROTH, 1985, The molecular response of cadmium resistant *Datura innoxia* cells to heavy metal stress. In: KEY and KOSUGE (eds.), 1985, pp. 145–160.

JACKSON, P. J., C. J. UNKEFER, J. A. DOOLEN, K. WATT, N. J. ROBINSON, 1987, Poly(gamma-glutamylcysteinyl)glycin: Its role in cadmium resistance in plant cells. Proc. Natl. Acad. Sci. USA 84: 6619–6623.

JAFFE, M. J., F. W. TELEWSKI, 1984, Thigmomorphogenesis: Callose and ethylene in the hardening of mechanically stressed plants. Rec. Advanc. Phytochem. 18: 79–96.

JAHNEN, W., K. HAHLBROCK, 1988a, Cellular localization of nonhost resistance reactions of parsley *(Petroselinum crispum)* to fungal infection. Planta 173: 197–204.

JAHNEN, W., K. HAHLBROCK, 1988b, Differential regulation and tissue-specific distribution of enzymes related to phenylpropanoid branch pathways in developing parsley seedlings. Planta 173: 453–458.

JAMET, E., B. FRITIG, 1986, Purification and characterization of 8 of the pathogenesis-related proteins in tobacco leaves reacting hypersensitively to tobacco mosaic virus. Plant Mol. Biol. 6: 69–80.

JENKINS, D. E., J. E. SCHULTZ, A. MATIN, 1988, Starvation-induced cross protection against heat or H_2O_2 challenge in *Escherichia coli.* J. Bacteriol. 170: 3910–3914.

JEROME, S. M. R., K. O. MÜLLER, 1958, Studies on phytoalexins. II. Influence of temperature on resistance of *Phaseolus vulgaris* towards *Sclerotinia fructicola* with reference to phytoalexin output. Austral. J. Biol. Sci. 11: 301–314.

JINDAL, S., A. K. DUDANI, B. SINGH, C. B. HARLEY, R. S. GUPTA, 1989, Primary structure of a human mitochondrial protein homologous to the bacterial and plant chaperonins and to the 65-kd mycobacterial antigen. Mol. Cell. Biol. 9: 2279–2283.

JOHNSON, C., G. N. CHANDRASEKHAR, C. GEORGOPOULOS, 1989, *Escherichia coli* DnaK and GrpE heat shock proteins interact both *in vivo* and *in vitro.* J. Bacteriol. 171: 1590–1596.

JOHNSON-FLANAGAN, A. M., J. SINGH, 1986, Membrane deletion during plasmolysis in hardened and non-hardened plant cells. Plant Cell Environ. 9: 299–305.

JOHNSON-FLANAGAN, A. M., J. SINGH, 1987, Alteration of gene expression during the induction of freezing tolerance in *Brassica napus* suspension cultures. Plant Physiol. 85: 699–705.

JOHNSTON, G. C., R. A. SINGER, 1980, Ribosomal precursor RNA metabolism and cell division in the yeast *Saccharomyces cerevisiae.* Mol. Gen. Genet. 178: 357–360.

JOHNSTON, R. N., B. L. KUCEY, 1988, Competitive inhibition of hsp70 gene expression causes thermosensitivity. Science 242: 1551–1554.

JONES, R. G. W., 1984, Phytochemical aspects of osmotic adaptation. Rec. Advanc. Phytochem. 18: 55–78.

JONES, D., R. H. RUSSNAK, R. J. KAY, E. P. M. CANDIDO, 1986, Structure, expression and evolution of a heat shock gene locus in *Caenorhabditis elegans* that is flanked by repetitive elements. J. Biol. Chem. 261: 12006–12015.

KAHL, G., 1978, Biochemistry of Wounded Plant. de Gruyter, Berlin.

KAHL, G., M. FURUTA, K. OBA, I. URITANI, 1982, Wounding-induced enhancement of the activity of chromatin-bound DNA-dependent RNA polymerases in sweet potato root. Agric. Biol. Chem. 46: 2457–2463.

KAUR-SAWHNEY, R., L. SHIH, T. CEGIELSKA, A. W. GALSTON, 1982, Inhibition of protease activity by polyamines. Relevance for control of leaf senescence. FEBS Letters 145: 345–349.

KAINA, B., 1982, Enhanced survival and reduced mutation and aberration frequencies induced in V79 Chinese hamster cells preexposed to low levels of methylating agents. Mutat. Res. 93: 195–211.

KANEKO, M., S. A. LEADON, T. ITO, 1988, Relationship between the induction of mitotic gene conversion and the formation of thymine glycols in yeast *S. cerevisiae* treated with hydrogen peroxide. Mutat. Res. 207: 17–22.

KAO, C. H., S. F. YANG, 1983, Role of ethylene in the senescence of detached rice leaves. Plant Physiol. 73: 881–885.

KAPISZEWSKA, M., L. E. HOOPWOOD, 1986, Changes in bleb-formation following hyperthermia treatment of CHO-cells. Radiat. Res. 105: 405–412.

KAPPEN, L., O. L. LANGE, 1968, Die Hitzeresistenz ausgetrockneter Blätter von *Commelina africana.* Ein Vergleich zwischen zwei Untersuchungsmethoden. Protoplasma 65: 119–132.

KARCH, F., I. TOEROEK, A. TISSIERES, 1981, Extensive regions of homology in front of the two hsp70 heat shock variant genes in *Drosophila melanogaster.* J. Mol. Biol. 148: 219–230.

KARIN, M, 1985, Metallothioneins: Proteins in search of function. Cell 41: 9–10.

KASSANIS, B., 1957, Effects of changing temperature on plant virus deseases. Advanc. Virus Res. 4: 221–241.

KASSENBROCK, C. K., P. D. GARCIA, P. WALTER, R. B. KELLY, 1988, Heavy-chain binding protein recognizes aberrant polypeptides translocated *in vitro.* Nature 333: 90–93.

KAUR-SAWHNEY, R., A.W.GALSTON, 1979, Interaction of polyamines and light on biochemical processes involved in leaf senescence. Plant Cell Environm. 2: 189–196.

KAY, R.J., R.J.BOISSY, R.H.RUSSNAK, E.P.M.CANDIDO, 1986, Efficient transcription of a *Caenorhabditis elegans* heat shock gene pair in mouse fibroblasts is dependent on multiple promoter elements which can function bidirectionally. Mol. Cell. Biol. 6: 3134–3143.

KAY, R.J., R.H.RUSSNAK, D.JONES, C.MATHIAS, E.P.M.CANDIDO, 1987, Expression of intron-containing *C. elegans* heat shock genes in mouse cells demonstrates divergence of 3' splice site recognition sequence between nematodes and vertebrates, and an inhibitory effect of heat shock on the mammalian splicing apparatus. Nucl. Acids Res. 15: 3723–3741.

KEE, S.C., P.S.NOBEL, 1986, Concomitant changes in high temperature tolerance and heat-shock proteins in desert succulents. Plant Phyisol. 80: 596–598.

KEEGSTRA, K., 1989, Transport and routing of proteins into chloroplasts. Cell 56: 247–253.

KEIL, M., J. SANCHEZ-SERRANO, J. SCHELL, L. WILLMITZER, 1986, Primary structure of a proteinase inhibitor II gene from potato *(Solanum tuberosum)*. Nucl. Acids Res. 14: 5641–5650.

KELLER, B., N.SAUER, C.J.LAMB, 1988, Glycine-rich cell wall proteins in bean: gene structure and association of the protein with the vascular system. EMBO J. 7: 3625–3633.

KELLER, B., M.D.TEMPLETON, C.J.LAMB, 1989, Specific localization of a plant cell wall glycine-rich protein in protoxylem cells of the vascular system. Proc. Natl Acad. Sci. USA 86: 1529–1533.

KELLEY, P.M., M.FREELING, 1982, A preliminary comparison of maize anaerobic and heat shock proteins. In: SCHLESINGER et al. (eds.), 1982, pp. 315–319.

KELLEY, P. M., M. FREELING, 1984a, Anaerobic expression of maize glucose phophate isomerase I. J. Biol. Chem. 259: 673–677.

KELLEY, P. M., M. FREELING, 1984b, Anaerobic expression of maize fructose 1,6-diphosphate aldolase. J. Biol. Chem. 259: 14180–14183.

KEMBLE, A.R., H.T.MACPHERSON, 1954, Liberation of amino acids in perennial rye grass during wilting. Biochem. J. 58: 46–49.

KEY, J.L., 1987, Auxin-regulated gene expression in a historical perspective and current status. In: J.E.Fox and M.JACOBS (eds.), Molecular Biology of Plant Growth Control. Alan R. Liss Inc., New York, pp. 1–21.

KEY, J.L., T.KOSUGE (eds.), 1985, Cellular and Molecular Biology of Plant Stress. Alan R.Liss, Inc., New York.

KEY, J. L., C. Y. LIN, Y. M. CHEN, 1981, Heat shock proteins of higher plants. Proc. Natl. Acad. Sci. USA 78: 3526–3530.

KEY, J. L., E. CZARNECKA, C.-Y. LIN, J. KIMPEL, C. MOTHERSHED, F. SCHÖFFEL, 1983, A comparative analysis of the heat shock response in crop plants. In: D. D. RANDALL, D. G. BLEVINS, R. L. LARSON, B. J. RAPP (eds.), Current Topics in Plant Biochemistry and Physiology. University of Missouri Press, Columbia, pp. 107–118.

KEY, J. L., J. A. KIMPEL, C. Y. LIN, R. T. NAGAO, E. VIERLING, E. CZARNECKA, W. B. GURLEY, J. K. ROBERTS, M. A. MANSFIELD, L. EDELMAN, 1985a, The heat shock response in soybean. In: KEY and KOSUGE (eds.), 1985, pp. 161–179.

KEY, J.L., J.KIMPEL, E.VIERLING, C.-Y.LIN, R.T.NAGAO, C.CZARNECKA, F.SCHOEFL, 1985b, Physiological and molecular analyses of the heat shock response in plants. In: ATKINSON and WALDEN (eds.), 1985, pp. 327–348.

KHOKHLOVA, V. A., S. A. PORFIROVA, A. B. FEDINA, E A. BURKHANOVA, 1987, Effect of heat shock on cell ultrastructure and protein synthesis in pumpkin roots. Fiziol. Rast 34: 869–878.

KIM, C.H., J.R.WARNER, 1983, Mild temperature shock alters transcription of a discrete class of *Saccharomyces cerevisiae* genes. Mol. Cell. Biol. 3: 457–465.

KIMPEL, J.A., J.L.KEY, 1985a, Heat shock in plants. Trends Biochem. Sci. 10: 353–357.

KIMPEL, J.A., J.L.KEY, 1985b, Presence of heat shock messenger RNA species in field grown soybeans *Glycine max.* Plant Physiol. 79 672–678.

KIRK, D.L., M.M.KIRK, 1986, Heat shock elicits production of sexual inducer in *Volvox.* Science 231: 51–54.

KISHORE, R., K. C. UPADYAYA, 1988, Heat shock proteins of pigeon pea *(Cajanus cajan).* Plant Cell Physiol. 29: 517–521.

KISTLER, H.C., H.D.VAN ETTEN, 1984, Regulation of pisatin demethylation in *Nectria haematococca* and its influence on pisatin tolerance and virulence. J. Gen. Microbiol. 130: 2605–2613.

KLEE, H.J., R.B.HORSCH, M.A.HINCHEE, M.B.HEIN, N.L.HOFFMANN, 1987, The effect of overproduction of two *Agrobacterium tumefaciens* T-DNA auxin biosynthetic gene products in transgenic petunia plants. Genes Devel. 1: 86–96.

KLEIN, J. D., I. B. FERGUSON, 1987, Effect of high temperature on calcium uptake by suspension-cultured pear fruit cells. Plant Physiol. 84: 153–156.

KLEINSCHMIDT, J.A., A.SEITER, 1988, Identification of domains involved in nuclear uptake and histone binding of protein N1 of *Xenopus leavis.* EMBO J. 7: 1605–1614.

KLOETZEL, P., E.K.F.BAUTZ, 1983, Heat-shock proteins are associated with hnRNP in *Drosophila melanogaster* tissue culture cells. EMBO J. 2: 705–710.

KLOPPSTECH, K., I. OHAD, 1986, Heat-shock protein synthesis in *Chlamydomonas reinhardii.* Translational control at the level of initiation of a poly(A)-rich-RNA coded 22-kDa protein in a cell-free system. Eur. J. Biochem. 154: 63–68.

KLOPPSTECH, K., G. MEYER, G. SCHUSTER, I. OHAD, 1985, Synthesis, transport and localization of a nuclear coded 22-kd heat-shock protein in the chloroplast membranes of peas and *Chlamydomonas reinhardii.* EMBO J 4: 1901–1909.

KLOPPSTECH, K., I. OHAD, H.-G. SCHWEIGER, 1986, Evidence for an extranuclear coding site for a heat-shock protein in *Acetabularia.* Eur. J. Cell Biol. 42: 239–245.

KLOPPSTECH, K., H. LORBERBOUM, N. DEGROOT, A. A. HOCHBERG, 1987, Translational control at the level of initiation between mRNAs for pre-existing proteins and a 22-kDa heat-shock protein of *Chlamydomonas* by small cytosolic RNAs. Eur. J. Biochem. 167: 501–505.

KLOTZ, F., W. J. HORST, 1988, Genotypic differences in aluminum tolerance of soybean (*Glycine max* L.) as affected by ammonium and nitrate nitrogen nutrition. J. Plant Physiol. 132: 702–707.

KLUGE, M., I. P. TING, 1978, Crassulacean Acid Metabolism. Analysis of an Ecological Adaptation. Springer Verlag, Berlin.

KNOGGE, W., E. KOMBRINK, E. SCHMELZER, K. HAHLBROCK, 1987, Occurrence of phytoalexins and other putative defense-related substances in infected parsley plants. Mol. Gen. Genet. 171: 279–287.

KODAMA, O., T. SUZUKI, J. MIYAKAWA, T. AKATSUKA, 1988, Ultraviolet-induced accumulation of phytoalexins in rice leaves. Agric. Biol. Chem. 52: 2469–2473.

KÖHLE, H., D. H. YOUNG, H. KAUSS, 1984, Physiological changes in suspension-cultured soybean cells elicited by treatment with chitosan. Plant Sci. Letters 33: 221–230.

KÖHLE, H., W. JEHLICK, F. POTEN, W. BLASCHEK, H. KAUSS, 1985, Chitosan-elicited callose synthesis in soybean cells as a Ca^{2+}-dependent process. Plant Physiol. 77: 544–551.

KOLATTUKUDY, P. E., C. L. SOLIDAY, 1985, Effects of stress on the defense barriers of plants. In: KEY and KOSUGE (eds.), 1985, pp. 381–400.

KOMBRINK, E., M. SCHROEDER, K. HAHLBROCK, 1988, Several 'pathogenesis-related' proteins in potato are 1,3-β-glucanases and chitinases. Proc. Natl. Acad. Sci. USA 85: 782–786.

KOSUGE, T., E. W. NESTER (eds.), 1987, Plant-Microbe Interactions, Molecular and Genetic Perspectives. MacMillan, New York.

KOZAK, M., 1988, Leader length and secondary structure modulate mRNA function under conditions of stress. Mol. Cell. Biol. 8: 2737–2744.

KRAUS, K. W., P. J. GOOD, R. L. HALLBERG, 1987, A heat shock-induced, polymerase III-transcribed RNA selectively associates with polysomal ribosomes in *Tetrahymena thermophila.* Proc. Natl. Acad. Sci. USA 84: 383–387.

KRISHNASAMY, S., R. MANNAR-MANNAN, M. KRISHNAN, A. GNANAM, 1988, Heat shock response of the chloroplast genome in *Vigna sinensis.* J. Biol. Chem. 263: 5104–5109.

KRONE, P. H., J. HEIKKILA, 1987, Analysis of hsp30, hsp70 and ubiquitin gene expression in *Xenopus laevis* tadpoles. Development 103: 59–67.

KRÜGER, C., B.-J. BENECKE, 1981, *In vitro* translation of *Drosophila* heat-shock and non-heat-shock mRNAs in heterologous and homologous cell-free systems. Cell 23: 595–603.

KUC, J., 1983, Induced systemic resistance in plants to diseases caused by fungi and bacteria. In: J. A. BAILEY and B. J. DEVERALL (eds.), The Dynamic of Host Defence. Acad. Press, Sydney, pp. 191–221.

KUHN, D. N., J. CHAPPELL, A. BOUDET, K. HAHLBROCK, 1984, Induction of phenylalanine ammonium lyase and 4-coumarate: CoA ligase mRNAs in cultured plant cells by UV light or fungal elicitor. Proc. Natl. Acad. Sci. USA 81: 1102–1106.

KULAEVA, O. N., E. A. BURKHANOVA, A. B. FEDINA, N. V. DANILOVA, G. ADAM, H. M. VORBRODT, V. A. KHRIPACH, 1989, Brassinosteroids in the regulation of protein synthesis of wheat leaves (russ.). Fiziol. Rast., in press.

KULOMAA, M. S., N. L. WEIGEL, O. A. KLEINSEK, W. G. BEATTIE, O. M. CONNELLY, C. MARCH, T. JARUCKI-SCHULZ, W. T. SCHRADER, B. W. O'MALLEY, 1986, Amino acid sequence of a chicken heat shock protein derived from the complementary DNA nucleotide sequence. Biochemistry 25: 6244–6252.

KURKELA, S., M. FRANCK, P. HEINO, V. LANG, E. T. PALVA, 1988, Cold induced gene expression in *Arabidopsis thaliana* L.. Plant Cell Rep. 7: 495–498.

KURTZ, S., J. ROSSI, L. PETKO, S. LINDQUIST, 1986, An ancient developmental induction: Heat-shock proteins induced in sporulation and oogenesis. Science 231: 1154–1157.

LALK, J., K. DÖRFFLING, 1985, Hardening, abscisic acid, proline and freezing resistance in two winter wheat varieties. Physiol. Plant. 63: 287–292.

LAMB, C. J., M. A. LAWTON, M. DRON, R. A. DIXON, 1989, Signals and transduction mechanisms for activation of plant defenses against microbial attack. Cell 56: 215–224.

LAMPORT, D. T. A., D. H. NORTHCOTE, 1960, Hydroxyproline in primary cell walls of higher plants. Nature 188: 665–666.

Lanford, R.E., P.Kanda, R.C.Kennedy, 1986, Induction of nuclear transport with a synthetic peptide homologous to the SV40 T antigen transport signal. Cell 46: 575–582.

Lange, O.L., 1955, Untersuchungen ueber die Hitzeresistenz der Moose in Beziehung zu ihrer Verbreitung I. Die Resistenz stark ausgetrockneter Moose. Flora 142: 381–399.

Lange, O.L., 1961, Die Hitzeresistenz einheimischer immer- und wintergrüner Pflanzen im Jahresverlauf. Planta 56: 666–683.

Lange, O.L., B.Schwemmle, 1960, Untersuchungen zur Hitzeresistenz vegetativer und blühender Pflanzen von *Kalanchoë blossfeldiana.* Planta 55: 208–225.

Langridge, J., 1963, Biochemical aspects of temperature response. Annu. Rev. Plant Physiol. 14: 441–462.

Lanks, K.W., 1986, Modulators of the enkaryotic heat shock response. Exper. Cell Res. 165: 1–10.

Laszlo, A., P.St. Lawrence, 1983, Parallel induction and synthesis of PDC and ADH in anoxic maize roots. Mol. Gen. Genet. 192: 110–117.

Laval, F., J.Laval, 1984, Adaptive response in mammalian cells: Cross-reactivity of different pretreatments on cytotoxicity as contrasted to mutagenicity. Proc. Natl. Acad. Sci. USA 81: 1062–1066.

Lawson, R., R. Mestril, P. Schiller, R. Voellmy, 1984, Expression of heat shock-beta-galactosidase hybrid genes in cultured *Drosophila* cells. Mol. Gen. Genet. 198: 116–124.

Lawton, M.A., C.J.Lamb, 1987, Transcriptional activation of plant defense genes by fungal elicitor, wounding and infection. Mol. Cell. Biol. 7: 335–341.

Lee, Ph. C., B. R. Bochner, B. N. Ames, 1983, ApppppA, heat-shock stress, and cell oxidation. Proc. Natl. Acad. Sci. USA 80: 7496–7500.

Lee, H., J.A. Simon, J.T. Lis, 1988, Structure and expression of ubiquitin genes of *Drosophila melanogaster.* Mol. Cell. Biol. 8: 4727–4735.

Legrand, M., S.Kaufmann, P.Geoffroy, B.Fritig, 1987, Biological function of pathogenesis-related proteins: Four tobacco pathogenesis-related proteins are chitinases. Proc. Natl. Acad. Sci. USA 84: 6750–6754.

Leicht, B.G., H.Biessmann, K.B.Palter, J.J.Bonner, 1986, Small heat shock proteins of *Drosophila* associate with the cytoskeleton. Proc. Natl. Acad. Sci. USA 83: 90–94.

LeJohn, H.B., E.E. Braithwaite, 1984, Heat and nutritional shock-induced proteins of the fungus *Achlya* are different and under independent transcriptional control. Can. J. Biochem. Cell Biol. 62: 837–846.

LeRudulier, D., R. C. Valentine, 1982, Genetic engineering in agriculture: Osmoregulation. Trends Biochem. Sci. 7: 431–433.

LeRudulier, D., A.R.Strom, A.M.Dandekar, L.Smith, R.C.Valentine, 1984, Molecular biology of osmoregulation. Science 224: 1064–1068.

Leshem, Y.Y., 1984, Interaction of cytokinins with lipid-associated oxygen free radicals during senescence: a prospective mode of cytokinin action. Canad. J. Bot. 62: 2943–2949.

Levinger, L., A. Varshavsky, 1981, Heat-shock proteins of *Drosophila* are associated with nuclease-resistant, high-salt resistant nuclear structures. J. Cell Biol. 90: 793–796.

Levitt, J., 1958, Frost, Drought, and Heat Resistance. Protoplasmatologia, Vol. VIII/6, Springer, Wien.

Levitt, J., 1980a, Responses of Plants to Environmental Stresses. Vol. I, Chilling, Freezing and High Temperature Stresses. Acad. Press, New York.

Levitt, J., 1980b, Responses of Plants to Environmental Stresses. Vol. II Water, Radiation, Salt, and Other Stresses. Academic Press, New York.

Lewis, M.J., H.R.B.Pelham, 1985, Involvement of ATP in the nuclear and nucleolar functions of the 70 kd heat shock protein. EMBO J. 4: 3137–3143.

Liberek, K., C.Georgopoulos, M.Zylicz, 1988, Role of the *Escherichia coli* DnaK and DnaJ heat shock proteins in the initiation of bacteriophage lambda DNA replication. Proc. Natl. Acad. Sci. USA 85: 6632–6636.

Lieberman, M., 1979, Biosynthesis and action of ethylene. Annu. Rev. Plant Physiol. 30: 533–591.

Lin, L.-S., T.D.Ho, 1986, Mode of action of abscisic acid in barley aleurone layers. Induction of new proteins by abscisic acid. Plant Physiol. 82: 289–297.

Lin, C.Y., J.L.Key, 1967, Dissociation and reassembly of polyribosomes in relation to protein synthesis in the soybean root. J. Mol. Biol. 26: 237–247.

Lin, C.-Y., J.K.Roberts, J.L.Key, 1984, Acquisition of thermotolerance in soybean seedlings. Plant Physiol. 74: 152–160.

Lin, C.-Y., Y.-M. Chen, J.L.Key, 1985, Solute leakage in soybean seedlings under various heat shock regimes. Plant Cell Physiol. 26: 1493–1498.

Lincoln, J.E., S.Cordes, E.Read, R.L.Fischer, 1987, Regulation of gene expression by ethylene during *Lycopersicon esculentum* (tomato) fruit development. Proc. Natl. Acad. Sci. USA 84: 2793–2797.

Lindquist, S., 1981, Regulation of protein synthesis during heat shock. Nature 293: 311–314.

Lindquist, S., 1986, The heat-shock response (review). Annu. Rev. Biochem. 55: 1151–1191.

Lindquist, S., E.A.Craig, 1988, The heat-shock proteins. Annu. Rev. Genet. 22: 631–677.

LIPETZ, J., 1965, Crown-gall tumorigenesis: Effect of temperature on wound healing and conditioning. Science 149: 865–867.

LIPPHARDT, S., R. BRETTSCHNEIDER, F. KREUZALER, J. SCHELL, J. L. DANGL, 1988, UV-inducible transient expression in parsley protoplasts identifies regulatory cis-elements of a chimeric *Antirrhinum majus* chalcone synthase gene. EMBO J. 7: 4027–4033.

LITTLEWOOD, T.D., D.C. HANCOCK, G.I. EVAN, 1987, Characterization of a heat shock-induced insoluble complex in the nuclei of cells. J. Cell Sci. 88: 65–72.

LJUNGDAHL, L. G., 1979, Physiology of thermophilic bacteria. Adv. Microb. Physiol. 19: 150–243.

LLEWELLYN, D.J., E.J. FINNEGAN, J.G. ELLIS, E.S. DENNIS, W.J. PEACOCK, 1987, Structure and expression of an alcohol dehydrogenase 1 gene from *Pisum sativum* (cv. 'Green feast'). J. Mol. Biol. 195: 115–123.

LOOMIS, W. F., 1979, Biochemistry of aggregation in *Dictyostelium*. Devel. Biol. 70: 1–12.

LOOMIS, W. F., S. WHEELER, 1980, Heat shock response of *Dictyostelium*. Dev. Biol. 79: 399–408.

LOVE, R., R.Z. SORIANO, R.J. WALSH, 1970, Effect of hyperthermia on normal and neoplastic cells *in vitro*. Cancer Res. 30: 1525–1533.

LOVEN, D.P., D.B. LEEPER, L.W. OBERLEY, 1985, Superoxide dismutase levels in Chinese hamster ovary cells and ovarian carcinoma cells after hyperthermia or exposure to cycloheximide. Cancer Res. 45: 3029–3033.

LUCAS, J., A. CAMACHO-HENRIQUEZ, F. LOTTSPEICH, A. HENSCHEN, H. L. SAENGER, 1985, Amino acid sequence of the 'pathogenesis-related' leaf protein p14 from viroid-infected tomato reveals a new type of structurally unfamiliar proteins. EMBO J. 4: 2745–2749.

LYNCH, D.V., P.L. STEPONKUS, 1987, Plasma membrane lipid alterations associated with cold acclimation of winter rye seedlings. Plant Physiol. 83: 761–767.

LYNCH, J., A. LAEUCHLI, 1988, Salinity affects intracellular calcium in corn root protoplasts. Plant Physiol. 87: 351–356.

MACDONALD T. L., R. B. MARTIN, 1988, Aluminum ion in biological systems. Trends Biochem. Sci. 13: 15–19.

MANDAVA, N. B., 1988, Plant growth-promoting brassinosteroids. Annu. Rev. Plant Physiol. 39: 23–52.

MANSFIELD, M. A., J. L. KEY, 1987, Synthesis of the low molecular weight heat shock proteins in plants. Plant Physiol. 84: 1007–1017.

MANSFIELD, M.A., J.L. KEY, 1988, Cytoplasmic distribution of heat shock proteins in soybean. Plant Physiol. 86: 1240–1246.

MANSFIELD, M.A., W.L. LINGLE, J.L. KEY, 1988, The effects of lethal heat shock on nonadapted and thermotolerant root cells of *Glycine max*. J. Ultrastruct. Res. 99: 96–105.

MARCOTTE, W.R., C.C. BAYLEY, R.S. QUATRANO, 1988, Regulation of wheat promoter by abscisic acid in rice protoplasts. Nature 335: 454–457.

MARINEAU, C., D.P. MATTON, N. BRISSON, 1987, Differential accumulation of potato tuber mRNAs during the hypersensitive response induced by arachidonic acid elicitor. Plant Mol. Biol. 9: 335–342.

MARMIROLI, N., F.M. RESTIVO, M. ODOARDI STANCA, V. TERZI, B. GIOVANELLI, F. TASSI, C. LORENZONI, 1986, Induction of heat shock proteins and acquisition of thermotolerance in barley seedlings. Genet. Argar. 40: 9–25.

MAROTO, F. G., J. M. SIERRA, 1988, Translational control in heat-shocked *Drosophila* embryos. Evidence for the inactivation of initiation factor(s) involved in the recognition of mRNA cap structure. J. Biol. Chem. 263: 15720–15725.

MASCARENHAS, J. P., M. ALTSCHULER, 1985, Responses to environmental heat stress in plant embryo. In: ATKINSON and WALDEN (eds.) 1985, 315–326.

MASON, P. J., L. M. C. HALL, J. GAUSZ, 1984, The expression of heat shock genes during normal development in *Drosophila melanogaster*. Mol. Gen. Genet. 194: 73–78.

MATSUMOTO, S., S. GUGG, J.G. LAFONTAINE, 1987, Ultrastructural investigation of the effects of supraoptimal temperature on late interphase nuclei of *Physarum polycephalum*. Biol. Cell 60: 87–96.

MATSUSHITA, K., I. URITANI, 1974, Change in invertase activity of sweet potato in response to wounding and purification and properties of its invertases. Plant Physiol. 54: 60–66.

MATTERS, G. L., J. G. SCANDALIOS, 1986a, Effect of the free radical-generating herbicide paraquat on the expression of the superoxide dismutase (SOD) genes in maize. Biochim. Biophys. Acta 882: 29–38.

MATTERS, G. L., J. G. SCANDALIOS, 1986b, Changes in plant gene expression during stress. Dev. Genet. 7: 167–176.

MAUCH, F., L. A. HADWIGER, T. BOLLER, 1984, Ethylene: Symptom, not signal for the induction of chitinase and beta-1,3-glucanase in pea pods by pathogens and elicitors. Plant Physiol. 76: 607–611.

MAUCH, F., L. A. HADWIGER, T. BOLLER, 1988a, Antifungal hydrolases in pea tissue. 1. Purification and characterization of 2 chitinases and 2 beta-1,3-glucanases differentially regulated during development and in response to fungal infection. Plant Physiol. 87: 325–333.

MAUCH, F., B. MAUCH-MANI, T. BOLLER, 1988b, Antifungal hydrolases in pea tissue. II. Inhibition of fungal growth by combinations of chitinase and β-1,3-glucanase. Plant Physiol. 88: 936–942.

MAXIMOV, N. A., 1912, Chemische Schutzmittel der Pflanzen gegen Erfrieren. Ber. Dtsch. Bot. Ges. 30: 52–65, 293–305, 504–516.

MAYAK, S., A. H. HALEVY, 1980, Flower senescence. In: K. V. THIMANN (ed.), Senescence in Plants. CRC Press, Boca Raton, pp. 131–156.

MAYRAND, S., T. PEDERSON, 1983, Heat shock alters nuclear ribonucleoprotein assembly in *Drosophila* cells. Mol. Cell. Biol. 3: 161–171.

MAZZARELLA, R. A., and M. GREEN, 1987, ERp 99, an abundant, conserved glycoprotein of the endoplasmic reticulum, is homologous to the 90-kDa heat shock protein (hsp90) and the 94-kDa glucose regulated protein (GRP94). J. Biol. Chem. 262: 8875–8883.

MCCLINTOCK, B., 1978, Mechanisms that rapidly reorganize the genome. Stadler Genetics Symp. 10: 24–47.

MCCLINTOCK B., 1984, The significance of responses of the genome to challenge. Science 226: 792–801.

MCCONELL, M., A. M. WHALEN, D. E. SMITH, P. A. FISHER, 1987, Heat shock-induced changes in the structural stability of proteinaceous karyoskeletal elements *in vivo* and morphological effects *in situ*. J. Cell Biol. 105: 1087–1098.

MCCORMICK, W., S. PENMAN, 1969, Regulation of protein synthesis in HeLa cells: Translation at elevated temperatures. J. Mol. Biol. 39: 315–333.

MCELFRESH, K. C., P. S. CHOUREY, 1988, Anaerobiosis induces transcription but not translation of sucrose synthase in maize. Plant Physiol. 87: 542–546.

MCGARRY, T. J., S. LINDQUIST, 1985, The preferential translation of *Drosophila* hsp70 mRNA requires sequences in the untranslated leader. Cell 42: 903–911.

MCMULLIN, T. W., R. L. HALLBERG, 1986, Effect of heat shock on ribosome structure: Appearance of a new ribosome-associated protein. Mol. Cell. Biol. 6: 2527–2535.

MCMULLIN, T. W., R. L. HALLBERG, 1987, A normal mitochondrial protein is selectively synthesized and accumulated during heat shock in *Tetrahymena thermophila*. Mol. Cell. Biol. 7: 4414–4423.

MCMULLIN, T. W., R. L. HALLBERG, 1988, A highly evolutionarily conserved mitochondrial protein is substructurally related to the protein encoded by the *E. coli* groEL gene. Mol. Cell. Biol. 8: 371–380.

MEHDY, M. C., C. J. Lamb, 1987, Chalcone isomerase cDNA cloning and mRNA induction by fungal elicitor, wounding and infection. EMBO J. 6: 1527–1533.

MEHRA, V., D. SWEETSER, R. A. YOUNG, 1986, Efficient mapping of the protein antigenic determinants. Proc. Natl. Acad. Sci. USA 83: 7013–7017.

MEINS, F., P. AHL, 1989, Induction of chitinase and β-1,3-glucanase in tobacco plants infected with *Pseudomonas tabaci* and *Phytophthora parasitica* var. *nicotianae*. Plant Sci. (in press).

MEINS, F., R. FORSTER, J. D. LUTZ, 1983, Evidence for a Medelian factor controlling the cytokinin requirement of cultured tobacco cells. Devel. Genetics 4: 129–141.

MEMELINK, J., J. H. C. HOGE, R. A. SCHILPEROORT, 1987a, Cytokinin stress changes the developmental regulation of several defence-related genes in tobacco. EMBO J. 6: 3579–3583.

MEMELINK, J., S. B. DEPATER, J. H. C. HOGE, R. A. SCHILPEROORT, 1987b, T-DNA hormone biosynthetic genes: Phytohormones and gene expression in plants. Devel. Genetics B: 321–337.

MEMELINK, J., S. C. DE VRIES, R. A. SCHILPEROORT, J. H. C. HOGE, 1988, Changes in the tissue-specific prevalence of translatable mRNAs in transgenic tobacco shoot containing the T-DNA cytokinin gene. Plant Mol. Biol. 11: 625–631.

MENDEL, D. B., J. E. BODWELL, B. GAMETCHU, R. W. HARRISON, A. MUNCK, 1986, Molybdate-stabilized nonactivated glucocorticoid-receptor complexes contain a 90-kDa non-steroid-binding phosphoprotein that is lost on activation. J. Biol. Chem. 261: 3758–3763.

MENDGEN, K., E. DRESSLER, 1983, Culturing *Puccinia coronata* on a cell monolayer of the *Avena* coleoptile. Phytopathol. Z. 108: 226–234.

METRAUX, J. P., L. STREIT, T. STRAUB, 1988, A pathogenesis-related protein in cucumber is a chitinase. Physiol. Mol. Plant Pathol. 33: 1–10.

MEYER, J., Y. CHARTIER, 1983, Long-lived and short-lived heat shock proteins in tobacco mesophyll protoplasts. Plant Physiol. 72: 26–32.

MEYER, A., P. MUELLER, G. SEMBDNER, 1987, Air pollution and plant hormones. Biochem. Physiol. Pflanzen 182: 1–21.

MEZA-BASSO, L., M. ALBERDI, M. RAYNAL, M. L. FERRARO-CADINANOS, M. DELSENY, 1986, Changes in protein synthesis in rape seed *Brassica napus* seedlings during a low temperature treatment. Plant Physiol. 82: 733–738.

MICHAELIS, A., S. TAKEHISA, R. RIEGER, O. AURICH, 1986, Ammonium chloride and zinc sulfate pretreatments reduce the yield of chromatid aberrations induced by triethylene amine and maleic hydrazide in *Vicia faba*. Mutat. Res. 173: 187–192.

MICHAELIS, A., R. RIEGER, H. NICOLOFF, 1988, Effects of ethidium bromide and nalidixic acid pretreatment on the induction of chromatid aberrations by TEM and maleic hydrazide in *Vicia faba* main root meristems. Mutat. Res. 208: 83–86.

MILARSKI, K. L., R. I. MORIMOTO, 1986, Expression of human HSP70 during the synthetic phase of the cell cycle. Proc. Natl. Acad. Sci. USA 83: 9517–9522.

MILLER, D., A. G. MCLENNAN, 1986, Changes in intracellular levels of Ap3A and Ap4A in cysts and larvae of *Artemia* do not correlate with changes in protein synthesis after heat-shock. Nucl. Acid Res. 14: 6031–6040.

MIRAULT, M. E., R. SOUTHGATE, E. DELWART, 1982, Regulation of heat-shock genes: A DNA sequence upstream of *Drosophila* hsp70 genes is essential for their induction in monkey cells. EMBO J. 1: 1279–1285.

MITCHELL, J. B., A. RUSSO, T. J. KINSELLA, E. GLATSTEIN, 1983, Glutathione elevation during thermotolerance induction and thermosensitization by glutathione depletion. Cancer Res. 43: 987–991.

MITCHELL, H. K., N. S. PETERSEN, C. H. BUZIN, 1985, Self-degradation of heat shock proteins. Proc. Natl. Acad. Sci. USA 82: 4969–4973.

MIZZEN, L. A., W. J. WELCH, 1988, Characterization of the thermotolerant cell: I. Effects on protein synthesis activity and the regulation of HSP70 expression. J. Cell Biol. 106: 1105–1116.

MOCQUOT, B., C. H. PRAT, C. MOUCHES, A. PRADET, 1981, Effect of anoxia on energy charge and protein synthesis in rice embryo, Plant Physiol. 68: 636–640.

MOCQUOT, B., B. RICARD, A. PRADET, 1987, Rice embryos can express heat-shock genes under anoxia. Biochimie 69: 677–681.

MOESTA, P., H. GRISEBACH, 1982, L-2-Aminooxy-3-phenylpropionic acid inhibits phytoalexin accumulation in soybean with concomitant loss of resistance against *Phytophthora megasperma* f. sp. *glycinea*. Physiol. Plant Pathol. 21: 65–70.

MOHANTY, N., S. D. S. MURTHY, P. MOHANTY, 1987, Reversal of heat induced alterations in photochemical activities in wheat primary leaves. Photosynth. Res. 14: 259–268.

MOHAPATRA S. S., R. J. POOLE, R. S. DHINDSA, 1987, Change in protein patterns and translatable mRNA populations during cold acclimation in alfalfa. Plant Physiol. 84: 1172–1176.

MOHAPATRA, S. S., R. J. POOLE, R. S. DHINDSA, 1988a, Abscisic acid-regulated gene expression in relation to freezing tolerance in alfalfa. Plant Physiol. 87: 468–473.

MOHAPATRA S. S., R. J. POOLE, R. S. DHINDSA, 1988b, Alterations in membrane protein-profile during cold treatment of alfalfa. Plant Physiol. 86: 1005–1007.

MOHNEN, D., H. SHINSHI, G. FELIX, F. MEINS, 1985, Hormonal regulation of β-1,3-glucanase messenger RNA levels in cultured tobacco tissues. EMBO J. 4: 1631–1635.

MOORE, S. K., C. KOZAK, E. A. ROBINSON, S. J. ULLRICH, , E. APPELLA, 1987, Cloning and nucleotide sequence of the murine hsp 84 cDNA and chromosome asssignment of related sequences. Gene 56: 29–40.

MORGAN, J. M., 1984, Osmoregulation and water stress in higher plants. Annu. Rev. Plant Physiol. 35: 299–319.

MORGAN, R. W., M. F. CHRISTMAN, F. S. JACOBSON, G. STORZ, B. N. AMES, 1986, Hydrogen peroxide-inducible proteins in *Salmonella typhimurium* overlap with heat shock and other stress proteins. Proc. Natl. Acad. Sci. USA 83: 8059–8063.

MORIMOTO, R., C. HUNT, S.-Y. HUANG, K. L. BERG, S. S. BANERJI, 1986, Organization, nucleotide sequence and transcription of the chicken HSP70 gene. J. Biol. Chem. 261: 12692–12699.

MORIMOTO, R., D. MOSSER, T. K. MCCLANAHAN, N. G. THEODORAKIS, G. WILLIAMS, 1989, Transcriptional regulation of the human hsp70 gene. In: PARDUE et al., (eds.) 1989, pp. 83–94.

MORRIS, R. O., 1986, Genes specifying auxin and cytokinin biosynthesis in phytopathogens. Annu. Rev. Plant Physiol. 37: 509–538.

MOSSER, D. D., N. G. THEODORAKIS, R. I. MORIMOTO, 1988, Coordinate changes in heat shock element-binding activity and hsp70 gene transcription rates in human cells. Mol. Cell. Biol. 8: 4736–4744.

MOTHES, K., 1928, Die Wirkung des Wassermangels auf den Eiweißumsatz in höheren Pflanzen. Ber. Dtsch. Bot. Ges. 46: 59–67.

MOTHES, K., 1960, Über das Altern der Blätter und die Möglichkeit ihrer Wiederverjüngung. Naturwissenschaften 47: 337–351.

MÜHLBACH, H. P., 1987, Viroide: Freie infektiöse RNA-Moleküle als Erreger von Pflanzenkrankheiten. Biol. uns. Zeit 27: 65–78.

MÜLLER-THURGAU, H., 1886, Über das Gefrieren und Erfrieren der Pflanzen. Landwirtsch. Jahrb. 15: 453–610.

MUELLER-URI, F., B. PARTHIER, L. NOVER, 1988, Jasmonate-induced alteration of gene expression in barley leaf segments analyzed by *in vivo* and *in vitro* protein synthesis. Planta 176: 241–247.

MUHICH, M. L., J. C. BOOTHROYD, 1988, Polycistronic transcripts in trypanosomes and their accumulation during heat shock: Evidence for a precursor role in mRNA synthesis. Mol. Cell Biol. 8: 3837–3846.

MUKOHATA, Y., T. YAGI, M. HIGASHIDA, K. SHINOZAKI, A. MATSUNO, 1973, Biophysical studies on subcellular particles. VI. Photosynthetic activities in isolated spinach chloroplasts after transient warming. Plant Cell Physiol. 14: 111–118.

MUNDY, J., N.-H. CHUA, 1988, Abscisic acid and water stress induce the expression of a novel rice gene. EMBO J. 7: 2279–2286.

MUNRO, S., H. R. B. PELHAM, 1984, Use of peptide tagging to detect proteins expressed from cloned genes: Deletion mapping functional domains of *Drosophila* hsp70. EMBO J. 3: 3087–3093.

MUNRO, S., H. R. B. PELHAM, 1986, An hsp70-like protein in the ER: Identity with the 78 kd glucose-regulated protein and immunoglobulin heavy chain binding protein. Cell 46: 291–300.

MUNRO, S., H. R. B. PELHAM, 1987, A C-terminal signal prevents secretion of luminal ER proteins. Cell 48: 899–907.

MURAKAMI, H., D. PAIN, G. BLOBEL, 1988, 70-kD heat shock-related protein is one of at least two distinct cytosolic factors stimulating protein import into mitochondria. J. Cell Biol. 107: 2051–2057.

MURPHY, T. M., G. B. MATSON, S. L. MORRISON, 1983, Ultraviolet-stimulated $KHCO_3$ efflux from rose cells. Regulation of cytoplasmic *p*H. Plant Physiol. 73: 20–24.

MURTI, K. G., SMITH, H. T., V. A. FRIED, 1988, Ubiquitin is a component of the microtubule network. Proc. Natl. Acad. Sci. USA 85: 3019–3023.

MUSGROVE, J. E., R. J. ELLIS, 1986, The Rubisco large subunit binding protein. Phil. Trans. R. Soc. Lnd. B 313: 419–428.

NAGAO, R. T., E. CZARNECKA, W. B. GURLEY, F. SCHÖFFL, J. L. KEY, 1985, Genes for low-molecular-weight heat shock proteins of soybeans: Sequence analysis of a multigene family. Mol. Cell. Biol. 5: 3417–3428.

NAGAO, R. T., J. A. KIMPEL, E. VIERLING, J. L. KEY, 1985, The heat shock response A comparative analysis. Oxford Surv. Plant Mol. Cell Biol. 3: 384–438.

NAKAGAWA, N., N. NAKAJIMA, H. IMASEKI, 1988, Immunochemical difference of wound-induced 1-aminocyclopropane-1-carboxylate synthase from the auxin-induced enzyme. Plant Cell Physiol. 29: 1255–1259.

NAPOLITANO, E. W., J. S. PACHTER, R. K. H. LIEM, 1987, Intracellular distribution of mammalian stress proteins. Effects of cytoskeletal-specific agents. J. Biol. Chem. 262: 1493–1504.

NEBIOLO, C. M., E. M. WHITE, 1985, Corn mitochondrial protein synthesis in response to heat shock. Plant Physiol. 79: 1129–1132.

NECCHI, A., N. E. POGNA, S. MAPELLI, 1987, Early and late heat shock proteins in wheats and other cereal species. Plant Physiol. 84: 1378–1384.

NEIDHARDT, F. C., R. A. VAN BOGELEN, V. VAUGHN, 1984, The genetics and regulation of heat-shock proteins. Annu. Rev. Genet. 18: 295–329.

NELSON, C. E., C. A. RYAN, 1980, *In vitro* synthesis of preproteins of vacuolar compartmented proteinase inhibitors that accumulate in leaves of wounded tomato plants. Proc. Natl. Acad. Sci. USA 77: 1975–1979.

NERLAND, A. H., A. S. MUSTAFA, D. SWEETSER, T. GODAL, R. A. YOUNG, 1988, A protein antigen of *Mycobacterium leprae* is related to a family of small heat shock proteins. J. Bacteriol. 170: 5919–5921.

NEUMANN, D., K.-D. SCHARF, L. NOVER, 1984, Heat shock induced changes of plant cell ultrastructure and autoradiographic localization of heat shock proteins. Eur. J. Cell Biol. 34: 254–264.

NEUMANN, D., U. zur NIEDEN, R. MANTEUFFEL, G. WALTER, K.-D. SCHARF, L. NOVER, 1987, Intracellular localization of heat shock proteins in tomato cells cultures. Eur. J. Cell Biol. 43: 71–81.

NICOLOFF, H., K. GECHEFF, R. RIEGER, A. MICHAELIS, 1985, "Clastogenic adaptation" in barley: Differential response of presoaked and dry seeds. Mutat. Res. 143: 83–95.

NIETO-SOTELO, J., T.-H. D. HO, 1986, Effect of heat shock on the metabolism of glutathione in maize roots. Plant Physiol. 82: 1031–1035.

NIETO-SOTELO, J., T.-H. D. HO, 1987, Absence of heat shock protein synthesis in isolated mitochondria and plastids from maize. J. Biol. Chem. 262: 12288–12292.

NISHIDA, E., S. KOYASU, H. SAKAI, I. YAHARA, 1986, Calmodulin-regulated binding of the 90 kDa heat shock protein to actin filaments. J. Biol. Chem. 261: 16033–16037.

NOLAN, R. C., T.-H. D. HO, 1988, Hormonal regulation of α-amylase expression in barley aleurone layers. The effect of gibberellic acid removal and abscisic acid and phaseic acid treatments. Plant Physiol. 88: 588–593.

NOLAN, R. C., L.-S. LIN, T.-H. D. HO, 1987, The effect of abscisic acid on the differential expression of α-amylase isozymes in barley aleurone layers. Plant Mol. Biol. 8: 13–22.

NOVER, L. (ed.), 1984, Heat Shock Response of Eukaryotic Cells. Springer Verlag, Berlin.

NOVER, L., 1987, Expression of heat shock genes in homologous and heterologous systems. Enzyme Microb. Technol. 9: 130–144.

NOVER, L., 1989, 125 years of experimental heat shock research: Historical roots of a discipline. Genome 31: (in press).

NOVER, L. (ed.), 1990, Heat Shock Response. CRC Press, Boca Raton (Fd.).

NOVER, L., K.-D. SCHARF, 1984, Synthesis, modification and structural binding of heat shock proteins in tomato cell cultures. Eur. J. Biochem. 139: 303–313.

NOVER, L., E. KRANZ, K.-D. SCHARF, 1982, Growth cycle of suspension cultures of *Lycopersicon esculentum* and *L. peruvianum*. Biochem. Physiol. Pflanzen 177: 483–499.

NOVER, L., K.-D. SCHARF, D. NEUMANN, 1983; Formation of cytoplasmic heat shock granules in tomato cell cultures and leaves. Mol. Cell. Biol. 3: 1648–1655.

NOVER, L., D. MUNSCHE, K. OHME, K.-D. SCHARF, 1986, Ribosome biosynthesis in heat shocked tomato cell cultures I. Ribosomal RNA. Eur. J. Biochem. 160: 297–304.

NOVER, L., K.-D. SCHARF, D. NEUMANN, 1989, Cytoplasmic heat shock granules are formed from precursor particles and contain a specific set of mRNAs. Mol. Cell. Biol. 9: 1298–1308.

OBA, K., N. MAKIMOTO, T. HATTORI, I. URITANI, 1982, Induction of polysome formation in sweet potato root tissue in response to wounding. Agric. Biol. Chem. 46: 1929–1931.

ÖZKAYNAK, E., D. FINLEY, M. J. SOLOMON, A. VARSHAVSKY, 1987, The yeast ubiquitin genes: A family of natural gene fusions. EMBO J. 6: 1429–1439.

OGREN, W. L., 1984, Photorespiration: Pathways, regulation and modification. Annu. Rev. Plant Physiol. 35: 415–442.

OHASHI, Y., M. MATSUOKA, 1985, Synthesis of stress proteins in tobacco leaves. Plant Cell Physiol. 26: 473–480.

OKIMOTO, R., M. M. SACHS, E. K. PORTER, M. FREELING, 1980, Patterns of polypeptide synthesis in various maize organs under anaerobiosis. Planta 150: 89–94.

OLIVIERI, G., J. BODYCOTE, S. WOLFF, 1984, Adaptive response of human lymphocytes to low concentrations of radioactive thymidine. Science 223: 594–597.

O'MALLEY, K., A. MAURON, J. D. BARCHAS, L. KEDES, 1985, Constitutively expressed rat mRNA encoding a 70-kilodalton heat-shock-like protein. Mol. Cell. Biol. 5: 3476–3483.

OMAR, R. A., S. YANO, Y. KIKKAWA, 1987, Antioxidant enzymes and survival of normal and Simian virus 40-transformed mouse embryo cells after hyperthermia. Cancer Res. 47: 3473–3476.

ORR, W., W. A. KELLER, J. SINGH, 1986, Induction of freezing tolerance in an embryonic cell suspension culture of *Brassica napus* by abscisic acid at room temperature. J. Plant Physiol. 126: 23–32.

ORT, D. R., J. S. BOYER, 1985, Plant productivity, photosynthesis, and environmental stress. In: ATKINSON and WALDEN (eds.) 1985: 279–313.

ORTIZ, W., C. J. WILSON, 1988, Induced changes in chloroplast protein accumulation during heat bleaching in *Euglena gracilis*. Plant Physiol. 86: 554–561.

OSTREM, J. A., S. W. OLSON, J. M. SCHMITT, H. J. BOHNERT, 1987, Salt stress increases the level of translatable mRNA for phosphoenolpyruvate carboxylase in *Mesembryanthemum crystallinum*. Plant Physiol. 84: 1270–1275.

OTA, K., 1988, Stimulation of CAM photosynthesis in *Kalanchoë blossfeldiana* by transferring to nitrogen-deficient conditions. Plant Physiol. 87: 454–457

OUGHAM, H. J., 1987, Gene expression during leaf development in *Lolium temulentum*: Pattern of protein synthesis in response to heat-shock and cold shock. Physiol. Plant. 70: 479–484.

OUGHAM, H. J., J. L. STODDART, 1986, Synthesis of heat-shock protein and acquisition of thermotolerance in high-temperature tolerant and high-temperature susceptible lines of *Sorghum*. Plant Sci. 44: 163–168.

OVERGAARD, J. (ed.), 1985, Hyperthermic Oncology 1984. Tylor and Francis, London, Vol. 1 and 2.

PAEK, K.-H., G. C. WALKER, 1987, *Escherichia coli* dnaK null mutants are inviable at high temperature. J. Bacteriol. 169: 283–290.

PALTER, K. B., M. WATANABE, L. STINSON, A. P. MAHOWALD, E. A. CRAIG, 1986, Expression and localization of *Drosophila melanogaster* hsp70 cognate proteins. Mol. Cell. Biol. 6: 1187–1203.

PANNIERS, R., E. B. STEWART, W. C. MERRICK, E. C. HENSHAW, 1985, Mechanism of inhibition of polypeptide chain initiation in heat-shocked Ehrlich cells involves reduction of eukaryotic initiation factor 4F activity. J. Biol. Chem. 260: 9648–9653.

PARADIES, I., J. R. KONZE, E. F. ELSTNER, J. PAXTON, 1980, Ethylene: Indicator but not inducer of phytoalexin synthesis in soybean. Plant Physiol. 66: 1106–1109.

PARDUE, M. L., J. R. FERAMISCO, S. LINDQUIST (eds.), 1989, Stress-Induced Proteins. Alan R. Liss, Inc., New York.

PARENT, J. G., A. ASSELIN, 1984, Detection of pathogenesis-related proteins (PR or b) and other proteins in the intercellular fluid of hypersensitive plants infected with tobacco mosaic virus. Can. J. Bot. 62: 564–569.

PARKER, J. E., K. HAHLBROCK, D. SCHEEL, 1989, Different cellwall components from *Phytophthora megasperma* f. sp. *glycinea* elicit phytoalexin production in soybean and parsley. Planta (in press).

PARTHIER, B., 1988, Gerontoplasts – the yellow end in the ontogenesis of chloroplasts. Endocytobios. Cell Res. 5: 163–190.

PARTHIER, B., 1989, Hormone-induced alterations in plant gene expression. Biochem. Physiol. Pflanzen 185: 289–314.

PATTERSON, B. D., D. GRAHAM, 1987, Temperature and metabolism. In: D. D. DAVIES (ed.), The Biochemistry of Plants. Vol. 12. Acad. Press, San Diego, pp. 153–199.

PAULI, D., C.-H. TONKA, 1987, A new *Drosophila* heat shock gene from locus 67B is expressed during embryogenesis and pupation. J. Mol. Biol. 198: 235–240.

PAULI, D., A. SPIERER, A. TISSIERES, 1986, Several hundred base pairs upstream of *Drosophila* hsp23 and 26 genes are required for their heat induction in transformed flies. EMBO J. 5: 755–761.

PEARCE, G., C.A.RYAN, D.LILJEGREN, 1988, Proteinase inhibitors I and II in fruit of wild tomato species: Transient components of a mechanism for defense and seed dispersal. Planta 175: 527–531.

PEARCY, R.W., J.A.BERRY, D.C.FORK, 1977, Effects of growth temperature on the thermal stability of the photosynthetic apparatus of *Atriplex lentiformis* (Torr.) Wats. Plant Physiol. 59: 873–878.

PEDERSON, T., 1983, Nuclear RNA-protein interactions and messenger RNA processing. J. Cell Biol. 97: 1321–1326.

PEGG, G.F., D.H.YOUNG, 1982, Purification and characterization of chitinase enzyme from healthy and *Verticilium albo-atrum*-infected tomato plants, and *V. albo-atrum*. Physiol. Plant Pathol. 21: 349–409.

PEKKALA, D., I.B.HEATH, J.C.SILVER, 1984, Changes in chromatin and the phosphorylation of nuclear proteins during heat shock of *Achlya ambisexualis*. Mol. Cell. Biol. 4: 1198–1205.

PELHAM, H.R.B., 1982, A regulatory upstream promoter element in the *Drosophila* hsp70 heat-shock gene. Cell 30: 517–528.

PELHAM, H.R.B., 1984, Hsp70 accelerates the recovery of nucleolar morphology after heat shock. EMBO J. 3: 3095–3100.

PELHAM, H.R.B., 1986, Speculations on the functions of the major heat shock and glucose-regulated proteins. Cell 46: 959–961.

PELHAM, H.R.B., 1988, Heat shock proteins. Coming in from the cold? Nature 332: 776–777.

PELHAM, H.R.B., M.BIENZ, 1982, A synthetic heat-shock promoter element confers heat-inducibility on the herpes simplex virus thymidine kinase gene. EMBO J. 1: 1473–1477.

PENA-CORTES, H., J. SANCHEZ-SERRANO, M. ROCHA-SOSA, L. WILLMITZER, 1988, Systemic induction of proteinaseinhibitor-II gene expression in potato plants by wounding. Planta 174: 84–89.

PENDHARKAR, M.B., P.M.NAIR, 1975, Induction of phenylalanine ammonia lyase (PAL) in gamma irradiated potatoes. Radiat. Bot. 15: 191–197.

PERDEW, G.H., 1988, Association of the Ah receptor with the 90-kDa heat shock protein. J. Biol. Chem. 263: 13802–13805.

PETKO, L., S.LINDQUIST, 1986, Hsp26 is not required for growth at high temperatures, nor for thermotolerance, spore development, or germination. Cell 45: 885–894.

PFITZNER, U.M., A.J.PFITZNER, H.M.GOODMAN, 1988, DNA sequence analysis of a PR-1a gene from tobacco: molecular relationship of heat shock and pathogen responses in plants. Mol. Gen. Genet. 211: 290–295.

PITT, D., M.GALPIN, 1971, Increase in RNase activity following mechanical damage of leaf and tuber tissues of *Solanum tuberosum* L. Planta 101: 317–322.

PITTO, L., F.LOSCHIAVO, G.GIOLIANO, T.TERZI, 1983, Analysis of heat-shock protein pattern during somatic embryogenesis of carrot. Plant Mol. Biol. 2: 231–237.

PLATEAU, P., M.FROMANT, S.BLANQUET, 1987, Heat shock and hydrogen peroxide responses of *Escherichia coli* are not changed by dinucleoside tetraphosphate hydrolase overproduction. J. Bacteriol. 169: 3817–3820.

PLESOFSKY-VIG, N., R.BRAMBL, 1985, Heat shock response of *Neurospora crassa*: Protein synthesis and induced thermotolerance. J. Bacteriol. 162: 1083–1091.

POPOVA, L.P., S.G.VAKLINOVA, 1988, Effect of jasmonic acid on the synthesis of ribulose-1,5-bisphosphate carboxylaseoxygenase in barley leaves. J. Plant Physiol. 133: 210–215.

POPOVA, L.P., T.D.TSONEV, S.G.VAKLINOVA, 1988, Changes in some photosynthetic and photorespiratory properties in barley leaves after treatment with jasmonic acid. J. Plant Physiol. 132: 257–261.

POWLES, S.B., 1984, Photoinhibition of photosynthesis induced by visible light. Annu. Rev. Plant Physiol. 35: 15–44.

PRADET, A., B.MOCQUOT, P.RAYMOND, C.MORISSET, L.ASPART, M.DELSENY, 1985, Energy metabolism and synthesis of nucleic acids and proteins under anoxic stress. In: KEY and KOSUGE (eds.), 1985, 227–245.

QUATRANO, R.S., 1986, Regulation of gene expression by abscisic acid during angiosperm embryo development. Oxford Surv. Plant Mol. Cell Biol. 3: 467–477.

RALHAN, R., G.S.JOHNSON, 1986, Destabilization of cytoplasmic mouse mammary tumor RNA by heat shock: Prevention by cycloheximide pretreatment. Biochem. Biophys. Res. Commun. 137: 1028–1033.

RAMAGOPAL, S., 1986, Protein synthesis in a maize callus exposed to NaCl and mannitol. Plant Cell Rep. 5: 430–434.

RAMAGOPAL, S., 1987, Differential mRNA transcription during salinity stress in barley. Proc. Natl. Acad. Sci. USA 84: 94–98.

RASCHKE, E., G. BAUMANN, P. SCHÖFFL, 1988, Nucleotide sequence analysis of soybean small heat shock protein genes belonging to two different multigene families. J. Mol. Biol. 199: 549–557.

READING, D.S., R.HALLBERG, A.M.MYERS, 1989, Characterization of the yeast HSP60 gene coding for a mitochondrial assembly factor. Nature 337: 655–659.

REBBE, N.F., J.WARE, R.M.BERTINA, P.MODRICH, D.W.STAFFORD, 1987, Nucleotide sequence of a cDNA for a member of the human 90 kDa heat-shock protein family. Gene 53: 235–245.

Rechsteiner, M., 1987, Ubiquitin-mediated pathways for intracellular proteolysis. Annu. Rev. Cell. Biol. 3: 1–30.

Rees, C.A.B., N.C. Hogan, D.B. Walden, B.G. Atkinson, 1987, Identification of mRNAs encoding low molecular mass heat shock protein in maize (*Zea mays* L.). Can. J. Genet. Cytol. 28: 1106–1114.

Reese, R.N., L.W. Roberts, 1984, Cadmium uptake and its effects on growth of tobacco cell suspension cultures. Plant Cell Rep. 3: 91–94.

Reinert, J., Y.P.S. Bajaj (eds.), 1977, Applied and Fundamental Aspects of Plant Cell, Tissue, and Organ Culture. Springer Verlag, Berlin.

Restivo, F.M., F. Tassi, E. Maestri, C. Lorenzoni, P.P. Puglisi, N. Marmiroli, 1986, Identification of chloroplast associated heat-shock proteins in *Nicotiana plumbaginifolia* protoplasts. Curr. Genet. 11: 145–151.

Rhoads, R.E., 1988, Cap recognition into the protein synthesis initiation cycle. Trends Biochem. Sci. 13: 52–56.

Rhodes, D., 1987, Metabolic responses to stress. In: E. Davies (ed.), The Biochemistry of Plants, Vol. 12. Acad. Press, San Diego, pp. 201–241.

Rhodes, D., S. Handa, R.A. Bressan, 1986, Metabolic changes associated with adaptation of plant cells to water stress. Plant Physiol. 82: 890–903.

Rice, G.C., K.A. Fisher, G.A. Fisher, G.M. Hahn, 1987, Correlation of mammalian cell killing by heat shock to intramembranous particle aggregation and lateral phase separation using fluorescence-activated cell sorting. Radiat. Res. 112: 351–364.

Rich, P.R., C.J. Lamb, 1977, Biophysical and enzymological studies upon the interaction of trans-cinnamic acid with higher plant microsomal cytochrome P-450. Europ. J. Biochem. 72: 353–360.

Rieger, R., A. Michaelis, 1965, Chemical induction of heat-reversible potential lesions in *Vicia faba* chromosomes. Nature 206: 741–742.

Rieger, R., A. Michaelis, 1988a, Heat shocks prior to mutagen treatment may result in protection against induction of chromatid structural changes in higher plants. Biol. Zent.bl. 107: 403–413.

Rieger, R., A. Michaelis, 1988b, Heat shock protection against induction of chromatid aberrations is dependent on the time span between heat shock and clastogen treatment of *Vicia faba* root tip meristem cells. Mutat. Res. 209: 141–144.

Rieger, R., A. Michaelis, H. Nicoloff, 1982, Inducible repair processes in plant root meristems? 'Below additivity effects' of unequally fractionated clastogen concentrations. Biol. Zent.bl. 101: 125–138.

Rieger, R., A. Michaelis, I. Schubert, 1985, Heat shock prior to treatment of *Vicia faba* root tip meristems with maleic hydrazide or TEM reduce the yield of chromatid aberrations. Mutation Res. 143: 79–82.

Rieger, R., A. Michaelis, H. Nicoloff, 1986a, Effects of stress factors on the clastogen response of *Vicia faba* root tip meristems – 'Clastogenic adaptation'. Biol. Zent.bl. 105: 19–28.

Rieger, R., A. Michaelis, I. Schubert, 1986b, Reduction by heat shock of maleic hydrazide-induced aberration yield is dependent on temperature and duration of heat pretreatment. Mutat. Res. 1744: 199–204.

Rigden, J., R. Coutts, 1988, Pathogenesis-related proteins in plants. Trends Genet. 4: 87–89.

Riggleman, R.C., B. Fristensky, L.A. Hadwiger, 1985, The disease resistance in pea is associated with increased levels of specific mRNAs. Plant Mol. Biol. 4: 81–86.

Risueno, M.C., J.C. Stockert, B. Gimenez-Martin, J.L. Diez, 1973, Effect of supraoptimal temperatures on meristematic cell nucleoli. J. Microscopie 16: 87–94.

Ritossa, F., 1962, A new puffing pattern induced by heat shock and DNP in *Drosophila*. Experientia 18: 571–573.

Roberts, J.A., G.A. Tucker (eds.), 1985, Ethylene and Plant Development. Butterworth, London.

Roberts, J.K.M., J. Callis, D. Wemmer, V. Walbot, D. Jardetzky, 1984a, Mechanism of cytoplasmic *p*H regulation in hypoxic maize root tips and its role in survival under hypoxia. Proc. Natl. Acad. Sci. USA 81: 3379–3383.

Roberts, J.K.M., J. Callis, O. Jardetzky, V. Walbot, M. Freeling, 1984b, Cytoplasmic acidosis as a determinant of flooding intolerance in plants. Proc. Natl. Acad. Sci. USA 81: 6029–6033.

Robertson, A.G., L.V. Gusta, M.J. Reaney, M. Ishikawa, 1987, Protein synthesis in bromegrass cultured cells during the induction of frost tolerance by ABA or low temperature. Plant Physiol. 84: 1331–1336.

Robertson, A.J., L.V. Gusta, M.J.T. Reaney, M. Ishikawa, 1988, Identification of proteins correlated with increased freezing tolerance in bromegrass (*Bromus inermis* Leyss. cv. Manchar) cell cultures. Plant Physiol. 86: 344–347.

Roby, D., M.T. Esquerré-Tugayé, 1987, Induction of chitinases and of translatable mRNA for these enzymes in melon plants infected with *Colletotrichum lagenarium*. Plant Sci. 52: 175–185.

Roby, D., A. Toppan, M.T. Esquerre-Tugaye, 1985, Cell surfaces in plant-microorganism interacions. V. Elicitors of fungal and plant origin trigger the synthesis of ethylene and of cell wall hydroxyproline-rich glycoprotein in plants. Plant Physiol. 77: 700–704.

Roby, D., A. Toppan, M. Esquerré-Tugayé, 1986, Cell surface in plant-microorganism interactions. VI. Elicitors of ethylene from *Colletotrichum* trigger chitinase activity in melon plants. Plant Physiol. 81: 228–233.

Rochester, D.E., J.A. Winter, D.M. Shah, 1986, The structure and expression of maize genes encoding the major heat shock protein, hsp70. EMBO J. 5: 451–458.

Rogers, K.R., F. Albert, A.J. Anderson, 1988, Lipid peroxidation is a consequence of elicitor activity. Plant Physiol. 86: 547–553.

Rohwer, F., K.-H. Fritzemeier, D. Scheel, K. Hahlbrock, 1987, Biochemical reactions of different tissues of potato (*Solanum tuberosum*) to zoospores or elicitor from *Phytophthora infestans*. Accumulation of sesquiterpenoid phytoalexins. Planta 170: 556–561.

Rossi, J.M., S. Lindquist, 1989, The intracellular location of yeast HSP26 varies with metabolism. J. Cell. Biol. 108. 425–439.

Rothman, J. E., S. L. Smid, 1986, Enzymatic recycling of clathrin form coated vesicles. Cell 46: 5–9.

Roti Roti, J. L., N. Uygur and R. Higashikubo, 1986, Nuclear protein following heat shock: Protein removal kinetics and cell cycle rearrangements. Radiat. Res. 107: 250–261.

Rowland, L. J., J. N. Strommer, 1986, Anaerobic treatment of maize roots affects transcription of Adh 1 and transcript stability. Mol. Cell. Biol. 6: 3368–3372.

Roy, H., S. Cannon, 1988, Ribulose bisphosphate carboxylase assembly: what is the role of the large subunit binding protein. Trends Biochem. Sci. 13: 163–165.

Roychowdhury, H. S., M. Kapoor, 1988, Ethanol and carbon-source starvation enhance the accumulation of hsp80 in *Neurospora crassa*. Can. J. Microbiol. 34: 162–168.

Russnak, R.H., E.P.M. Candido, 1985, Locus encoding a family of small heat shock genes in *Caenorhabditis elegans*: Two genes duplicated to form a 3.8-kilobase inverted repeat. Mol. Cell. Biol. 5: 1268–1278.

Ryan, C.A., 1974, Assay and biochemical properties of the proteinase inhibitor-inducing factor, a wound hormone. Plant Physiol. 54: 328–332.

Ryan, C.A., 1987, Oligosaccharide signalling in plants. Annu. Rev. Cell Biol. 3: 295–317.

Ryan, C.A., 1988, Oligosaccharines as recognition signals for the expression of defensive genes in plants. Biochemistry 27: 8879–8883.

Ryan, C.A., W. Huisman, 1970, The regulation of synthesis and storage of chymotrypsin inhibitor I in leaves of potato and tomato plants. Plant Physiol. 45: 484–489.

Ryan, C.A., P. Bishop, G. Pearce, A.G. Darvill, M. McNeil, P. Albersheim, 1981, A sycamore cell wall polysaccharide and a chemically related tomato leaf polysaccharide possess similar proteinase inhibitor-inducing activities. Plant Physiol. 68: 616–618.

Ryder, T.B., C.L. Cramer, J.N. Bell, M.P. Robbins, R.A. Dixon, 1984, Elicitor rapidly induces chalcone synthase mRNA in *Phaseolus vulgaris* cells at the onset of the phytoalexin defense response. Proc. Natl. Acad. Sci. USA 81: 5724–5728.

Ryder, T. B., S. A. Hedrick, J. N. Bell, X. Liang, S. D. Clouse, C. J. Lamb, 1987, Organization and differential activation of a gene family encoding the plant defense enzyme chalcone synthase in *Phaseolus vulgaris*. Mol. Gen. Genet. 210: 219–233.

Sabbah, M., G. Redeulich, C. Secco, E.-E. Baulieu, 1987, The binding activity of estrogen receptor to DNA and heat shock protein (Mr 90.000) is dependent on receptor bound metal. J. Biol. Chem. 262: 8631–8635.

Sacher, J. A., J. Tseng, R. Williams, A. Cabello, 1982, Wound-induced RNase activity in sweat potato. Evidence for regulation at transcription. Plant Physiol. 69: 1060–1065.

Sachs, J., 1864, Über die obere Temperatur-Grenze der Vegetation. Flora 47: 5–12, 24–29, 33–39, 64–75.

Sachs, M. M., M. Freeling, 1978, Selective synthesis of alcohol dehydrogenase during anaerobic treatment of maize. Mol. Gen. Genet. 161: 111–115.

Sachs, M. M., T.-H. D. Ho, 1986, Alteration of gene expression during environmental stress in plants. Annu. Rev. Plant Physiol. 37: 363–376.

Sachs, M. M., M. Freeling, R. Okimoto, 1980, The anaerobic proteins of maize. Cell 20: 761–767.

Saglio, P. H., M. C. Drew, A. Pradet, 1988, Metabolic acclimation to anoxia induced by low (2–4 kPa partial pressure) oxygen pretreatment (hypoxia) in root tips of *Zea mays*. Plant Physiol. 86: 61–66.

Sanchez, E. R., S. Meshinchi, W. Tienrungroj, M. J. Schlesinger, D. O. Toft, W. B. Pratt, 1987, Relationship of the 90-kDa murine heat-shock protein to the untransformed and transformed status of the L cell glucocorticoid receptor. J. Biol. Chem. 262: 6986–6991.

Sanchez-Martinez, D., P. Pugdomenech, M. Pages, 1986, Regulation of gene expression in developing *Zea mays* embryos. Protein synthesis during embryogenesis and early germination. Plant Physiol 82: 543–549.

Sanders, M. M., D. F. Triemer, A. S. Olsen, 1986, Regulation of protein synthesis in heat-shocked *Drosophila* cells. Soluble factors control translation *in vitro*. J. Biol. Chem. 261: 2189–2196.

Santarius, K. A., 1975, Sites of heat sensitivity in chloroplasts and differential inactivation of cyclic and noncyclic photophosphorylation by heating. J. Therm. Biol. 1: 101–107.

Sapper, I., 1935, Versuche zur Hitzeresistenz der Pflanzen. Planta 23: 518–556.

SARHAN, F., M. PERRAS, 1987, Accumulation of a high molecular weight protein during cold hardening of wheat (*Triticum aestivum*). Plant Cell Physiol. 28: 1173–1179.

SATLER, S. O., K. V. THIMANN, 1981, Le jasmonate de methyle: nouveau et puissant promoteur de la senescence des feuilles. Compt. Rend. Acad. Sci. Paris, Ser. III 293: 735–740.

SCHAFER, W., G. KAHL, 1981, Auxin-induced changes in chromosomal protein phosphorylation in wounded potato tuber parenchyma. Plant Molec. Biol. 1, 5–17.

SCHAFFER, M. A., R. L. FISCHER, 1988, Analysis of messenger RNAs that accumulate in response to low temperature identifies a thiol protease gene in tomato. Plant Physiol. 87: 431–436.

SCHARF, K.-D., L. NOVER, 1982, Heat shock induced alterations of ribosomal protein phosphorylation in plant cell cultures. Cell 30: 427–437.

SCHARF, K.-D., L. NOVER, 1987, Control of ribosome biosynthesis in plant cell cultures under heat shock conditions II. Ribosomal proteins. Biochim. Biophys. Acta 909: 44–57.

SCHEEL, D., K. HAHLBROCK, 1989, Physiology and molecular biology of phenylpropanoid metabolism, Annu. Rev. Plant Physiol. 40: 347–369

SCHLESINGER, M. J., 1986, Heat shock proteins: the search for functions. J. Cell Biol. 103: 321–325.

SCHLESINGER, M. J., U. BOND, 1987, Ubiquitin genes. In: Oxford Survey of Eukaryotic Genes. Vol. 4: pp. 77.

SCHLESINGER, M. J., M. ASHBURNER, A. TISSIERES (eds.), 1982, Heat Shock. From Bacteria to Man. Cold Spring Harbor Lab., Cold Spring Harbor, N. Y.

SCHLUMBAUM, A., F. MAUCH, U. VOEGELI, T. BOLLER, 1986, Plant chitinases are potent inhibitors of fungal growth. Nature 324: 365–367.

SCHMELZER, E., W. JAHNEN, K. HAHLBROCK, 1988, *In situ* localization of light-induced chalcone synthase mRNA, chalcone synthase, and flavonoid end products in epidermal cells of parsley leaves. Proc. Natl. Acad. Sci. USA 85: 2989–2993.

SCHMITT, J. M., C. MICHALOWSKI, H. J. BOHNERT, 1988, Gene expression during CAM induction under salt stress in *Mesembryanthemum*: cDNA library and increased levels of mRNA for phosphoenolpyruvate carboxylase and pyruvate orthophosphate dikinase. Photosynth. Res. 17: 159–171.

SCHNEPF, E., U. SCHMITT, 1981, Destruction and reconstitution of the dictyosome in the chrysophycean flagellate, *Poterioochromonas malhamensis*, after heat-shock, and other heat-shock effects. Protoplasma 106: 261–271.

SCHÖFFL, F., J. L. KEY, 1982, An analysis of mRNAs for a group of heat shock proteins of soybean using cloned cDNAs. J. Mol. Appl. Genet. 1: 301–314.

SCHÖFFL, F., J. L. KEY, 1983, Identification of a multigene family for small heat shock proteins in soybean and physical characterization of one individual gene coding region. Plant Mol. Biol. 2: 269–278.

SCHÖFFL, F., G. BAUMANN, 1985, Thermo-induced transcripts of a soybean heat shock gene after transfer into sunflower using a Ti plasmid vector. EMBO J. 4: 1119–1124.

SCHÖFFL, F., E. RASCHKE, R. T. NAGAO, 1984a, The DNA sequence analysis of soybean heat shock genes and identification of possible regulatory promoter elements. EMBO J. 3: 2491–2497.

SCHÖFFL, F., C.-Y. LIN, J. L. KEY, 1984b, Soybean heat shock proteins: Temperature regulated gene expression and the development of thermotolerance. In: G. R. STEWART and P. J. LEA (eds.), Genetic Manipulation of Plants and its Application to Agriculture. Oxford Univers. Press, p. 129–140.

SCHÖFFL, F., G. BAUMANN, E. RASCHKE, M. BEVAN, 1986, The expression of heat-shock genes in higher plants. Philos. Trans. R. Soc. Lnd. B 314: 453–468.

SCHÖFFL, F., M. RIEPING, G. BAUMANN, 1987a, Constitutive transcription of a soybean heat-shock gene by a cauliflower mosaic virus promoter in transgenic tobacco plants. Dev. Genet. 8: 365–374.

SCHÖFFL, F., I. ROSSOL, S. ANGERMUELLER, 1987b, Regulation of the transcription of heat shock genes in nuclei from soybean *(Glycine max)* seedlings. Plant. Cell Envir. 10: 113–119.

SCHÖFFL, F., G. BAUMANN, E. RASCHKE, 1988, The expression of heat shock genes. A model for environmental stress response. In: B. GOLDBERG and D. P. S. VERMA (eds.), Plant Gene Research. Vol. V. Springer Verlag Wien, pp. 235–273.

SCHÖFFL, F., M. RIEPING, G. BAUMANN, M. BEVAN, S. ANGERMUELLER, 1989, The function of plant heat shock promoter elements in the regulated expression of chimaeric genes in tansgenic tobacco. Mol. Gen. Genetics 217: 246–253.

SCHREIBER, U., P. A. ARMOND, 1978, Heat induced changes of chlorophyll fluorescence in isolated chloroplasts and related heat-damage at the pigment level. Biochim. Biophys. Acta 502: 138–151.

SCHRÖDER, G., J. W. S. BROWN, J. SCHROEDER, 1988, Molecular analysis of resveratrol synthase, cDNA, genomic clones and relationship with chalcone synthase. Eur. J. Biochem. 172: 161–169.

SCHUH, S., W. YONEMOTO, J. BRUGGE, V. J. BAUER, R. M. RIEHL, W. F. SULLIVAN, D. O. TOFT, 1985, A 90,000-dalton binding protein common to both steroid receptors and the Rous sarcoma virus transforming protein, pp60v-sarc. J. Biol. Chem. 260: 14292–14296.

Schulze-Lefert, P., J. L. Dangl, M. Becker-Andre, K. Hahlbrock, W. Schulz, 1989, Inducible *in vivo* DNA footprints define sequences necessary for light activation of the parsley chalcone synthase gene. EMBO J. 8: 651–656.

Schuster, A., E. Davies, 1983, Ribonucleic acid and protein metabolism in pea epicotyls. II. Response to wounding in aged tissue. Plant Physiol. 73: 817–821.

Schuster, G., D. Even, K. Kloppstech, I. Ohad, 1988, Evidence for protection by heat-shock proteins against photoinhibition during heat-shock. EMBO J. 7: 1–6.

Schwartz, D., 1969, An example of gene fixation resulting from selective advantage in suboptimal conditions. Amer. Nat. 103: 479–481.

Schwemmle, B., O. L. Lange, 1959, Endogen-tagesperiodische Schwankungen der Hitzeresistenz bei *Kalanchoë blossfeldiana.* Planta 53: 134–144.

Segal, E., J.-B. Le Pecq, 1986, Relationship between cellular diadenosine 5′, 5‴-P1,P4-tetraphosphate level, cell density, cell growth stimulation and toxic stresses Exp. Cell Res. 167: 119–126.

Sembdner, G., C. Klose, 1985, (-)-Jasmonsäure—ein neues Phytohormon?. Biol. Rundsch. 23: 29–40.

Shall, S., 1984, ADP-ribose in DNA repair. Advanc. Radiat. Biol. 11: 1–69.

Sharma, S., L. Rodgers, J. Brandsma, M.-J. Gething, J. Sambrook, 1985, SV40 T antigen and the exocytotic pathway. EMBO J. 4: 1479–1489.

Shin, D.-Y., I. Uno, T. Ishikawa, 1987, Control of the G1–G0 transition and GO protein synthesis by cyclic AMP in *Saccharomyces cerevisiae.* Curr. Genetics 12: 577–582.

Shinnick, T. M., D. Sweetser, J. E. R. Thole, J. D. A. van Embden, R. A. Young, 1987, The etiologic agents of leprosy and tuberculosis share immunoreactive protein antigen with the vaccine strain *Mycobacterium bovis* BCG. Infect. Immun. 55: 1932–1935.

Shinshi, H., D. Mohnen, F. Meins, 1987, Regulation of a plant pathogenesis-related enzyme: Inhibition of chitinase and chitinase mRNA accumulation in cultured tobacco tissues by auxin and cytokinin. Proc. Natl. Acad. Sci. USA 84: 89–93.

Shinshi, H., H. Wenzler, J.-M. Neuhaus, G. Felix, J. Hofsteenge, F. Meins, 1988, Evidence for N- and C-terminal processing of a plant defense-related enzyme: Primary structure of tobacco prepro-β-1,3-glucanase. Proc. Natl. Acad. Sci. USA 85: 5541–5545.

Shirras, A. D., D. H. Northcote, 1984, Molecular cloning and characterization of cDNAs complementary to mRNAs from wounded potato *(Solanum tuberosum)* tuber tissue. Planta 162: 353–360.

Showalter, A. M., J. N. Bell, C. L. Cramer, J. A. Bailey, J. E. Varner, C. J. Lamb, 1985, Accumulation of hydroxyproline-rich glycoprotein mRNAs in response to fungal elicitor and infection. Proc. Natl. Acad. Sci. USA 82: 6551–6555.

Shuey, D. J., C. S. Parker, 1986, Binding of *Drosophila* heat-shock gene transcription factor to the hsp70 promoter. J. Biol. Chem. 261: 7934–7940.

Simard, R., W. Bernhard, 1967, A heat-sensitive cellular function located in the nucleolus. J. Cell Biol. 34: 61–76.

Simon, M. C., K. Kitchener, H.-T. Kao, E. Hickey, L. Weber, R. Voellmy, N. Heintz, J. R. Nevins, 1987, Selective induction of human heat shock gene transcription by the adenovirus E1A gene products, including the 12S E1A product. Mol. Cell. Biol. 7: 2884–2890.

Singh, J., A. Laroche, 1988, Freezing tolerance in plants: A biochemical overview. Biochem. Cell Biol. 66: 650–657.

Singh, N. K., A. K. Handa, P. M. Hasegawa, R. A. Bressan, 1985, Proteins associated with adaptation of cultured tobacco cells to NaCl. Plant Physiol. 79: 126–137.

Singh, N. K., P. C. La Rosa, A. K. Handa, P. M. Hasegawa, R. A. Bressan, 1987a, Hormonal regulation of protein synthesis associated with salt tolerance in plant cells. Proc. Natl. Acad. Sci. USA 84: 739–743.

Singh, J., B. Iu, A. M. Johnson-Flanagan, 1987b, Membrane alterations in winter rye and *Brassica napus* cells during lethal freezing and plasmolysis. Plant Cell Environ. 10: 163–168.

Sinibaldi, R. M., P. W. Morris, 1981, Putative function of *Drosophila melanogaster* heat shock proteins in the nucleo-skeleton. J. Biol. Chem. 256: 10735–10738.

Sinibaldi, R. M., T. Turpen, 1985, A heat shock protein is encoded within mitochondria of higher plants. J. Biol. Chem. 260: 15382–15385.

Skelly, S., C.-F. Fu, B. Dalie, B. Redfield, T. Coleman, N. Brot, H. Weissbach, 1988, Antibody to sigma 32 cross-reacts with DnaK: Association of DnaK protein with Escherichia coli RNA polymerase. Proc. Natl. Acad. Sci. USA 85: 5497–5501.

Slocum, R. D., R. Kaur-Sawhney, A. W. Galston, 1984, The physiology and biochemistry of polyamines in plants. Arch. Biochem. Biophys. 235: 283–303.

Smillie, R. M., S. E. Hetherington, J. Hie, R. Nott, 1988, Photoinhibition at chilling temperatures. Austral. J. Plant Physiol. 15: 207–222.

Smith, I. K., A. C. Kendall, A. J. Keys, J. C. Turner, P. J. Lea, 1985, The regulation of the biosynthesis of glutathione in leaves of barley (*Hordeum vulgare* L.). Plant Sci. 41: 11–17.

SMITH, J. J., E. P. MULDOON, J. J. WILLARD, D. T. A. LAMPORT, 1986, Tomato extensin precursors P1 and P2 are highly periodic structures. Phytochemistry 25: 1021–1030.

SMITH, L. T., J.-A. POCARD, T. BERNARD, D. LERUDULIER, 1988, Osmotic control of glycine betaine biosynthesis and degradation in *Rhizobium meliloti.* J. Bacteriol. 170: 3142–3149.

SMITH, M. A., 1981, Characterization of carrot cell wall protein. I. Effect of $\alpha_1\alpha'$-dipyridyl on cell wall protein synthesis and secretion in incubated carrot discs. Plant Physiol. 68: 956–963.

SMITH, T. A., 1984, Putrescine and inorganic icons. Rec. Advanc. Phytochem. 18: 7–54.

SMITH, T. A., 1985, Polyamines. Annu. Rev. Plant Physiol. 36: 117–143.

SNUTCH, T. P., M. F. P. HESCHL, D. L. BAILLIE, 1988, The *Caenorhabditis elegans* hsp70 gene family: A molecular genetic characterization. Gene 64: 241–255.

SOLLNER-WEBB, B., J. TOWER, 1986, Transcription of cloned eukaryotic ribosomal RNA genes. Annu. Rev. Biochem. 55: 801–830.

SOMMERVILLE, J., 1986, Nucleolar structure and ribosome biogenesis. Trends Biochem. Sci. 11: 438–442.

SOMSSICH, I. E., E. SCHMELZER, J. BOLLMANN, K. HAHLBROCK, 1986, Rapid activation by fungal elicitor of genes encoding "pathogenesis-related" proteins in cultured parsley cells. Proc. Natl. Acad. Sci. USA 83: 2427–2430.

SOMSSICH, I. E., E. SCHMELZER, P. KAWALLECK, K. HAHLBROCK, 1988, Gene structure and *in situ* transcript localization of pathogenesis-related protein 1 in parsley. Mol. Gen. Genet. 213: 93–98.

SOMSSICH, I. E., J. BOLLMANN, K. HAHLBROCK, E. KOMBRINK, W. SCHULZ, 1989, Differential early activation of defense-related genes in elicitor-treated parsley cells. Plant Mol. Biol. (subm.).

SORGER, P. K., H. R. B. PELHAM, 1987a, Cloning and expression of a gene encoding hsc 73, the major hsp 70—like protein in unstressed rat cells. EMBO J. 6: 993–998.

SORGER, P. K., H. R. B. PELHAM, 1987b, Purification and characterization of a heat-shock element binding protein from yeast. EMBO J. 6: 3035–3042.

SORGER, P. K., H. R. B. PELHAM, 1988, Yeast heat shock factor is an essential DNA-binding protein that exhibits temperature-dependent phosphorylation. Cell 54: 855–864.

SORGER, P. K., M. J. LEWIS, H. R. B. PELHAM, 1987, Heat shock factor is regulated differently in yeast and HeLa cells. Nature 329: 81–84.

SOUTHGATE, R., A. AYME, R. VOELLMY, 1983, Nucleotide sequence analysis of the *Drosophila* small heat shock gene cluster at locus 67B. J. Mol. Biol. 165: 35–57.

SPADORO, J. P., D. W. COPERTINO, L. D. STRAUSBAUGH, 1986, Differential expression of histone sequences in *Drosophila* following heat shock. Dev. Genet. 7: 133–148.

SPECTOR, M. P., Z. ALIABADI, T. GONZALES, J. W. FORSTER, 1986, Global control in *Salmonella typhimurium*: Two-dimensional electrophoretic anaysis of starvation-, anaerobiosis-, and heat shock-inducible proteins. J. Bacteriol. 168: 420–424.

SPENA, A., J. SCHELL, 1987, The expression of a heat-inducible chimeric gene in transgenic tobacco plants. Mol. Gen. Genet. 206: 436–440.

SPENA, A., R. HAIN, U. ZIERVOGEL, H. SAEDLER, J. SCHELL, 1985, Construction of a heat-inducible gene for plants. Demonstration of heat-inducible activity of the *Drosophila* hsp70 promoter in plants. EMBO J. 4: 2739–2743.

SPINELLI, G., C. CASANO, F. GIANGUZZA, M. CIACCIO, F. PALLA, 1982, Transcription of sea urchin mesenchyme blastula histone genes after heat shock. Eur. J. Biochem. 128: 509–513.

SPRADLING, A., M. L. PARDUE, S. PENMAN, 1977, Messenger RNA in heat-shocked *Drosophila* cells. J. Mol. Biol. 109: 559–587.

SPRANG, G. K., I. R. BROWN, 1987, Selective induction of a heat shock gene in fibre tracts and cerebellar neurons of the rabbit brain detected by *in situ* hybridization. Mol. Brain Res. 3: 89–93.

SPRINGER, B., W. WERR, P. STARLINGER, D. C. BENNETT, M. ZOKOLICA, M. FREELING, 1986, The *Shrunken* gene on chromosome 9 of *Zea mays* L is expressed in various plant tissues and encodes an anaerobic protein. Mol. Gen. Genet. 205: 461–468.

STACHEL, S. E., E. MESSEUS, M. VAN MONTAGU, P. ZAMBRYSKI, 1985, Identification of the signal molecules produced by wounded plant cells that activate T-DNA transfer in *Agrobacterium tumefaciens.* Nature 318: 624–629.

STAEB, M. R., J. EBEL, 1987, Effects of Ca^{2+} on phytoalexin induction by fungal elicitor in soybean cells. Arch. Biochem. Biophys. 257: 416–423.

STAPLES, R. C., H. J. GRAMBOW, H. C. HOCH, 1983, Contact with membrane grooves induces wheat stem rust uredospore germlings to differentiate appressoria but not vesicles. Phytopathology 73: 1436–1439.

STEFFENS, J. C., D. F. HUNT, B. G. WILLIAMS, 1986, Accumulation of non-protein metal-binding polypeptides (gammaglutamyl-cysteinyl)-glycine in selected cadmium resistant tomato cells. J. Biol. Chem. 261: 13878–13882.

STEFFENS, M., F. ETTL, D. KRANZ, H. KINDL, 1989, Vanadate mimics effects of fungal cell wall in eliciting gene activation in plant cell cultures. Planta 177: 160–163.

STEKOLL, M., C. A. WEST, 1978, Purification and properties of an elicitor of castor bean phytoalexin from culture filtrates of the fungus *Rhizopus stolonifer.* Plant Physiol. 61: 38–45.

STELZIG, D. A., R. D. ALLEN, S. K. BHATIA, 1983, Inhibition of phytoalexin synthesis in arachidonic acid-stressed potatoe tissue by inhibitors of lipoxygenase and cyanide-resistant respiration. Plant Physiol. 72: 746–749.

STEPANYAN, N. S., G. F. KRUPNOVA, V. D. ZHESTYANIKOV, 1983, Reparation of cytogenetical damage in plant cells upon an unequal fractionation of injurious action 1. Treatment by nitrogen mustard (russ.). Citologija 25: 958–964.

STEPONKUS, P. L., 1984, Role of the plasma membrane in freezing injury and cold acclimation. Annu. Rev. Plant Physiol. 35: 543–548.

STEPONKUS, P. L., M. UEMURA, R. A. BALSAMO, T. ARVINTE, D. V. LYNCH, 1988, Transformation of the cryobehaviour of rye protoplasts by modification of the plasma membrane lipid composition. Proc. Natl. Acad. Sci. USA 85: 9026–9030.

STERMER, B. A., R. HAMMERSCHMIDT, 1984, Heat shock induces resistance to *Cladosporium cucumerinum* and enhances peroxidase activity in cucumbers. Physiol. Plant Pathol. 25: 239–249.

STERMER, B. A., R. HAMMERSCHMIDT, 1985, The induction of disease resistance by heat shock. In: KEY and KOSUGE (eds.), 1985, 291–302.

STERMER, B. A., R. HAMMERSCHMIDT, 1987, Association of heat shock induced resistance to disease with increased accumulation of insoluble extension and ethylene synthesis. Physiol. Mol. Plant. Pathol. 31: 453–462.

STEWART, C. R., G. VOETBERG, 1985, Relationship between stress-induced ABA and proline accumulations and ABA-induced proline accumulation in excised barley leaves. Plant Physiol. 79: 24–27.

STEWART, C. R., G. VOETBERG, P. J. RAYAPATI, 1986, The effects of benzyladenine, cycloheximide, and cordycepin on wilting-induced abscisic acid and proline accumulations and ABA- and salt-induced proline accumulation in barley leaves. Plant Physiol. 82: 703–707.

STOECKLE, M. Y., S. SUGANO, A. HAMPE, A. VASHISTHA, D. PELLMAN, H. HANAFUSA, 1988, 78-Kilodalton glucose-regulated protein is induced in Rous sarcoma virus-transformed cells independently of glucose deprivation. Mol. Cell. Biol. 8: 2675–2680.

STORTI, R. V., M. P. SCOTT, A. RICH, M. L. PARDUE, 1980, Translational control of protein synthesis in response to heat shock in *D. melanogaster* cells. Cell 22: 825–834.

STORZ, G., F. C. JACOBSON, L. A. TARTAGLIA, R. W. MORGAN, L. A. SILVEIRA, B. N. AMES, 1989, An alkyl hydroperoxide reductase induced by oxidative stress in *Salmonella typhimurium* and *Escherichia coli*: Genetic charaterization and cloning of ahp. J. Bacteriol. 171: 2049–2055.

STRAUS, D. B., W. A. WALTER, C. A. GROSS, 1988, *Escherichia coli* heat shock gene mutants are defective in proteolysis. Genes Devel. 2: 1851–1858.

STRITTMATTER, G., N. H. CHUA, 1987, Artificial combination of two cis-regulatory elements generates a unique pattern of expression in transgenic plants. Proc. Natl. Acad. Sci. USA 84: 8986–8990.

STÜRZBECHER, H.-W., P. CHUMAKOV, W. J. WELCH, J. R. JENKINS, 1987, Mutant p53 proteins bind hsp72/73 cellular heat shock-related proteins in SV40-transformed monkey cells. Oncogene 1: 201–211.

SÜSS, K. H., I. T. YORDANOV, 1986, Biosynthetic cause of *in vivo* acquired thermotolerance of photosynthetic light reactions and metabolic responses of chloroplasts to heat stress. Plant physiol. 81: 192–199.

SUSSMAN, M. D., P. SETLOW, 1987, Nucleotide sequence of a *Bacillus megaterium* gene homologous to the dnak gene of *Escherichia coli.* Nucl. Acids Res. 15: 3923.

SUTTLE, J. C., 1981, Effect of polyamines on ethylene production. Phytochemistry 20: 1477–1480.

SWANSON, C. P., C. S. RUPERT, H. T. YOST, 1953, Infrared absorption and temperature studies' on the buds and chromosomes of *Tradescantia paludosa.* Amer. J. Bot. 40: 557–565.

SWINDLE, J., J. AJIOKA, H. EISEN, B. SANWAL, C. JACQUEMOT, Z. BROWDER, G. BUCK, 1988, The genomic organization and transcription of the ubquitin genes of *Trypanosoma cruzi.* EMBO J. 7: 1121–1127.

SYBENGA, J., G. KLEIJER, 1976, 'Below-aditivity' and 'protective effects' of dose fractionation in *Crotolaria intermedia.* Mutat. Res. 34: 131–140.

TANAKA, K., K. MATSUMOTO, A. TOH-E, 1988, Dual regulation of the expression of the polyubiquitin gene by cyclic AMP and heat shock in yeast. EMBO J. 7: 495–502.

TANGUAY, R. M., 1988, Transcriptional activation of heat shock genes in eukaryotes. Biochem. Cell Biol. 66: 584–593.

TANKSLEY, S. D., R. A. JONES, 1981, Effects of O_2 stress on tomato alcohol dehydrogenase activity: Description of a second ADH coding gene. Biochem. Genet. 19: 397–409.

TAYLOR, S. Y., R. HOLLIDAY, 1984, Induction of thermotolerance and mitotic recombination by heat-shock in *Ustilago maydis.* Curr. Genet. 9: 59–64.

TEO, J., B. SEDGWICK, B. DEMPLE, B. LI, T. LINDAHL, 1984, Induction of resistance to alkylating agents in *E. coli*:

the ada gene product serves both as a regulatory protein and as an enzyme for repair of mutagenic damage. EMBO J. 3: 2151–2157.

TEO, J., B. SEDGWICK, M. W. KILPATRICK, T. V. MCCARTHY, T. LINDAHL, 1986, The intracellular signal for induction of resistance to alkylating agents in *E. coli.* Cell 45: 315–324.

TEWES, A., A. WAPPLER, E.-M. PESCHKE, R. GARVE, L. NOVER, 1982, Morphogenesis and embryogenesis in long-term cultures of *Digitalis.* Z. Pflanzenphysiol. 106: 311–324.

THEBUD, R., K. A. SANTARIUS, 1982, Effects of high temperature stress on various biomembranes of leaf cells *in situ* and *in vitro.* Plant Physiol. 70: 200–206.

THEILLET, C., F. DEPEYROUX, M. FISZMAN, R. REIGNER, R. ESNAULT, 1982, Influence of the excision shock on the protein metabolism of *Vicia faba* L. meristematic root cells. Planta 155: 478–485.

THEODORAKIS, N. G., R. I. MORIMOTO, 1987, Posttranscriptional regulation of HSP 70 expression in human cells: Effects of heat shock, inhibition of protein synthesis, and mRNA stability. Mol. Cell. Biol. 7: 4357–4368.

THIMANN, K. V., 1985, The interaction of hormonal and environmental factors in leaf senescence. Biol. Plant. 27: 83–91.

THORNBURG, R. W., G. AN, T. E. CLEVELAND, R. JOHNSON, C. A. RYAN, 1987, Wound-inducible expression of a potato inhibitor II-chloramphenicol acetyltransferase gene fusion in transgenic tobacco plants. Proc. Natl. Acad. Sci. USA 84: 744–748.

TILLY, K., N. MCKITTRICK, M. ZYLICZ, C. GEORGOPOULOS, 1983, The dnaK protein modulates the heat-shock response of *Escherichia coli.* Cell 34: 641–646.

TIMMERMANN, B. N., C. STEELINK, F. A. LOEWUS (eds.), 1984, Phytochemical Adaptations to Stress. Series: Rec. Advanc. Phytochem., Vol. 18, Plenum Press, New York.

TISCHNER, R., H. LORENZEN, 1975, Physiologische Auswirkungen von Hitzeschocks auf synchrone Chlorellen im empfindlichsten Entwicklungsstadium. Biochem. Physiol. Pflanzen 168: 233–245.

TISSIERES, A., H. K. MITCHELL, U. M. TRACY, 1974, Protein synthesis in salivary glands of *D. melanogaster.* Relation to chromosome puffs. J. Mol. Biol. 84: 389–398.

TÖRÖK, I., R. KARCH, 1980, Nucleotide sequences of heat shock activated genes in *Drosophila melanogaster.* I. Sequences in the regions of the 5′ and 3′ ends of the hsp70 gene in the hybrid plasmid 56H8. Nucl. Acids Res. 8: 3105–3123.

TOMASOVIC, S. P., G. N. TURNER, W. C. DEWEY, 1978, Effect of hyperthermia on nonhistone proteins isolated with DNA. Radiat. Res. 73: 535–552.

TOMIYAMA, K., 1967, Further observations on the time requirement for hypersensitive cell death of potatoes infected by *Phytophthora infestans* and its relation to metabolic activity. Phytopathol. Ztschr. 58: 367–378.

TON-THAT, T. C., G. TURIAN, 1984, High-resolution autoradiography of nuclear modifications during and after heat treatment of *Neurospora crassa.* Protoplasma 120: 165–171.

TON-THAT, T. C., G. TURIAN, J. FAKAN, A. GAUTIER, 1981, Ultrastructural cytochemistry of perinucleolar dense spots in heat-treated macroconidia of *Neurospora crassa.* Eur. J. Cell Biol. 24: 317–319.

TRAYLOR, E. A., S. H. SHORE, R. F. RANSOM, L. D. DUNKLE, 1987, Pathotoxin effects in sorghum are also produced by mercuric chloride treatment. Plant Physiol. 84: 975–978.

TRIPLETT, B. A., R. S. QUATRANO, 1982, Timing, localization and control of wheat germ agglutinin synthesis in developing wheat embryos. Devel. Biol. 91: 491–496.

TSENG, M. J., P. H. LI, 1986, Alterations of gene expression in tuber-bearing *Solanum* species during cold acclimation and ABA treatment. Plant Physiol. 86: 109.

UEDA, J., J. KATO, 1980, Isolation and identification of a senescence-promoting substance from wormwood (*Artemisia absinthium* L.) Plant physiol. 66: 246–249.

UEDA, J., J. KATO, 1982, Inhibition of cytokinin-induced plant growth by jasmonic acid and its methyl ester. Physiol. Plant. 54: 249–252.

UEDA, J., J. KATO, H. YAMANE, N. TAKAHASHI, 1981, Inhibitory effect of methyl jasmonate and its related compounds on kinetin-induced retardation of oat leaf senescence. Physiol. Plant. 52: 305–309.

UEMURA, M., S. YOSHIDA, 1984, Involvement of plasma membrane alterations in cold acclimation of winter rye seedlings *Secale cereale* L. cv. Puma. Plant Physiol. 75: 818–826.

UKNES, S. J., T. D. HO, 1984, Mode of action of ABA in barley aleurone layers. ABA induces its own conversion to phaseic acid. Plant Physiol. 75: 1126–1132.

UNGEWICKELL, E., 1985, The 70-kd mammalian heat shock proteins are structurally and functionally related to the uncoating protein that releases clathrin triskelia from coated vesicles. EMBO J. 4: 3385–3391.

VALLIAMMAI, T., A. GNANAM, K. DHARMALINGAM, 1987a, Heat shock response in *Chlorella protothecoides.* Plant Cell Physiol. 28: 975–986.

VALLIAMMAI, T., A. GNANAM, R. MANAR MANNAN, 1987b, Heat shock response of *Chlorella protothecoides* during greening. J. Biosciences 12: 219–228.

VAN BERGEN EN HENEGOUWEN, P. M. P., W. A. M. LINNEMANS, 1987, Heat shock gene expression and cytoskeletal alterations in mouse neuroblastoma cells. Exp. Cell Res. 171: 367–375.

van Bergen en Henegouwen, P. M. P., G. Berbers, W. A. M. Linnemans, R. van Wijk, 1987, Cytoplasmic and nuclear localization of the 84,000 dalton heat shock protein in mouse neuroblastoma cells. Eur. J. Cell Biol. 43: 469–478.

van Bogelen, R. A., P. M. Kelley, F. C. Neidhardt, 1987, Differential induction of heat shock, SOS, and oxidation stress regulons and accumulation of nucleotides in *Escherichia coli.* J. Bacteriol. 169: 26–32.

Vance, C. P., T. K. Kirk, R. T. Sherwood, 1980, Lignification as a mechanism of disease resistance. Annu. Rev. Phytopathol. 18: 259–288.

van Holst, G. J., J. E. Varner, 1984, Reinforced polyproline II conformation in a hydroxproline-rich cell wall glycoprotein from carrot root. Plant Physiol. 74: 247–251.

van Loon, L. C., 1983, Mechanisms of resistance in virus-infected plants. In: J. A. Bailey and B. J. Deverall (eds.), The Dynamic of Host Defence. Acad. Press, Sydney, pp. 123–190.

van Loon, L. C., 1985, Pathogenesis-related proteins Plant Mol. Biol. 4: 11–116.

Varian, A., J. A. Sacher, 1978, Wound-induced RNase in bean pod tissue: II. Auxin regulation of RNase synthesis at transcription. Plant Cell Physiol. 19: 1185–1193.

Vayda, M. E., H. J. Schaeffer, 1988, Hypoxic stress inhibits the appearance of wound-response proteins in potato tubers. Plant Physiol. 88: 805–809.

Velazquez, J. M., S. Lindquist, 1984, Hsp70: Nuclear concentration during environmental stress and cytoplasmic storage during recovery. Cell 36: 655–662.

Velazquez, J. M., B. J. DiDomenico, S. Lindquist, 1980, Intracellular localization of heat shock proteins in *Drosophila.* Cell 20: 679–689.

Veleminsky, J., T. Gichner, T. J. Satava, 1983, Reduction in the frequency of N-methyl-N-nitrosourea-induced somatic mutations in *Tradescantia* by pretreatment with low doses of alkylating agents. Mutat. Res. 122: 229–234.

Vera, P., V. Conejero, 1988, Pathogenesis-related proteins of tomato. P-69 as an alkaline endoproteinase. Plant Physiol. 87: 58–63.

Vera, P., J. H. Yago, V. Conejero, 1988, Immunocytochemical localization of the major "pathogenesis-related" (PR) protein of tomato plants. Plant Sci. 55: 223–230.

Vernon, D. M., J. A. Ostrem, J. M. Schmitt, H. J. Bohnert, 1988, PEPCase transcript levels in *Mesembryanthemum crystallinum* decline rapidly upon relief from salt stress. Plant Physiol. 86: 1002–1004.

Vierling, E., J. L. Key, 1985, Ribulose 1,5-bisphosphate carboxylase synthesis during heat shock. Plant Physiol. 78: 155–162.

Vierling, E., M. L. Mishkind, G. W. Schmidt, J. L. Key, 1986, Specific heat shock proteins are transported into chloroplasts. Proc. Natl. Acad. Sci. USA 83: 361–365.

Vierling, E., R. T. Nagao, A. E. DeRocher, L. M. Harris, 1988, A heat shock protein localized to chloroplasts is a member of a eukaryotic superfamily of heat shock proteins. EMBO J. 7: 575–582.

Vierling, E., L. M. Harris, Q. Chen, 1989, The major low molecular weight heat shock protein in chloroplasts shows antigenic conservation among diverse higher plant species. Mol. Cell. Biol. 9: 461–468.

Vodkin, M. H., J. C. Williams, 1988, A heat shock operon in *Coxiella burnetii* produces a major antigen homologous to a protein in both mycobacteria and *Escherichia coli.* J. Bacteriol. 170: 1227–1234.

Voegeli, U., F. Meins, T. Boller, 1988, Co-ordinated regulation of chitinase and β-1,3-glucanase in bean leaves. Planta 174: 364–372.

Voegeli-Lange, R., A. Hansen-Gehri, T. Boller, F. Meins, 1988, Induction of the defense-related glucanohydrolases, beta-1,3-glucanase and chitinase, by tobacco mosaic virus infection of tobacco leaves. Plant Sci. 54: 171–176.

Vogt, P. K., T. J. Bos, R. F. Doolittle, 1987, Homology between the DNA-binding domain of the GCN4 regulatory protein of yeast and the carboxyl-terminal region of a protein coded for by the oncogen jun. Proc. Natl. Acad. Sci. USA 84: 3316–3319.

Volger, H., K. A. Santarius, 1981, Release of membrane proteins in relation to heat injury of spinach chloroplasts. Physiol. Plant 51: 195–200.

Wagih, E. E., R. H. A. Coutts, 1981, Similarities in the soluble protein profiles of leaf tissue following either a hypersensitive reaction to virus infection or plasmolysis. Plant Sci. Letters 21: 61–69.

Walker, G. C., 1985, Inducible DNA repair systems. Annu. Rev. Biochem. 54: 425–457.

Walker, J. C., J. L. Key, 1982, Isolation of cloned cDNAs to auxin-responsive poly(A+)RNAs of elongating soybean hypocotyl./Proc. Natl. Acad. Sci. USA 79: 7185–7189.

Walker, G. C., L. Marsh, L. O. Dodson, 1985, Genetic analysis of DNA repair: Inference and extrapolation. Annu. Rev. Genetics 19: 103–126.

Walker, J. C., E. A. Howard, E. S. Dennis, W. J. Peacock, 1987, DNA sequences required for anaerobic expression of the maize alcohol dehydrogenase 1 gene. Proc. Natl. Acad. Sci. USA 84: 6624–6628.

Walker-Simmons, M., C. A. Ryan, 1977, Wound-induced accumulation of trypsin inhibitor activities in plant leaves. Survey of several plant genera. Plant Physiol. 59: 437–439.

WALKER-SIMMONS, M., C. A. RYAN, 1986, Proteinase inhibitor I accumulation in tomato suspension cultures. Induction by plant and fungal cell wall fragments and an extracellular polysaccharide secreted into the medium. Plant Physiol. 80: 68–71.

WALKER-SIMMONS, M., H. HOLLAENDER-CZYTKO, J. K. ANDERSEN, C. A. RYAN, 1984, Wound signals in plants: A systemic plant wound signal alters plasma membrane integrity. Proc. Natl. Acad. Sci. USA 81: 3737–3741.

WALKEY, D. G. A., V. C. COOPER, 1975, Effect of temperature on virus eradication and growth of infected tissue cultures. Ann. Appl. Biol. 80: 185–190.

WALKUP, L. K. B., T. KOGOMA, 1989, *Escherichia coli* proteins inducible by oxidative stress mediated by the superoxide radical. J. Bacteriol. 171: 1476–1484.

WALTER, H., 1955, The water economy and the hydrature of plants. Annu. Rev. Plant Physiol. 6: 239–252.

WALTER, M., 1989, The induction of phenylpropanoid biosynthetic enzymes by ultraviolet light or fungal elicitor in cultured parsley cells is overriden by a heat shock treatm ent. Planta 177: 1–8.

WALTER, M. H., K. HAHLBROCK, 1985, Synthesis of characteristic proteins in nutrient-depleted cell suspension cultures of parsley. Planta 166: 194–200.

WANG, P. J., 1985, Producing pathogen-free plants using tissue culture. Plant Tissue Cult. Newsletter No 46: 2–8.

WANNER, R., H. FOERSTER, K. MENDGEN, R. C. STAPLES, 1985, Synthesis of differentiation-specific proteins in germlings of the wheat stem rust fungus after heat shock. Exp. Mycol. 9: 279–283.

WARNER, J. R., 1982, The yeast ribosome. In: J. V. STRATHERN, E. W. JONES and J. R. BROACH (eds.), 1982, The Molecular Biology of the Yeast *Saccharomyces.* Cold Spring Harbor Lab., pp. 529–560.

WAROCQUIER, R., K. SCHERRER, 1969, RNA metabolism in mammalian cells at elevated temperature. Eur. J. Biochem. 10: 362–370.

WATOWICH, S. S., R. I. MORIMOTO, 1988, Complex regulation of heat shock and glucose responsive genes in human cells. Mol. Cell. Biol. 8: 393–405.

WEBSTER, P. L., 1980, 'Stress' protein synthesis in pea root meristem cells?. Plant Sci. Letters 20: 141–145.

WEBSTER, P. L., J. VAN'T HOF 1973, Polyribosomes in proliferating and non-proliferating root meristem cells. Amer. J. Bot. 60: 117–121.

WEIDHASE, R. A., J. LEHMANN, H. KRAMELL, G. SEMBDNER, B. PARTHIER, 1987a, Degradation of ribulose-1,5-bisphosphate carboxylase and chlorophyll in senescing barley leaf segments triggered by jasmonic acid methylester, and counteraction by cytokinin. Physiol. Plant. 69: 161–166.

WEIDHASE, R. A., H. M. KRAMELL, J. LEHMANN, H.-W. LIEBISCH, W. LERBS, B. PARTHIER, 1987b, Methyljasmonate-induced changes in the polypeptide pattern of senescing barley leaf segments. Plant Sci. 51: 177–186.

WEIGEL, H.-J., 1983, The effect of high temperatures on leaf cells of *Valerianella*: Relative heat stability of the tonoplast membrane of mesophyll vacuoles. Planta 159: 398–403.

WEIGEL, P., E. A. WERETILNYK, A. D. HANSON, 1986, Betaine aldehyde oxidation by spinach chloroplasts. Plant Physiol. 82: 753–759.

WEIS, E., 1981, Reversible heat-inactivation of the Calvin cycle: A possible mechanism of the temperature regulation of photosynthesis. Planta 151: 33–39.

WELCH, W. J., J. R. FERAMISCO, 1985, Rapid purification of mammalian 70,000-dalton stress proteins: Affinity of the proteins for nucleotides. Mol. Cell. Biol. 5: 1229–1237.

WELCH, W. J., J. P. SUHAN, 1985, Morphological study of the mammalian stress response: Characterization of changes in cytoplasmic organelles, cytoskeleton, and nucleoli, and appearance of intranuclear actin filaments in rat fibroblasts. J. Cell. Biol. 101: 1198–1211.

WERNER-WASHBURNE, M., D. E. STONE, E. A. CRAIG, 1987, Complex interactions among members of an essential subfamily of hsp70 genes in *Saccharomyces cerevisiae.* Mol. Cell. Biol. 7: 2568–2577.

WESTGATE, M. E., J. S. BOYER, 1985, Osmotic adjustment and the inhibition of leaf, root, stem and silk growth at low water potentials in maize. Planta 164: 540–549.

WHITE, F. P., S. R. WHITE, 1986, Isoproterenol induced myocardial necrosis is associated with stress protein synthesis in rat heart and thoracic aorta. Cardiovasc. Res. 20: 512–515.

WIEDERRECHT, G., D. SETO, C. S. PARKER, 1988, Isolation of the gene encoding the *S. cerevisiae* heat shock transcription factor. Cell 54: 841–853.

WIEGAND, K. M., 1906, The occurrence of ice in plant tissue. Plant World 9: 25–39.

WIEGANT, F. A. C., M. TUYL, W. A. M. LINNEMANS, 1985, Calmodulin inhibitors potentiate hyperthermic cell killing. Int. J. Hyperthermia 1: 157–170.

WIEGANT, F. A. C., P. M. P. VAN BERGEN EN HENEGOUWEN, A. A. M. S. VAN DONGEN, W. A. M. LINNEMANS, 1987a, Stress-induced thermotolerance of the cytoskeleton. Cancer Res. 47: 1674–1680.

WIEGANT, F. A. C., F. J. BLOK, P. M. P. VAN BERGEN EN HENEGOUVEN, W. A. M. LINNEMANS, 1987b, Heat shock induced cytoskeletal reorganization is related to cell survival in mammalian cells. Thesis, Rijksuniversiteit Utrecht, pp. 79–94.

WIELGAT, B., G. KAHL, 1979, Gibberellic acid activates chromatin-bound DNA-dependent RNA polymerase in wounded potato tuber tissue. Plant Physiol. 64: 863–866.

WILKISON, W. O., R. M. BELL, 1988, sn-Glycerol-3-phosphate acyltransferase tubule formation is dependent upon heat shock proteins (htpR). J. Biol. Chem. 263: 14505–14510.

WILLIAMS, J. P., M. U. KHAN, K. MITCHELL, G. JOHNSON, 1988, The effect of temperature on the level and biosynthesis of unsaturated fatty acids in diacylglycerols of *Brassica napus* leaves. Plant Physiol. 87: 904–910.

WILLIAMSON, J. D., R. S. QUATRANO, 1988, ABA regulation of two classes of embryo-specific sequences in mature wheat embryos. Plant Physiol. 86: 208–215.

WILLIAMSON, D., R. S. QUATRANO, A. C. CUMMING, 1985, Em polypeptide and its mRNA levels are modulated by abscisic acid during embryogenesis in wheat. Eur. J. Biochem. 152: 501–507.

WINGATE, V. P. M., M. A. LAWTON, C. J. LAMB, 1988, Glutathione causes a masive and selective induction of plant devense genes. Plant Physiol. 87: 206–210.

WINTER, K., D. J. WILLERT, 1972, NaCl induzierter Crassulaceen-Säurestoffwechsel bei *Mesembryanthemum crystallinum.* Z. Pflanzenphysiol. 67: 166–170.

WINTER, J., R. WRIGHT, N. DUCK, C. GASSER, R. FRALEY, D. SHAH, 1988, The inhibition of *Petunia* hsp70 mRNA processing during CdCl stress. Mol. Gen. Genet. 211: 315–319.

WISE, R. R., A. W. NAYLOR, 1988, Stress ethylene does not originate directly from lipid peroxidation during chilling-enhanced photooxidation. J. Plant Physiol. 133: 62–66.

WOLFFE, A. P., A. J. PERLMAN, J. R. TATA, 1984, Transient paralysis by heat shock of hormonal regulation of gene expression. EMBO J. 3: 2763–2770.

WU, C., 1984, Activating protein factor binds *in vitro* to upstream control sequences in heat shock gene chromatin. Nature 311: 81–84.

WU, M.-T., S. J. WALLNER, 1983, Heat stress responses in cultured plant cells. Development and comparison of viability tests. Plant Physiol. 72: 817–820.

WU, M.-T., S. J. WALLNER, 1984, Heat stress responses in cultured plant cells. Heat tolerance induced by heat shock versus elevated growing temperature. Plant Physiol. 75: 778–780.

WU, M.-T., S. J. WALLNER, J. W. WADDELL, 1984, Heat stress responses in cultured plant cells. Effect of culture handling and age. Plant Physiol. 74: 944–946.

WU, B. J., T. W. GREGG, R. I. MORIMOTO, 1987a, Detection of three protein binding sites in the serum-regulated promoter of human gene encoding the 70-kDa heat shock protein. Proc. Natl. Acad. Sci. USA 84: 2203–2207.

WU, C., S. WILSON, B. WALKER, I. DAVID, T. PAISLEY, V. ZIMARINO, H. UEDA, 1987b, Purification and properties of *Drosophila* heat shock activator protein. Science 238: 1247–1253.

WU, C. H., T. CASPAR, J. BROWSE, S. LINDQUIST, C. SOMERVILLE, 1988a, Characterization of an HSP70 cognate gene family in *Arabidopsis.* Plant Physiol. 88: 731–740.

WU, C. H., H. L. WARREN, K. SITARAMAN, C. Y. TSAI, 1988b, Translational alterations in maize leaves responding to pathogen infection, paraquat treatment, or heat shock. Polysome dissociation and accumulation of a 57 kilodalton protein. Plant Physiol. 86: 1323–1329.

XIAO, C. M., J. P. MASCARENHAS, 1985, High temperature-induced thermotolerance in pollen tubes of *Tradescantia* and heat-shock proteins. Plant Physiol. 78: 887–890.

XIAO, H., J. T. LIS, 1986, A consensus sequence polymer inhibits *in vivo* expression of heat shock genes. Mol. Cell. Biol. 6: 3200–3206.

XIAO, H., LIS, J. T., 1988, Germline transformation used to define key features of heat-shock response elements. Science 239: 1139–1142.

XIAO, H., J. T. LIS, 1989, Heat shock and developmental regulation of the *Drosophila melanogaster* hsp83 gene. Mol. Cell. Biol. 9: 1746–1753.

YACOOB, R. K., W. G. FILION, 1986, Temperature-stress response in maize: A comparison of several cultivars. Can. J. Genet. Cytol. 28: 1125–1131.

YACOOB, R., W. G. FILION, 1987, The effects of cold-temperature stress on gene expression in maize. Biochem. Cell Biol. 65: 112–119.

YANG, S. F., N. E. HOFFMAN, 1984, Ethylene biosynthesis and its regulation in higher plants. Annu. Rev. Plant Physiol. 35: 155–189.

YAROSH, D. B., 1985, The role of 06-methylguanine-DNA methyl-transferase in cell survival, mutagenesis and carcinogenesis. Mutat. Res. 145: 1–16.

YARWOOD, C. E., 1956, Heat-induced susceptibility of beans to some viruses and fungi. Phytopathology 46: 523–525.

YARWOOD, C. E., 1961, Acquired tolerance of leaves to heat. Science 134: 941–942.

YARWOOD, C. E., 1964, Adaptation and sensitization of bean leaves to heat. Phytopathology 54: 936–940.

YOST, H. J., S. LINDQUIST, 1986, RNA splicing is interrupted by heat shock and is rescued by heat shock protein synthesis. Cell 45: 185–193.

YOUNG, N.D., A.W. GALSTON, 1983, Putrescine and acid stress. Induction of arginine decarboxylase activity and putrescine accumulation by low *p*H. Plant Physiol. 71: 767–771.

YUYAMA, S., A.M. ZIMMERMAN, 1972, RNA synthesis in *Tetrahymena*: Temperature-pressure studies. Exp. Cell Res. 71: 193–203.

ZAKERI, Z.F., D.J. WOLGEMUTH, C.R. HUNT, 1988, Identification and sequence analysis of a new member of the mouse HSP70 gene family and characterization of its unique cellular and developmental pattern of expression in the male germ line. Mol. Cell. Biol. 8: 2925–2932.

ZAMBRYSKI, P., J. TEMPE, J. SCHELL, 1989, Transfer and function of T-DNA genes from *Agrobacterium* Ti and Ri plasmids in plants. Cell 56: 193–201.

ZIEMIECKI, A., M.-G. CATELLI, I. JOAB, B. MONCHARMONT, 1986, Association of the heat shock protein HSP90 with steroid hormone receptors and tyrosine kinase oncogene products. Biochem. Biophys. Res. Commun. 138: 1298–1307.

ZIMARINO, V., C. WU, 1987, Induction of sequence-specific binding of *Drosophila* heat shock activator protein without protein synthesis. Nature 327: 727–730.

ZIMMERMANN, R., M. SAGSTETTER, M.J. LEWIS, H.R.B. PELHAM, 1988, Seventy-kilodalton heat shock proteins and an additional component from reticulocyte lysate stimulate import of M13 procoat protein into microsomes. EMBO J. 7: 2875–2880.

ZIVY, M., 1987, Genetic variability for heat shock proteins in common wheat. Theor. Appl. Genet. 74: 209–213.

ZUKER, C., J. CAPPELLO, R.L. CHISHOLM, H.F. LODISH, 1983, A repetitive *Dictyostelium* gene family that is induced during differentiation and by heat shock. Cell 34: 997–1005.

ZYLICZ, M., J.H. LEBOWITZ, R. MCMACKEN, C. GEORGOPOULOS, 1983, The dnaK protein of *Escherichia coli* pocesses an ATPase and autophosphorylating activity and is essential in an *in vitro* DNA replication system. Proc. Natl. Acad. Sci. USA 80: 6431–6435.

Received May 22, accepted July 5, 1989

Authors' address: DIETER NEUMANN, LUTZ NOVER, BENNO PARTHIER,
KLAUS-DIETER SCHARF, REINHOLD WOLLGIEHN, UTA ZUR NIEDEN
Institut für Biochemie der Pflanzen
der AdW der DDR
Weinberg 3
Halle
DDR – 4050

RIGOMAR RIEGER
Zentralinstitut für Genetik und Kulturpflanzenforschung
der AdW der DDR
Corrensstr. 3
Gatersleben
DDR-4325

Subject index

H

T